·精 选·
汉英科技语篇翻译

/

王　欣　刘晓琳·主编

/

Jingxuan

Hanying Kejiyupian

Fanyi

黑龙江教育出版社

图书在版编目（ＣＩＰ）数据

精选汉英科技语篇翻译 / 王欣，刘晓琳主编. -- 哈
尔滨 ： 黑龙江教育出版社，2020.6
　ISBN 978-7-5709-1513-2

　Ⅰ．①精… Ⅱ．①王… ②刘… Ⅲ．①科学技术一英
语一翻译一高等学校一教材 Ⅳ．①N

中国版本图书馆CIP数据核字(2020)第112451号

精选汉英科技语篇翻译
Jingxuan Hanying Keji Yupian Fanyi

王　欣　刘晓琳　**主编**

责任编辑　徐永进
封面设计　孙维福
责任校对　程　丽
出版发行　黑龙江教育出版社
　　　　　　（哈尔滨市道里区群力第六大道 1305 号）
印　　刷　哈尔滨市强生印务有限责任公司
开　　本　787 毫米×1092 毫米　1/16
印　　张　18.5
字　　数　470 千
版　　次　2020 年 6 月第 1 版
印　　次　2020 年 6 月第 1 次印刷

书　　号　ISBN 978 - 7 - 5709 - 1513 - 2　　**定　　价**　38.00 元

黑龙江教育出版社网址:www.hljep.com.cn
如需订购图书,请与我社发行中心联系。联系电话:0451 - 82533097　82534665
如有印装质量问题,影响阅读,请与我公司联系调换。联系电话:13796681795
如发现盗版图书,请向我社举报。举报电话:0451 - 82533087

前　言

随着中国经济的发展,中国与世界的交流日益频繁,为了更好地展示中国的科技发展,向外宣传自己,扩大国际合作与交流,需要科技英语翻译者能够忠实准确地传递原文内容,因此科技英语翻译的重要性逐渐凸显出来,对译者的要求也不断提高,不仅需要译者有良好的语言能力,掌握一定的翻译理论基础,包括对翻译标准、原则、目的和方法的认识;而且要有一定的科学素质和很强的学习能力,不断努力增加自身的知识储备、拓宽翻译视野和领域。同时,译者还必须进行大量翻译实践来提高自身的语言素质和翻译水平。鉴于此,我们编写了此教材,目的是为翻译者提供丰富的翻译实践材料和实用的科技翻译方法。

本书共分为十五章,每一章包括汉语原文、英语参考译文、单词释义和译文解析四个板块。在选取文章时,我们注意了科技文章题材的多样性、前沿性和趣味性。同时,也关注了原文的难度系数,谨防过难或过易的文章入选,以保证本书的读者在练习翻译时在内容上具有可读性,在练习中具有一定的挑战性。在单词释义中,我们主要将原文的难词、专有词汇和语言难点进行了注释,这样有利于读者在翻译时无需再查找专业词典。在译文解析部分,主要是通过文章例句解释了汉英翻译技巧和翻译方法的使用。如果教师将此书用于课堂教学,建议教师可以先让学生自学单词释义,然后课堂上着重讲解译文解析部分,最后选取部分或全部原文进行练习。

本书适合英语语言文学专业翻译方向本科生、研究生、翻译专业硕士和各界热爱翻译的仁人志士作为教材或者辅助材料之用,也可作为高校英语教师的翻译授课教材或者补充教材使用。因编者的水平有限,书中谬误之处在所难免,请广大读者批评指正。本教材是中央高校基本业务科研经费资助项目《生态话语的及物性分析模式构建研究》(项目编号:3072020CF1202)的研究成果;2020 年黑龙江省高等教育教学改革研究项目《基于课程思政育人理念的"一体两翼式"大学英语教学模式研究与实践》(项目编号:SJGY20200160)的阶段性成果。

编　者

2020 年 11 月于哈尔滨

目　　录

第一章　柔性电子

让我们先来发挥一下想象力吧！电脑显示屏可以像手帕一样折叠起来塞进口袋，要两个壮汉才能搬动的大屏幕电视，一个人便可以像卷海报那样轻松地拎回家；遭遇不测而无法恢复的皮肤，可以换成全新的人造皮肤，不仅和原来一样健康美观，还能感受到极其微小的压力；将一块小小的薄片，贴在皮肤上，不仅可以 24 小时检测自己身体的健康状况，而且日常生活中的每一个微小动作都可以转换成电能，例如走路、开车、打字，甚至心脏的跳动。这些看似非同寻常的憧憬其实都可以通过柔性电子学变成触手可及的现实。柔性电子学，又称柔性电路，是一种制备轻薄、可折叠的电子器件及其形成的电路技术。

柔性显示来了

在介绍柔性显示屏前，我们先了解一下显示屏的两个基本电子器件：晶体管（transistor）和发光二极管（light emitting diode）。晶体管就像一道神奇的阀门，可以根据输入电压控制输出电流。我们平时所用的开关就是晶体管的应用之一，晶体管是所有集成电路最基本的组成部分。

很多人或许没有听过发光二极管，可是提起它的另一个名字 LED，相信大家一定不会觉得陌生。相比传统的钨丝灯泡，LED 光源明亮稳定，节约能耗，而且通过加入不同的物质，可以呈现出不同的色彩，完全颠覆了我们对照明的认知。北京奥运会的标志性建筑水立方，就是利用 LED 光源变幻出各种色彩和图案。奥妙在于 LED 通过电压控制器件发出强弱不同的光，而特定的材料会吸收特定的光，从而通过材料的选择改变发出光的颜色（波长）。

显示屏有两种驱动方法：被动矩阵型和主动矩阵型。被动矩阵型是采用 X 和 Y 轴的交叉方式来驱动发光二极管，这种方式驱动的屏幕越大，需要的线路就越多，速度也就越慢，而且显示点（pixel）之间会有电信号干扰，从而影响画面质量。主动矩阵型的显示屏是由之前提到的晶体管和发光二极管组成，显示区域的每一个显示点都由一个晶体管控制，几根线路就可以快速控制非常庞大的屏幕，显示点之间的电信号干扰

1

会大幅减少,能耗也会更低。

传统的晶体管和发光二极管都是由无机半导体材料组成的,而柔性电路采用的器件则主要是有机半导体材料。这种材料于20世纪60年代被发现,以碳元素为主,也包括氢、氮、氧等元素。自80年代第一个用有机材料做出的发光二极管(OLED)和晶体管(Organic transistor)诞生起,研发人员就产生了用有机材料做柔性电子器件的想法。OLED显示主要依靠透明电极,这些电极不但导电,还能让显示屏发出的光透出去。传统的透明电极材料叫作ITO,是一种陶瓷材料,一弯曲就很容易破碎。硅谷的C3Nano公司则选用导电纳米材料来实现这个目标,由于纳米材料可通过溶液处理和打印,所以制备成本也会低了不少。研究的碳纳米材料主要包括:碳纳米管(carbon nanotubes)、富勒烯(C60)和石墨烯(graphene)。2015年,C3Nano收购了韩国最大的银纳米线公司(Aid-en Co. Ltd),开始使用银纳米线制作透明电极。2015年6月,日立公司(Hita-chi)已经开始使用C3Nano的透明电极材料,用于研发大面积柔性触摸显示屏。

除了Hitachi,不少公司都开始展示柔性显示屏产品,包括Plastic logic、三星、LG等。其中柔性显示屏做得最轻最薄的是一家叫作柔宇科技(Royole)的公司,是由斯坦福毕业生刘自鸿博士领导的团队创建的。这家公司做出的显示屏只有0.01毫米,卷曲半径可以达到1毫米,刷新了世界纪录。在一次展示中,由手机控制的录像就在这个薄如蝉翼的显示屏中播放,没有一丝缺陷。2015年10月,李克强总理还亲自参观了他们的研发中心。这家公司最近一次融资高达1.7亿美元,估值超过10亿美元,发展十分迅猛。

柔性器材:医学监测、诊断和治疗

柔性器件在生物医学方面也有非常广泛的应用。对于人体外部检测而言,刺激大脑电信号或者心脏律动的传统方法是将电极接合在导电凝胶之中,然后贴在体表。这种方法有许多不足,比如,由于凝胶会逐渐干燥,失去黏性,所以不能进行长时间、连续的检测;与人体体表接合较差,易滑落;舒适度偏低,尤其对于低龄化患者使用较为困难;将电极接合的过程也相对烦琐费时。

而柔性器件可以解决这些问题。柔软轻薄的柔性器件可以非常紧密地贴合在皮肤心脏或者大脑上面,从而通过对电学或压力信号的检测获取更准确的身体信息。并且人体佩戴的体感较好,不会有明显的不适。人们甚至可以24小时佩戴这此薄如蝉翼的器件,随时监测人体的心跳、脉搏、血压、脑电波、心电图等各种身体指标。长时间随时监测,对于心脑血管疾病的诊断、预防和及时治疗意义重大。如果在器件中安装药物缓释材料,甚至可以有效地针对疾病进行治疗,比如糖尿病病人可以佩戴定期释放胰岛素的柔性材料。

伊利诺伊大学厄巴纳－香槟分校的教授约翰·罗杰斯（John Rogers）是这个领域的翘楚。他运用柔性电路的制备方法，将各种传感器制备在一个可以贴在皮肤上的超薄贴纸上，用于测量身体各项指标。他还在 2008 年成立了坐落于波士顿的 MC10 公司，致力于研发推广舒适、安全和与人体皮肤紧密接合的检测性产品。

MC10 公司的其中一项研究是与欧莱雅公司合作研制的高灵敏度可穿戴皮肤检测贴片，其电路由金属细丝制成，与柔性材料交织，佩戴时甚至不会感觉到它的存在，可以佩带几个星期，洗澡、游泳也不会脱落，信号的准确性也不会受到影响。所有检测信号可以通过蓝牙传送到电脑或者智能手机上，譬如血液流动引起的肌肤温度的变化——充足的血液流通是肌肤健康的标志性指标，还有皮肤的含水量——这可是检验绝大多数美容产品的标志性指标。这种检测贴片可以帮助欧莱雅公司研发深度补水乳液和清爽型润肤霜等诸多产品。这对于广大爱美的女性同胞来说绝对是个福音。想象一下，你可以 24 小时跟踪自己的皮肤含水量，用直观的数据来检测不同美容产品的使用效果，针对自己的皮肤状况选择最适合自己的产品和使用量。它还可以检测皮肤炎症，这对于敏感性皮肤的顾客更是救星。

对于植入人体的器件来说，柔性器件具有更大的优越性。为了治疗一些严重的神经系统的疾病，如帕金森病、癫痫或者抑郁症，医生需要用外部电信号来刺激大脑神经元，使其恢复功能，也就是令人闻风丧胆的电疗。另一种方法是将检测电路运送到大脑深处的目标细胞进行深脑刺激。这些传统方法使用的是金属制成的电极，但坚硬的金属刺穿皮质会导致正常细胞受损或死亡，副作用非常大。与此相比，柔性材料的优越性就体现出来了，柔软材料制成的电极在刺穿过程中会极大地减少对正常脑细胞的损伤。柔性电极也可运用到瘫痪或者失去手臂的病人中，如果将柔性电极植入脑中读取信号就可以控制机械肢体。

由于柔性电极非常软，如何将电极运送到目标细胞成为科学家的研究重点。目前有三种运输方法：第一种是把柔性器件用可溶解的胶，粘在一个坚硬的物体表面将其运送到目标细胞（比如注射用的针头），待胶完全溶解后取出坚硬的物体即可。第二种是把极软的柔性电路放进注射器中，像打针那样将柔性电路注射到目标细胞，针管还可以同时传输液态药物或者细胞，使其共同作用。第三种方法则是将柔性电路冷冻，使其变得坚硬，就可以直接刺穿人体细胞到达目标细胞，再通过人体体温溶解继而发挥作用，整个过程并不影响其灵敏度和测量的准确度。第二种和第三种方法都是由纳米界领军人物—哈佛大学的查尔斯·利伯（Charles Lieber）教授的团队在 2015 年发表报道的。

此外，使用表面积更大的纳米材料做成的电极，电阻抗更小，所以电极能记录更微小的信号。表面积大的材料还拥有更高电容值，这样用更小的电压就达到刺激神经的效果。传输到深脑的电极通常需要一个和外部仪器连接的电线,最新的技术可以通过无线电的方法在体外刺激神经或者接收检测到的信号。

近几年,随着柔性材料的不断发展,许多柔性器件都使用可降解材料制作,还可以通过材料控制器件的降解时间,信号收集完毕之后在人体内自行降解,使病人免受二次手术之苦。

人造皮肤:可以用于人,用于机器人

在 20 世纪 70 年代的科幻电影《无敌金刚》里,主人公因飞机事故失去了自己的肢体和一只眼睛,机缘巧合地装上了人造皮肤覆盖的机械肢体和用摄像头做成的眼睛,由此获得了非同寻常的速度和力量,开始他危险刺激的特工生涯。20 世纪 80 年代的《星球大战》中,卢克·天行者(Luke Skywalker)的机械手甚至拥有完整的感觉能力,能够非常灵敏地感觉到外界压力。这些丰富的想象大概也是推动电子皮肤研究的原动力之一吧。

手指接近火苗感到灼热刺痛的瞬间快速缩回手指,这看似简单的动作却包含大量人体神经细胞的精准运作。我们的皮肤是人体感知系统的第一道大门,皮肤下遍布神经细胞网络,皮肤感受到的压力有任何细微的变动,都会通过神经细胞树突和轴突间的联系将信号传入神经中枢,送达大脑,让我们迅速做出反应。

皮肤的这种感知功能可以用传感器模拟。从 20 世纪 90 年代起科学家们就开始研发柔性器件做传感器,近几年,柔性传感器更是有了非常迅速的发展。对于制作电子人造皮肤的材料来说,有两点非常重要:一是要求材料对压力的敏感度很高,即便是非常细小的变化也可以被感知;二是要求材料具有较高的弹性度,不但能经受弯曲,还能够被拉伸。在这个领域,斯坦福大学的鲍哲楠教授和其所带领的团队取得了多项突破。他们发现用一列金字塔形结构的介电层(dielectric)可以大大提高电子皮肤的敏感度。即使只有很小的压力也能改变介电层的厚度,从而改变晶体管的电流。应用这项技术制成的压力传感器的敏感度达到了历史新高,即使一只蝴蝶或者一只苍蝇落在上面,该压力传感器也能感受得到。同时这种传感器还能贴在手腕上检测脉搏,有望以后在疾病诊断领域有所应用。此外这种压力传感器的反应速度也极其快(<10 毫秒),比没有用这种金字塔结构的同种材料快 100 倍。为了提高电子皮肤的拉伸能力,他们把碳纳米管提前拉伸,这样回归原位时会形成弯曲的结构,再次拉伸时该材料的导电能力不会有太大的变化。

有了感知功能的人造皮肤越来越逼真,然而还缺少人体皮肤的自我修复能力。即使是柔性电子器件也很难实现这个自然界的奇迹。鲍教授的团队却发现了一种能够自我修复的有机材料,加入导电的纳米材料改造后,不仅可以当作导电的压力传感器,还有自我修复的能力。材料被切断后,完全丧失了导电能力,但过一段时间后,显微镜下被切断的材料外表不但能够完全修复,导电能力也基本上能够恢复如初。这种材料受到压力之后,纳米颗粒之间的距离会减小,导电性发生变化,因此对压力也有很高的敏感度。

人造皮肤的研究并不止步于让皮肤承受的压力转化为电流,终极目标是让这种感知上传到人的中枢神经,让使用假肢的人们真正恢复触觉。我们的皮肤之所以能感受到压力,是因为一种感受神经元上有机械性刺激感受器,而那些使用假肢的人群早已失去了这些感受神经元。2015 年,鲍教授的团队联合斯坦福著名的神经学家卡尔·代塞尔罗思(Karl Deisseroth)课题组在《科学》杂志上发表了利用有机电子材料和纳米材料制成的有机械性刺激感受器功能的电子器件,能够把压力转化成不同频率的电子信号和光学信号,并且让神经元真正地感受到。

柔性电子器件还可以应用于制作人造器官。无论对失去某些器官的人类还是机器人而言,这一点都非常重要。比如,人造眼其实就是一个柔性感光传感器的阵列,人造鼻就是一个非常精巧的气体传感器,而人造舌头则是一个能感觉出酸甜苦辣的液体传感器。通过应用不同材料,这些传感器甚至可以感受到人体器官感知不到的元素。而拥有这些超级器官的机器人,甚至能感受到空气中的污染物质、食物饮料中的营养成分,还能"看到"除了可见光以外的其他波长。

能 量 转 化 器 件

如今能源问题日渐凸显,研究如何利用新能源、如何进行能量采集极其重要。如果使用柔性材料制作能源转化器件,可以极大提高能源利用率,颠覆我们的生活。

相信大家对摩擦生电都不会陌生,一种材料比较容易失去电子,另一种材料比较容易得到电子,有这两种特性的材料互相摩擦就可以带电。当两种材料接触在一起的时候,正负电荷中心在同一个平面,处于中和状态,则对外不显电性。如果施加一个外界机械力,使得两种材料分离,正负电荷就会永久地保留在材料上面,两种材料之间就会形成电场。如果在两种材料背后分别镀上一个金属电极,电场就会使得金属电极之间的电子发生转移,从而对外电路形成一个电信号。该原理应用到柔性电路,可以将电信号用来进行各种机械传感运动。比如佐治亚理工学院的王中林教授团队研发的智能键盘就是其中一种代表性应用。由于人的皮肤容易失去电子,如果用容易得到电子的柔性材料制作智能键盘的表面,那么当手指接触键盘时,就会产生电子转移——手指带正电,键盘带负电。如果手指离开键盘,就会产生电场变化。这会使智能键盘内部安装的金属电极产生电势差。电势差会使电极之间发生电子转移,而转移的数量与手指敲击键盘的力度和手指与键盘的接触面有着密切的关系,因而可以识别使用键盘的对象,若有非授权人士使用这个机器就会触发警报系统。由于自然界大部分材料都具有容易得到或者失去电子的特性,这种器材选材非常广泛,因而成本低廉,而且工序并不复杂,易于推广。

除了摩擦生电,还有许多原本被浪费的能量也可以借助柔性器件转化为电能。臂如柔性材料制作的太阳能电池,基于柔性材料质量轻、贴合度高的特点,可以把它铺在高低不平的表面上将太阳能转换成电能,不仅可以大幅度提高太阳能利用率,而且便于工人携带作业,减轻表面承重。如果我们在书包或者野营帐篷上包裹一层柔性太阳能电池,就可以利用太阳能给手机电脑充电了。还有柔性热电器件,能够通过温度差把热能转化成电能,柔性器件包裹在正在发热的工厂机器上,或者包裹在家家户户都使用的厨具上,不仅可以保温,还获取了这些原本会被浪费的热能。柔性器件还可以安装在生活中所有有机械运动的地方, 公路、车轮、运动器材, 甚至鞋子,充分利用一切零散的能量。

甚至人体也有许多能量可以被利用,心脏的跳动、运动时体温的升高都可以通过柔性材料转化成能量。穿一种超级衣服,在人体运动时造成材料分离,摩擦生电。未来移动电源可能会被淘汰,因为,人体本身将有望变成一个小型移动发电站,每个人都可以为自己的电子设备供电。

柔性材料制备方法

这一系列神奇的功能究竟是如何实现的? 柔性材料究竟是如何制成的? 这个部分将揭开柔性材料的神秘面纱。传统的电子器件和电路都是用坚硬的半导体材料硅以及金属导体制成,如果我们想使它们拥有柔性,有四种方法:

第一种方法是选用承受应变(strain)能力强的电子材料。所谓应变,是指物体在外力作用下发生的形变。这里所说的承受应变能力强是指材料受到拉伸以后,不但自身的结构没有被破坏,而且电学性能也不会变化。目前,承受应变能力强的新型电子材料包括有机的小分子、高分子、碳纳米管、硅纳米线等。这些新型电子材料还有一个特点,就是可以通过溶液进行加工处理,然后用打印的方法把它们制成器件。与传统电路的生产工艺相比,打印的方法不仅省去了好多烦琐的生产步骤,极大地提高了生产效率,而且生产过程中不需要高温,还降低了成本。但大规模生产应用这些新型材料仍有一些技术问题需要克服,比如说在非真空空气中和潮湿环境下的稳定性不够好,合成的过程比较复杂,重复性不高等。

第二种方法是把电子器件做薄。这个器件所能够承受的应变主要与卷曲半径(bending radius)和器件的厚度有关。所以如果材料厚度比折叠半径薄很多的话,这个材料所承受的应变也就不会那么大了。比如说,一个厚度为1毫米的材料,如果把它弯曲到半径为 1 厘米,那么这个材料所受到的应变应该是 5%。而如果把这个材料的厚度减小 1000 倍,只有 1 微米的话,把它弯曲到 1 厘米所受到的应变应该就只有 0.005% 了。但我们把器件做薄会影响材料和器件本身的电学性能,工艺十分复杂,成本也会随之增高。另外,大部分传统半导体材料(硅、砷化镓、氧化物等)即便受到很小的应变,也很容易折断。

第三种方法是将一些易碎的电学材料放在塑料基底的中间。一般在基底弯曲的情况下,上表面所受到的伸张应变(tensile strain)是最大的,下表面会受到压缩应变(compressive strain),所以塑料基底中间的那一部分所受到的应变是最小的。通过这种三明治结构的设计,把关键的易碎材料放在塑料基底的中间,可以极大地提高材料的卷曲半径,从而达到任意折叠、弯曲的设计目的。

第四种方法则是制作成网眼结构(mesh structure)。比起连续的线性或者平面结构,网眼设计可以在材料弯曲的时候让中间的空洞部分承受大部分的应变。此外,整个结构也会更柔软,更容易被弯曲。在横向的连接部分设计一些弯曲结构,甚至是一些凸出来的结构,这样连接部分底下的塑料可以承担基底弯曲所引起的应变。但这种弯曲的结构会使制作工艺变得更加复杂。而且,在同一个空间内,采用网眼结构制备出的器件密度会小很多,所以会影响大量器件的最终集成。

以上着重介绍的柔性显示屏和人造电子皮肤等技术只是柔性电子学或柔性电路的一部分。此类技术在能源、医学、环境、可穿戴设备等领域都有非常广阔的应用前景。电影中的超级英雄拥有的超凡脱俗的能力正在从荧屏步入现实。梦想实现的方法,正是无尽的想象力和不懈的科学探索。柔性电子学如同科学星空中冉冉升起的一颗璀璨明星,闪耀着人类探索未知、突破极限的智慧光芒。

参考译文:

Flexible Electronics

Let's play our imagination first! The computer screen can be folded like a handkerchief and stuffed into a pocket. A large-screen TV that requires two strong men to move can be easily carried home like a poster roll. The irrecoverable skin caused by unexpected accidents can be replaced with new artificial skin, which is not only healthy and beautiful as before, but also able to feel extremely tiny pressure. Applying a small piece of thin sheet to the skin can not only check your health for 24 hours a day, but also convert every tiny action in daily life into electric energy, such as walking, driving, typing, even the beating of your heart. These seemingly extraordinary visions can be turned into

accessible reality through flexible electronics. Flexible electronics, also known as flexible circuit, is a kind of circuit technology for the preparation of light, thin and foldable electronic devices.

Here Comes The Flexible Display

Before introducing the flexible display, let's take a look at the two basic electronic devices of the display: transistor and light emitting diode. The transistor is like a magic valve, which can control the output current according to the input voltage. The switch we usually use is one of the applications of transistor, which is the most basic component of all integrated circuits.

Maybe a lot of people have never heard of light emitting diode, but when it comes to LED, I'm sure that you won't feel strange. Compared with traditional tungsten light bulbs, LED light source is bright, stable and energy efficient. Furthermore, it can show different colors by adding various substances, which completely subverts our understanding of lighting. The LED light source is used in Water Cube, the landmark building of Beijing Olympic Games, to change various colors and patterns. The secret lies in that LED emits different light through voltage controller, and specific materials will absorb specific light, thus changing the color (wavelength) of the light emitted through the selection of materials.

There are two driving methods for display screen: passive matrix and active matrix. Passive matrix type drives LED through the cross mode of X-axis and Y-axis. The larger the screen driven in this mode, the more lines are needed and the slower the speed is. In addition, there will be electrical signal interference between the pixels, which will affect the picture quality. The active matrix display screen is composed of transistors and light emitting diodes above-mentioned. Each pixel in the display area is controlled by a transistor. Several lines can quickly control a very large screen. The electrical signal interference between pixels will be greatly reduced, and the energy consumption will be lower.

Traditional transistors and light emitting diodes are made of inorganic semiconductor materials, while the devices used in flexible circuits are mainly organic semiconductor materials. This kind of material was found in 1960s. It is mainly composed of carbon element, including hydrogen, nitrogen, oxygen and other elements. Since the first OLED and organic transistor made of organic materials were born in the 1980s, researchers have come up with the idea of using organic materials as flexible electronic devices. OLED display mainly relies on transparent electrodes, which not only conduct electricity, but also let out the light emitted from the display. The traditional transparent electrode material, named ITO, is a kind of ceramic material, which is easily broken when bent. The C3 Nano company in Silicon Valley chooses conductive nanomaterials to achieve this goal. Nanomaterials can be processed and printed by solution, so the preparation cost will be much lower. Carbon nanomaterials studied mainly include carbon nanotubes, fullerenes (C60) and graphene. In 2015, C3 Nano acquired Aid-en Co. Ltd, the largest company in South Korea, and began to use silver nanowires to make transparent electrodes. In June 2015, Hitachi has begun to use transparent electrode materials made by C3 Nano for the development of large-area flexible touch screen.

In addition to Hitachi, many companies began to exhibit flexible display products, including Plastic logic, Samsung, LG, etc. The lightest and thinnest flexible display was made by a company named Royole, which was founded by a team led by Stanford graduate Dr. Liu Zihong. The display made by the company is only 0.01mm, and the curling radius 1mm, which is a new world record. At an exhibition, the video controlled by the mobile phone was played in the thin display screen without any defect. In October 2015, Premier Li Keqiang visited their R&D center in person. The company has grown very fast. With the latest financing reached $ 170 million, its estimated valuation has exceeded $ 1 billion.

Flexible Equipment: Medical Monitoring, Diagnosis And Treatment

Flexible devices are also widely applied in biomedical field. In terms of external detection of human body, the traditional way to stimulate brain electrical signals or cardiac rhythm is to bond electrodes into electrographic gel, and then attach them to the body surface. However, there are many shortcomings in this method. For example, since the gel will gradually dry out and lose viscosity, it can not be detected for a long time and continuously. Besides, it is poorly bonded to the human body surface and easy to slip. Moreover, the comfort level is low, especially for younger patients and the process of electrode bonding is relatively cumbersome and time-consuming.

Flexible devices can be applied to address these problems. Soft, light and thin, the flexible devices fit snugly over the skin, heart or brain, so as to obtain more accurate body information through the detection of electrical or pressure signals. And the body feeling to wear flexible devices is better and there will be no obvious discomfort. People can even wear these gossamer-thin devices for 24 hours a day to monitor the body's heartbeat, pulse, blood pressure, brain waves, electrocardiogram and other physical indicators

anytime. Long-term monitoring at any time is of great significance for the diagnosis, prevention and timely treatment of cardiovascular and cerebrovascular diseases. If the device is equipped with drug delivery materials, it can even treat diseases effectively. For instance, diabetes patients can wear flexible materials that release insulin regularly.

John Rogers, a professor at the University of Illinois at Urbana-Champaign, is a leader in the field. He used the flexible circuit preparation methods to prepare various sensors on an ultra-thin sticker that can be attached to the skin to measure various indicators of the body. He also founded Boston-based MC10 company in 2008, which is dedicated to research, development and promotion of the detection products that are comfortable, safe and tightly bonded to human skin.

One of the research of MC10 company is a wearable skin-detection patch with high sensitivity developed in cooperation with L'Oreal company. Its circuit is made of metal filaments and interwoven with flexible materials. You will not even feel its existence when wearing. It can be worn for several weeks and will not fall off even when you are bathing or swimming. Furthermore, the accuracy of the signal will not be affected. All detection signals can be transmitted to computers or smartphones via Bluetooth, such as the changes of skin temperature caused by blood flow—adequate blood circulation is a hallmark of skin health, as well as the moisture content of the skin—which is a signature indicator of testing most beauty products. This kind of detection patch can help L'OREAL develop a wide range of products such as deep-moisturizing lotions and refreshing moisturizers. This is definitely a good news for the majority of women in pursuit of beauty. Imagine that you can track the moisture content of your skin for 24 hours a day, use visual data to test the effects of different beauty products, and select the most suitable products and usage amount according to your skin condition with the detection patch. Additionally, it can detect skin inflammation, which is a lifesaver for customers with sensitive skin .

As for devices implanted in human body, flexible devices are more advantageous. In order to treat some serious neurological diseases, such as Parkinson's disease, epilepsy or depression, doctors need to use external electrical signals to stimulate brain neurons to restore their functions, that is, electric therapy, which is frightening. Another method is to transport the detection circuit to the target cell deep in the brain for deep-brain stimulation. These traditional methods use electrodes made of metal, but hard metal piercing the cortex will cause damage or death to normal cells, accompanied with very serious side effects. Compared with this, the superiority of flexible materials is manifested in that the electrodes made of soft materials can greatly reduce the damage to normal brain cells in the process of piercing. Flexible electrodes can also be used for patients with paralysis or lossing of arms. The mechanical limbs can be controlled if the flexible electrodes are implanted into the brain to read signals.

Since the flexible electrode is very soft, the way to transport the electrode to the target cell has become the focus of scientists' research. At present, there are three methods of transportation: the first is to use soluble glue to attach the flexible device to the surface of a hard object and transport it to the target cell (such as the needle for injection) and then take out the hard object after the glue is completely dissolved. The second is to put the extremely soft flexible circuit into the syringe and then inject the flexible circuit into the target cell like injection. The syringe can also transport liquid drugs or cells at the same time, so as to make them work together. The third method is to freeze the flexible circuit to make it hard, so that it can directly penetrate human cells to reach the target cell and dissolves through body temperature and then works. The whole process does not affect its sensitivity and measurement accuracy. The second and third methods were both reported in 2015 by a team led by Professor Charles Lieber of Harvard University.

In addition, electrodes made of nanomaterials with a larger surface area have smaller electrical impedance, so they are able to record even smaller signals. The materials with large surface area also have higher capacitance value, so that the effect of nerve stimulation can be achieved with a smaller voltage. Electrodes that are transmitted to the deep brain usually require a wire connected to an external instrument. The latest technology can stimulate nerves in vitro or receive detected signals by radio.

In recent years, with the continuous development of flexible materials, many flexible devices are made of degradable materials and the degradation time of the devices can also be controlled by the materials. The flexible devices are self-degraded in the human body after the signal collection finished, so that patients can be protected from the pain of the second operation.

Artificial Skin: For People, For Robots

In the science fiction movie The Six Million Dollar Man in the 1970s, the protagonist lost his limbs and an eye due to an airplane crash. By chance, he was fitted with mechanical limbs covered with artificial skin and an eye made with a camera so that extraordinary speed and strength were gained and then his dangerous and exciting spy career started. In Star Wars in the 1980s, Luke Skywalker's machine hands even boasted full sensory abilities and were able to sense external pressure very sensitively. The rich

imagination is probably one of the driving forces behind the development of electronic skin research.

The moment your fingers approach the flame, you can feel the burning and tingling and then quickly retract your fingers. This seemingly simple action contains the precise operation of a large number of human nerve cells. Our skin is the first door to the human perception system. Under the skin, there is a network of nerve cells. Any slight changes in the pressure felt by the skin will send signals to the nerve center through the connection between the dendrites and axons of nerve cells, and then to the brain, so that we can react quickly.

This sensory function of the skin can be simulated by sensors. Since the 1990s, scientists have been developing flexible devices for sensors and in recent years, flexible sensors have developed very rapidly. For the materials of electronic artificial skin, two points are very important: first, the material is required to be highly sensitive to pressure and even very small changes should be perceived; second, the material is required to boast a high degree of elasticity, which can not only withstand bending, but also be stretched. Professor Bao Zhenan of Stanford University and his team have made many breakthroughs in this field. They found that the use of a column of pyramidal dielectric layers can greatly improve the sensitivity of electronic skin. Even a small pressure can change the thickness of the dielectric layer, thus changing the current of the transistor. The sensitivity of the pressure sensor made with this technology has reached a record high and even if a butterfly or a fly falls on it, the pressure sensor can perceive. At the same time, the sensor can be attached to the wrist to detect the pulse, which is expected to be applied in the field of disease diagnosis in the future. In addition, the response speed of the pressure sensor is extremely fast (< 10 ms), 100 times faster than that of the same material without the pyramid structure. In order to improve the stretching ability of the electronic skin, they stretch the carbon nanotubes in advance, so that when they return to the original position, a curved structure will be formed, and the conductivity of the material will not change much when stretched again.

With the function of perception, artificial skin becomes increasingly lifelike, but it lacks the self-healing ability of human skin. Even flexible electronic devices can hardly achieve this miracle of nature. But Professor Bao's team discovered an organic material that can be self-repairing. After being modified with conductive nanomaterials, it can not only be used as a conductive pressure sensor, but also repair itself. After the material is cut off, its conductivity is completely lost. However, after a period of time, the surface of the material cut off under the microscope can be completely repaired and the conductivity can also be restored basically. When the material is subjected to pressure, the distance between the nanoparticles is reduced and the conductivity changes, so it has a high sensitivity to pressure.

The study of artificial skin does not stop at converting the pressure on skin into electricity. The ultimate goal is to transmit this perception to the human central nervous system, so that people with artificial limbs can truly restore the sense of touch. The reason why our skin can feel pressure is that there are mechanical stimulating receptors on one kind of sensory neurons, which have been lost for those with artificial limbs. In 2015, Professor Bao's team, together with the famous Stanford neuroscientist Karl Deisseroth's research group, published in the journal Science that electronic devices with mechanical stimulating receptors' function made of organic electronic materials and nanomaterials can transform pressure into electronic and optical signals at different frequencies and then allow neurons to truly feel them.

Flexible electronic devices can also be used to make artificial organs. This is very important for both robots and humans who have lost certain organs. For example, the artificial eye is actually an array of flexible photosensitive sensors, the artificial nose is a very delicate gas sensor, and the artificial tongue is a liquid sensor that can sense various flavors. By applying different materials, these sensors can even sense elements that can not be perceived by human organs. Robots with these super organs can even feel pollutants in the air, nutrients in food and beverages, and see wavelengths other than visible light.

Energy Conversion Device

Nowadays, the problems of energy are becoming increasingly prominent. It is of great significance to study how to use new energy and collect energy. If we use flexible materials to make energy conversion devices, the energy efficiency can be greatly improved and our lives subverted .

It is believed that everyone is familiar with triboelectrification. One material is more likely to lose electrons, while the other material is easier to get electrons. Materials with these two characteristics can be charged by rubbing against each other. When the two materials are in contact, the positive and negative charge centers are on the same plane and in a neutral state, then there is no apparent electricity. If an external mechanical force is applied to separate the two materials, the positive and negative charges will be permanently retained on the material, then an electric field will be formed between the two materials. If

a metal electrode is plated on the back of the two materials separately, the electric field will transfer the electrons between the metal electrodes, thus forming an electric signal to the external circuit. The principle is applied to flexible circuits, and electrical signals can be used for various mechanical sensing motions. For example, the smart keyboard developed by the team led by Professor Wang Zhonglin of Georgia Institute of technology is one of the representative applications. Since human skin tends to lose electrons, so if the surface of smart keyboard is made of flexible materials that can easily obtain electrons, there will be electron transfer when the fingers touch the keyboard—fingers positively charged and keyboard negatively charged. If the finger leaves the keyboard, the electric field will change. This will cause a potential difference between the metal electrodes installed inside the smart keyboard, which will cause electronic transfer between the electrodes, and the number of transferred electrons is closely related to the strength of the fingers tapping the keyboard and the contact surface between the fingers and the keyboard. Therefore, those using the keyboard can be identified and the alarm system will be triggered if an unauthorized person uses the machine. Since most materials in nature are easy to obtain or lose electrons, this kind of equipment has a wide range of materials to select, so that the cost is low. Additionally, the process is not complex and easy to promote.

In addition to triboelectrification, many of the originally wasted energy can also be converted into electrical energy by means of flexible devices. For example, the solar cell made of flexible materials can be spread on uneven surfaces to convert solar energy into electrical energy based on the characteristics of light weight and high adhesion of flexible materials, which can not only greatly improve the utilization of solar energy, but also facilitate workers to carry and reduce the surface load. If we wrap a layer of flexible solar cells around a schoolbag or camping tent, the solar energy can be used to charge the mobile phones and computers. There are also flexible thermoelectric devices that convert heat energy into electrical energy through the temperature difference. The flexible devices are wrapped on hot-burning factory machines, or on the kitchenware used by households, which can not only preserve heat, but also obtain the heat energy that would have been wasted. Flexible devices can also be installed in all areas of life where there is mechanical movement, including roads, wheels, sports equipment and even shoes, to make full use of all scattered energy.

Even the human body boasts a lot of energy that can be used. The beating of heart and the rise of body temperature during exercise can be converted into energy through flexible materials. We can wear a kind of super clothes, which can separate materials and generate

14

electricity through friction when the human body moves. In the future, mobile power may be obsoleted because the human body itself is expected to become a small mobile power station, and everyone can power their own electronic devices.

Flexible Material Preparation Method

How exactly are these magical functions achieved? How are the flexible materials made? This part will uncover the mystery of flexible materials. Traditional electronic devices and circuits are made of hard semiconductor material—silicon and metal conductors. If we want to make them flexible, there are four ways:

The first method is to select electronic materials with strong strain resistance. The so-called strain refers to the deformation of an object under the action of external forces. The strong strain resistance above-mentioned refers to the fact that after the material being stretched, not only does its structure remain undamaged, but the electrical properties will also remain unchanged. At present, new electronic materials with strong strain resistance include organic small molecules, polymers, carbon nanotubes, silicon nanowires, etc. Another feature of these new electronic materials is that they can be processed by solution and then made into devices by printing. Compared with the traditional circuit production process, the printing method not only saves a lot of cumbersome production steps and greatly improves the production efficiency but also reduces the cost for the reason that high temperature is unnecessary in the production process. However, there are still some technical problems to be overcome in the large-scale production and application of these new materials, such as poor stability in non-vacuum air and humid environment, complex synthesis process and low repeatability.

The second way is to thin the electronic devices. The strain that the device can bear is primarily related to the bending radius and the thickness of the device. So if the thickness of the material is much thinner than the folding radius, the strain borne by the material will not be so large. For example, if a material with a thickness of 1 mm is bent to a radius of 1 cm, it should be subjected to a strain of 5%. If the thickness of this material is reduced by a factor of 1000, that is, only 1 μm, the strain subjected to bending it to 1 cm should be only 0.005%. But if we thin the device, the electrical properties of the material and the device itself will be affected. The process is very complicated and the cost will increase correspondingly. In addition, most of the traditional semiconductor materials (silicon, gallium arsenide, oxide, etc.) are easily broken even under small strain.

The third method is to place some fragile electrical materials in the middle of the

15

plastic substrate. Generally, in the case where the substrate is bent, the tensile strain on the upper surface is the largest, and the lower surface is subjected to the compressive strain, so the strain on the middle part of the plastic substrate is the smallest. Through the design of the sandwich structure, the key fragile materials are placed in the middle of the plastic substrate, which can greatly increase the bending radius of the material, so as to achieve the design purpose of arbitrary folding and bending.

The fourth method is to make a mesh structure. Compared with continuous linear or planar structures, the mesh design allows the intermediate cavity to bear most of the strain when the material is bent. Moreover, the whole structure will be softer and more easily bent. In the transverse connection part, some bending structures, even some protruding structures are designed, so that the plastic under the connection part can bear the strain caused by the bending of the substrate. But the bending structure will make the manufacturing process even more complicated. Furthermore, in the same space, the density of devices fabricated with the mesh structure will be much smaller, which will affect the final integration of a large number of devices.

The techniques such as flexible display and artificial electronic skin highlighted above are only parts of flexible electronics or flexible circuits. such technologies boast an extremely broad application prospect in the fields of energy, medicine, environment, wearable devices and so on. The extraordinary abilities owned by the superheroes in movies are being realized from the screen into reality. The way to realize our dreams is exactly through endless imagination and unremitting scientific exploration. Flexible electronics, like a bright star rising in the scientific sky, shines with the wisdom of human beings to explore the unknown and push the limits.

单词释义：

1. 柔性电子　　　　flexible electronics
2. 晶体管　　　　　transistor
3. 发光二极管　　　light emitting diode
4. 集成电路　　　　integrated circuit
5. 显示点　　　　　pixel
6. 碳纳米材料　　　carbon nanomaterials
7. 导电凝胶　　　　electrographic gel
8. 神经元　　　　　neuron
9. 人造皮肤　　　　artificial skin

10. 树突　　　　　　　dendrite

11. 轴突　　　　　　　axon

12. 介电层　　　　　　dielectric

13. 摩擦生电　　　　　triboelectrification

14. 电势差　　　　　　potential difference

15. 卷曲半径　　　　　bending radius

16. 伸张应变　　　　　tensile strain

17. 压缩应变　　　　　compressive strain

18. 阀门　　　　　　　valve

19. 电子器件　　　　　electronic devices

20. 钨丝灯泡　　　　　tungsten light bulb

21. 被动矩阵型　　　　passive matrix type

22. 主动矩阵型　　　　active matrix type

23. 石墨烯　　　　　　graphene

24. 心脏律动　　　　　cardiac rhythm

25. 心电图　　　　　　electrocardiogram

26. 心脑血管疾病　　　cardiovascular and cerebrovascular disease

27. 药物缓释材料　　　drug delivery material

28. 胰岛素　　　　　　insulin

29. 金属细丝　　　　　metal filament

30. 帕金森病　　　　　parkinson's disease

31. 癫痫　　　　　　　epilepsy

32. 皮质　　　　　　　cortex

33. 电阻抗　　　　　　electrical impedance

34. 电容值　　　　　　capacitance value

35. 神经细胞　　　　　nerve cells

36. 神经中枢　　　　　nerve center

37. 高分子　　　　　　polymer

38. 砷化镓　　　　　　gallium arsenide

39. 塑料基底　　　　　plastic substrate

40. 网眼结构　　　　　mesh structure

译文解析：

1. 汉语无主句的翻译

汉语无主句是汉语中特有的一种句型,即只有谓语部分而没有主语的句子。因为汉语注重意合,只要意思表达正确即可;而英语注重形合,句子一般都不能缺少主语。汉语无主句一般有两种处理办法:(1)增补主语;(2)转换句型。

原文:手指接近火苗感到灼热刺痛的瞬间快速缩回手指,这看似简单的动作却包含大量人体神经细胞的精准运作。

译文:The moment your fingers approach the flame, you can feel the burning and tingling and then quickly retract your fingers. This seemingly simple action contains the precise operation of a large number of human nerve cells.

分析:这里采用的是增补主语的方法,原文中没有主语,翻译时添加了隐含主语 you,既符合英语的表达习惯,又拉近了与读者之间的距离。

原文:如果施加一个外界机械力,使得两种材料分离,正负电荷就会永久地保留在材料上面,两种材料之间就会形成电场。

译文:If an external mechanical force is applied to separate the two materials, the positive and negative charges will be permanently retained on the material, then an electric field will be formed between the two materials.

分析:这里采用的是语态转换的方法,将原文的主动语态译为被动语态,这也符合科技英语常用被动语态的特点,使得译文更加客观、说服力更强。在译文中,我增加了一个副词 then,从而使译文更加连贯,可读性更强。

2. 汉译英时词性的灵活转换

汉译英时的词性转换,即将汉语中的某一词类译成英语中的另一词类。由于汉语和英语的表达方式不同,我们在翻译时不能逐字逐句的套用结构。适当的改变原文中某些词语的词性才能有效地传达原文的意思,也能使译文符合译入语的特点与表达习惯。

原文:对于植入人体的器件来说,柔性器件具有更大的优越性。

译文:As for devices implanted in human body, flexible devices are more advantageous.

分析:原文中"更大的优越性"是形容词修饰名词,翻译时较灵活地处理成"more advantageous",形容词转换为副词,名词转换为形容词,表达非常简洁。

原文:但大规模生产应用这些新型材料仍有一些技术问题需要克服,比如说在非真空空气中和潮湿环境下的稳定性不够好,合成的过程比较复杂,重复性不高等。

译文:However, there are still some technical problems to be overcome in the large-scale production and application of these new materials, such as poor stability in non-vacuum air and humid environment, complex synthesis process and low repeatability.

分析：原文中"大规模生产应用"是副词修饰动词，翻译时较灵活地处理成
"large-scale production and application of"，副词转换为形容词，动词转换为名词，表达
非常地道，符合英语是静态语言，常用名词这一特点。原文的"不够好"译为"poor"，
"不高"译为"low"，原文从反面表达而译文从正面表达，反说正译，思维转换非常灵
活，译文更简洁、地道。

3. 定语翻译的后置处理

汉语中定语是修饰和限制名词或代词的，一般放在名词或代词前，一个句子甚至
多个句子修饰一个词时也都是前置定语；而英语中，词组和句子在修饰一个单词时往
往后置定语。

原文：在20世纪70年代的科幻电影《无敌金刚》里，主人公因飞机事故失去了
自己的肢体和一只眼睛，机缘巧合地装上了人造皮肤覆盖的机械肢体和用摄像头做成
的眼睛

译文：In the science fiction movie- The Six Million Dollar Man in the 1970s, the
protagonist lost his limbs and an eye due to an airplane crash. By chance, he was fitted
with mechanical limbs covered with artificial skin and an eye made with a camera.

分析：本句定语较长，后置翻译更符合英语的表达习惯，可读性更强。

4. 汉译英时注意找出隐含的逻辑关系

因为汉语是意合语言，句子之间往往是隐性连接。句子各成分之间不通过语言形
式手段连接，而是通过语言内在的逻辑关系和叙事的时间顺序紧密结合。汉语注重语
义连贯，句子各成分之间不用或少用关联词，句子结构较为松散；而英语是形合语言，
句子之间往往呈显性连接。即用我们看得见的语言形式手段来连接词语或句子。这
种构句特点往往注重句子形式，以形显意。因此，英语大量使用关系词和连接词。汉
译英时，要注意找出隐含的逻辑关系。

原文：皮肤下遍布神经细胞网络，皮肤感受到的压力有任何细微的变动，都会通过
神经细胞树突和轴突间的联系将信号传入神经中枢，送达大脑，让我们迅速做出反应。

译文：Under the skin, there is a network of nerve cells. Any slight changes in the
pressure felt by the skin will send signals to the nerve center through the connection
between the dendrites and axons of nerve cells, and then to the brain, so that we can react
quickly.

分析：翻译时增加了时间副词 then，增加了译文的连贯性。分析原文可以看出
"信号传入神经中枢，送达大脑"的结果是"我们迅速做出反应"，因此译文中增加了 so
that 来引导结果状语从句，译文逻辑关系得以凸显。原文"皮肤感受到的压力"翻译为

19

" the pressure felt by the skin ",也是定语后置翻译,符合英语的表达习惯。

原文: 柔性器件还可以安装在生活中所有有机械运动的地方,公路、车轮、运动器材,甚至鞋子,充分利用一切零散的能量。

译文: Flexible devices can also be installed in all areas of life where there is mechanical movement, including roads, wheels, sports equipment and even shoes, to make full use of all scattered energy.

分析: 原文"生活中所有有机械运动的地方"译为 also be installed in all areas of,也是定语后置翻译。这些地方与"公路、车轮、运动器材,甚至鞋子"是包含关系,而原文中没有体现出来,因此汉译英时增加了"including"。一词。"柔性器件还可以安装在生活中所有有机械运动的地方"是为了"充分利用一切零散的能量",因此汉译英时增加了介词"to"来体现原文隐含的目的关系。

5. 汉译英时的语态转换

汉语常用主动语态而英语常用被动语态,这一点在科技英语中体现得尤为明显。一方面使得叙述更加客观,避免作者个人的主观臆想;另一方面避免了不必要的人称代词,使结构更加严密紧凑。

原文: 柔性器件在生物医学方面也有非常广泛的应用。

译文: Flexible devices are also widely applied in biomedical field.

分析: 汉语的主动语态译为英语的被动语态,符合英语的表达习惯。

6. 汉译英时的去形象化处理

汉语表达多用比喻,十分形象。但有时按照中文的表达译成英语不免会使目的语读者感到困惑,这种情况下可以去形象化处理,只要翻译出中文想要表达的意思即可。

原文: 在一次展示中,由手机控制的录像就在这个薄如蝉翼的显示屏中播放,没有一丝缺陷。

译文: At an exhibition, the video controlled by the mobile phone was played in the thin display screen without any defect.

分析: 若直译"薄如蝉翼"会使读者感到困惑,因为英语国家文化中没有这种表达,因此这里我只翻译成"thin"便于读者理解。

原文: 这个部分将揭开柔性材料的神秘面纱。

译文: This part will uncover the mystery of flexible materials.

分析: 中文表达十分形象优美,但若把面纱翻译出来不免有些突兀,造成目标语读者的理解障碍,因此我将"神秘面纱"简译为"mystery"以便于目标语读者理解。

第二章　雷达照进商业

说起雷达,人们一般会联想到巨型天线,或者各类军用地笨重装置,似乎和日常生活关系不大。然而 2015 年在谷歌 I/O 中亮相的 Project Soli 迷你雷达,令人眼前一亮。该雷达芯片以及全部天线合成在一起也不过指甲般大小,这样的尺寸,使得它完全可以嵌入可穿戴设备以及其它各种微型装置中,其商业应用也让人充满了想象。说到雷达的应用,涵盖军事、科研、家居、娱乐等多个领域,各种新功能也是层出不穷,令人眼花缭乱叹为观止。本文将分几节给大家详细聊聊雷达背后的原理及其商用前景。

雷达的前世今生

雷达,英文叫 Radar(RAdio Detection And Ranging),其基本原理是利用发射"无线电磁波"得到反射波来探测目标物体的距离、角度和瞬时速度。雷达的雏形在自然界早已存在:比如蝙蝠或者海豚便是利用声音的反射波(也称声纳)定位。

19 世纪中叶,麦克斯韦建立了电磁场方程,为整个无线通信以及雷达应用奠定了理论基础。该方程完整地描述了电场的变化如何导致磁场的变化,磁场的变化又如何导致电场的变化,从而产生了所谓电磁波概念。随后不久,赫兹就通过实验证实了电磁波的真实存在。1904 年,Christian Huelsmeyer 首先提出将电磁波应用于雷达,利用电磁波反射来探测海面上的船只。1922 年,无线电之父马可尼也将雷达的概念完整地表述了出来,他们都可以算作现代雷达的开山鼻祖。

雷达技术真正突飞猛进,是在第二次世界大战时期。这个时期无论英美还是德国都在积极研制更精准的雷达用以实时定位对方的飞机船只。德国的 Freya 雷达以及英国的 CHAIN HOME 雷达阵列都是比较早期投入军事侦察的应用示例。这些军事雷达对于第二次世界大战的走势和战局都起了关键性的作用。战争的较量,在很大程度上即是战争背后各国军事科技的较量。能够提前掌握对方的军事动态并且做出预判,从而有效干预,是战场上的制胜法宝之一,所谓知己知彼方能百战不殆。

二战后,原本只是用于发现和跟踪导弹的雷达就没有了太多用武之地。于是许多雷达技术就逐步从军用转为科研和民用。比如卫星遥感雷达、气象雷达、深空探测雷达、警察在高速公路旁经常使用的测速雷达、生命体征检测雷达、探底金属雷达、穿墙透视雷达等，甚至专门用来接收外星人讯号的雷达，不胜枚举，各类应用简直可以汇总成一个雷达"百货店"了。随着天线尺寸和芯片的极度缩小，在可预见的未来，更多的雷达设备将会以卫星器件面世。就比如前文提到的 Project Soli 项目，它们能嵌入可穿戴设备，成为物联网的一类重要传感器。随着技术的普及，也将逐渐走入寻常百姓家，为人们的生活起居带来方便。这种改变时革命性的改变，原因在于雷达具有许多其他技术无法替代的功能。

雷达基本特性

相比于其他隔空检测或者体感技术，例如体感相机、超声波等，雷达有着一些天然优势：首先时稳定性强，无论白天黑夜、暴晒寒风，雷达皆可正常工作；其次是制造起来相对容易且硬件成本低；最后是功能强大，高频雷达测量物体距离通常可以精确到毫米级别；而低频雷达则可以做到"穿墙而过"，完全无视遮挡物的存在。这些特性让雷达，尤其是微型雷达，在未来都有着广阔的应用前景。

首先要解释的是电磁波频段本身。一般雷达工作的频段从 3MHz 到 300GHz 不等。不同频率的电磁波易受到大气环境的影响。大气中的水蒸气和氧是电磁波衰减的主要原因，当电磁波频率小于 1GHz 时，大气衰减可忽略。一般规律是：频率越高，传输损耗受天气影响越大。所以低频波段比较适合远距离物体探测，但是精度不高；高频波段定位精度较好，但是作用距离较短。需要按照不同应用场景来选择相应的频段。

就拿谷歌的 Project Soli 来说，它的中心频率选择再 61.25GHz 左右。如此选择的好处是该频率可以捕捉到细微的手指动作，精度可以达到 mm（毫米）左右。但是由于低功率的需求，Project Soli 的作用范围不超过 1 米。同时该频段（61～61.5GHz）属于 ISM（Industrial, Scientific, and Medical）频段，不需要特殊执照可以免费使用。关于频段的使用，各个国家都有着严格的规定，对于商业用途而言，购买某一个特殊频段的使用权通常要花费巨大资金参与竞标，动辄数十甚至上百亿美元。所以免费范围的 ISM 频段通常是商业雷达的第一选择。频段选择是一个非常复杂的话题，这里就不详细叙述了。有兴趣的朋友可以参考相关的规定，比如美国的 FCC Regulation（联邦通信规则）。

透视眼与多路径效应

很多人认为雷达可以轻易地越过障碍物，穿透云层、墙壁和人体。这点并不完全正确。雷达是否能穿墙隔空探测物体，取决于墙本身的材料以及雷达频率的选择。首先是频率：3GHz 地电磁波能穿透 10cm 厚的墙，而 60GHz 雷达如 Project Soli 雷达恐怕连一张薄薄的纸都无法穿透。此外墙本身的材料也很重要,同样是 10cm 厚的墙,如果是一般的土砖或者木头制成的,就很容易穿过,而贴墙就难以逾越。

相对于穿墙,雷达波有时却可以"绕"过墙看到墙背后的物体,这其实是利用了电磁波的多次反射(Multipath,也称作多路径效应)。在某些特定的场景中,它可以成为雷达的一类特殊应用,比如利用多路径效应来检测视线不可及之处有否藏有异物。当然有时也会出现"Ghost",也就是噪音,雷达会探测到一些根本就不存在的物体(Ghost Object),这往往也是由于多路径效应造成的。

电磁波有时可以"绕墙而过"看到隐藏在背后的物体;多路径效应有时会令雷达误以为识别了不存在的物体;解决的方法通常是在信号处理曾检查返回信号的强度以及相位差来判断该反射信号是否来自真实存在的物体。

天线与发射信号图样

一般电磁波是以球面波或者至少在某一个平面上均匀地向外辐射出去的(omnidirectional),这对于一般通信而言是极好的,因为它可以保证通信在各个方向上都畅通无阻。然而对于雷达的特定功能来说就显得不够了,通常雷达需要能够电磁波朝某个方向上发射出去,这样才能"有的放矢",而这就需要特殊的有向天线设计了。

在给定发射频率情况下,天线的有向性(即波束发散角度)同天线的面积成反比。这也就能解释为什么深空探测的雷达天线要做得那么大。天线设计本身是一门非常精深的学问,我们这里的介绍知识抛砖引玉。它通常有一个突起的主轴(main lope)和周围的一些小突起(side lobe)。主轴的宽度决定了天线的波束宽度,而周围的小突起则一般作为噪音来处理。

一般有向天线的三维电磁辐射图样。基本的图样是一个主轴附带一些效地突起。主轴的宽度决定了天线的有向性。比如该图的波束宽度大约在 60 度左右。可以通过增大天线面积,增多发射器的个数,或者提高发射频率来减小波束的宽度。

雷达的组成

雷达一般由发射器、接收器、发射/接收天线、信号处理单元,以及终端设备组成。发射器通过天线将经过调频或调幅的电磁波发射出去;部分电磁波触碰物体后被反射回接收器,这就好比声音碰到墙壁被反射回来一样。信号处理单元分析接收到的信号并从中提取有用的信息,诸如物体的距离、角度以及行进速度,这些结果最终被实时地显示在终端设备上。传统的军事雷达还常配有机械控制地旋转装置用以调整天线地朝向,而新型雷达则更多通过电子方式做调整。

为节省材料和空间,通常发射器和接收器共享同一天先,方法是交替开关发射或接收器避免冲突。终端设备通常是一个可以显示物体位置地屏幕,但在迷你雷达的应用中,更多是将雷达提取的物理信息作为输入信号传送给诸如手表等电子设备。信号处理单元才是雷达真正的创意和灵魂所在,主要利用数学物理分析以及计算机算法对雷达信号过滤、筛选,并计算出物体的方位。在这基础之上,还可以利用前沿的机器学习算法对捕捉到的信号做体感手势识别等。

测距与测速

目前雷达的基本功能仍然是测距和测速。例如警察执法中通常会使用测速雷达来判断车辆是否超速。测距和测速背后的基本原理并不难理解。就拿测距来说吧,最简单的做法就是发射一个脉冲波,并等待其返回接收器。因为电磁波是以光速行进的,那么通过测量等待时间就可以间接地获取距离啦,是不是很简单呢?

电磁波遇到障碍物后,大部分能量散射到空间各个角落,小部分能量被反射回接收天线。通过精确测量发射到接受回波地时间,就可以推断雷达到物体地距离。

当然,发射脉冲对于发射机的峰值功率有较高的要求,并且电路实现相对复杂。比较普遍的低功耗获取距离信息的方法,是对发射信号的频率做调剂。此类雷达的术语叫做 FMCW(Frequency Modulation Continuous Wave,调频连续波),操作方法是发射一个线性调频信号(chirp)。因为频率与距离的关系是线性的,通过检测反射波与发射波当前的频率差异即可推断物体的距离。笔者估计谷歌 I/O 发布的 Project Soli 就是一款基于 FMCW 的微型雷达。FMCW 在目前的商用中是极其普遍的,主要源自于它对带宽要求低、功耗较低,以及电路设计相对容易实现。除此之外还有超宽频(UWB)雷达,在此就不多介绍了。

雷达的另一项优势是可以测量物体的瞬时速度,这就要提到物理中鼎鼎大名的"多普勒效应"了。其大意是说,反射波的频率会因为物体进行的速度改变而改变。一个经典的例子是有关声波的传播。远方疾驶过来的火车鸣笛声因为火车速度变快而变得尖细(即频率变高),而远去的火车鸣笛声因为火车速度变慢而变得低沉。那么,利用此规律,只需洞悉了频率变化就可以推断物体的速度了!事实上,上面提到的 FMCW 雷达可以同时提取物体的距离和速度,可谓一箭双雕。具体的算法细节可以参考文献。

手势识别

前面所讲的测距或者测速都把物体想象成一个抽象的点。而真实的物体,如人的手则可以认为是一堆三维点的集合。将雷达用于近距离识别各类手势是一个较新的研究领域。在这之前很多人都尝试过使用相机来做手势识别,问题是相机成本较高,需要一个较好的镜头才有可能实现,同时耗电量较大,并不适合放置在可穿戴设备上。而微型雷达在理论上可以做到低功耗、低成本,镜头也不会突兀在设备外面。

从 Project Soli 公开的资料来看,它主要是通过分析雷达反射信号在时间轴上的变化来区分不同的手势,这些手势可以是微笑的手指舒张缩放、手掌的张开合拢,或者是手指的前后位置摆放。雷达的反射波中已然蕴藏了手上许多个点的距离与速度信号。同时呈现这些信息的一个好方法叫做距离—多普勒映射(Range-Dopler Map),简称 RDM。RDM 中的横轴是速度,纵轴是距离。它可以认为是一张反射波能量的分布图或概率图,每一个单元的数值都代表了反射波从某个特定距离到达以某个特定速度的运动的物体,所得到的反射波能量。利用 FMCW 雷达构建 RDM 是极其容易的,只需要通过二维的傅里叶变换即可。RDM 中已然可以窥见探测物体的特征运动身形。基于 RDM 及其时间序列,我们可以采用机器学习的方法识别特定的能量模式变化,进而识别手势及动作。其实在 Project Soli 推出之前,Nvidia(英伟达)也做过和 Soli 十分类似的研究。

相位阵列与定位

除了简单的手势识别之外,雷达还可以用来定位。无论测距、测速,或者手势识别,都不能精准地指出物体所在的三维位置。要实现定位也不难,最简单粗暴的做法就是利用一个有向天线和一个机械旋转装置,通过不停地旋转天线来扫描天空地各个位置。

这种通过机械方式旋转天线的方法,对于移动产品来说显得很笨重,耗电量大且不方便。一个聪明又有趣的解决办法是通过“相位阵列”以电子的方式调控天线的合成方向,也被称为波束成形。其主要原理是使用多个发射器,通过调整波形的相位和波形间的向长和相消干涉(constructive and destructive interference),来控制合成发射波的朝向。更简单地说,就是“打时间差”。为便于理解,不妨想象一下水波之间的干涉条纹。如果可以自由任性的调整天线朝向,再配合上测距的原理,雷达就可以实现自动定位啦!

定位的另一个常用方法是使用多个接收器!因为多个接收器受到的反射波的相位略有不同,通过测量它们之间的相位差即可做定位。从介绍上看,谷歌的新款的迷你雷达 Project Soli 拥有 2 个发射器和 4 个接收器,这样就可以同时利用波束成形和相位差的方法做手掌的定位、跟踪和手势识别。

雷达的其他各种神奇应用

照妖镜

Xethru 公司提供的利用雷达隔空探测呼吸节律的方法,和 Project Soli 雷达有着本质的不同。Xethru 雷达基于的是超宽频技术,而 Soli 雷达使用的是窄频技术。超宽频雷达通过发送与接收非常短的脉冲,可以探测极其细微的动作。一般而言,雷达能够感知的动作细微度与使用的宽带成反比,Xethru 使用的 3.1~10.6GHz 的频段,将感知精度又带入了一个新的层次。该频段的电磁波可以轻易穿透墙壁或者衣服,甚至可以隔空检测人的心跳。假想不久的将来,警察局里将用上新型雷达测谎仪,隔空测心跳来判断嫌疑犯是不是在撒谎;相亲派对上的技术宅男们,也能通过雷达判断对面的美女是不是对其有意而"怦然心动",采取行动猛烈追求……目前 Xethru 的最大问题是如何在人移动的情况下检测呼吸状况。根据笔者推算,它目前的应用场景可能还是在假定人在静止不动的情况下。当然,即使如此也不错了,至少可以做到像巫师的"照妖镜",隔空区分"是人是妖"吧。

掘地机

探地雷达也是一项很有意思的发明,它可以利用电磁波穿透土壤的特性,窥探泥土底下隐藏着什么不可告人的秘密,电影《侏罗纪公园》中就又这样一个应用场景。除了发现地底下的管道、化石,据说还有人发现了金子哦!如果你家里有一个小院子,不妨试试看,说不定有惊喜。

探地雷达:电磁波穿透岩土遇到不同的物体比如水管、岩石、化石,或是金子会反射回来。改图选自《侏罗纪公园》,男主人公在领导一个小组勘探恐龙化石。

参考译文:

Radar in Business

When it comes to radar, it will usually reminds people of giant aerial or different kinds of heavy devices and it seems to have little relation with our daily life. However, the mini radar, Project Soli showed up in the 2005 Google I/O conference and enlightened us. The combination of the radar chip and all the antennas is no larger than one fingernail. This size enables it to be embedded in wearable devices or other various miniature devices and the commercial application is also imaginable. When it comes to the application of radar, it contains military, scientific research, house furnishing, entertaining and etc. Different kinds of new function has continuously come into being, making us dazzled and gasp in admiration. This text will elaborate the principles behind radar and its commercial prospect.

The Past and Present of theRadar

Radar, with its English name Radar (Radio Detection And Ranging), the basic principle is receiving reflected wave by sending radio electromagnetic to detect the distance, angle and instantaneous velocity of the object. The rudiment of the radar already existed in nature: Such as bats and dolphins, they locate by the reflected wave of the sound (aka sonar). Different from sonar, radar uses electromagnetic. It can work in vacuum smoothly without the existence of any medium.

In the mid-nineteenth century, Maxwell has set up the electromagnetic equation, laying the theoretical foundation for the whole radio communication and the application of radar. That equation entirely describes how the change of the electric field leads to change of the magnetic field and how the change of the magnetic field leads to the change of the electric field, thus generating the so-called electromagnetic concept. Soon after, Hertz has verified the actual existence of electromagnetic wave through experiment. In 1904, Christian Huelsmeyer has initially proposed to apply electromagnetic wave into radar and use the reflection of electromagnetic wave to explore the sea vessels. In 1922, Guglielmo Marchese Marconi , the initiator of the radio, has also stated the concept of the radar entirely. All of them can be regarded as the initiators of contemporary radar.

Outstanding progress in radar technique has been made during the Second World War. At the time, Britain, America or Germany were actively developing the radar which is more precise to locate others' aircraft and ships simultaneously. Both the German Freya radar array and the British CHAIN HOME radar array are the example of the application of the military scout at the relatively early stage. To a great extent, the contest of war is the contest of the military technology of each country behind the war. The capability of getting hold of the military dynamic and make judgement of your opponent in advance to intervene is one of the key to success in the battlefield, and that's what is called "Know yourself and know your enemy, you will win every war."

After the Second World War, the radar only used for discovering and tracking missile originally has seldom been used. Therefore, the usage of many of the radar technologies have been changed form military-used to civil use. Such as satellite remote sensing radar, meteorological radar, deep space detection radar, the velocity radar are often used by the police by the freeway, vital signs detection radar, ground-penetrating metal detection radar, wall-penetrating radar and so on. Moreover, there is even the radar specially used for receiving the signal from alien. With numerous types of its usages, they can be summarized

as a "radar store". With the extreme shrinking of the antenna and chip in size, more radar devices will emerge in the form of satellite devices. Like the Project Soli mentioned before, they can be embedded into wearable devices and become one kind of important sensor in the Internet of Things. With popularization of technology, these devices have been widely used and bring much convenience for our daily life. These changes are the revolutionary ones, for which radar contains the functions which can not replaced by many other technologies.

Characteristics of Radar

Compared to other technology of detection through interval or motion-sensing technology like motion-sensing camera, supersonic and ect., radar has some inherent advantages: First of all, it has strong stability as it can work normally all day in all weather conditions. Besides, its manufacture process is comparatively easy and the cost of its hardware is low. Lastly, with its advanced function, measurement with HF radar of the distance between objects can be accurate to millimeter while low frequency radar can receive and send signal through walls, which is entirely free from obstacles. These characteristics bring radar, especially mini-radar a future with broad application.

The Frequency Band and Alternatives of Electromagnetic Wave

What should be explained first is the frequency band of the electromagnetic wave itself. The working frequency band of common radar ranges from about 3 MHz to 300 GHz. Electromagnetic waves with different frequency are easy to be affected by atmosphere environment. The water vapor and oxygen in the atmosphere are the main cause of the decay of electromagnetic waves. When the frequency of the electromagnetic wave is lower than 1 GHz, the decay of the atmosphere can be omitted. The regular rule is that when the frequency is higher, the more weather will affect on the loss in the transmitting process. Therefore, low frequency band is more suitable for detecting objects between long distance, but the accuracy is not high while high frequency band is more accurate with shorter operating distance. Different frequency band needs to be adopted according to different applying occasions.

Take Google's Project Soli as an example, the chosen central frequency for it is about 61.25 GHz. The advantage of this choice is that it can capture subtle finger movement with this frequency. However, because of the requirement of low power state, the operation range of Project Soli is shorter than 1 meter. At the same time, as the frequency band (61 ~ 61.

5GHz) belongs to ISM frequency band, it is for free without special license. There are strict regulation on the use of that frequency band in each country. For commercial purpose, purchasing the right to use of a special frequency band needs to cost a huge sum of money which can be several billion or more than 10 billion to participate in a biding. Therefore, ISM frequency band for free is the first choice of business radar. As the choice of frequency band is a complicated topic , elaboration is to spare here. Those who are interested in it can refer to the relevant stipulation, like American FCC Regulation (Federal Communication Commission).

X-ray Vision and M-Path Effect

Many people think that radar can pass through obstacles like cloud, wall and human body. It is not entirely true. Whether radar can detect objects through wall depends on the material of the wall and the choice of the frequency of the radar. The frequency is to come first: the electromagnetic wave of 3GHz can pass through 10cm-thick wall while radar of 60GHz like Project Soli can hardly pass through a comparatively thin paper. Besides, the material of the wall is also important. For the same 10cm-thick wall, if it is made of common adobe admixture or wood, it is easy to be passed through, but an iron wall is hard to be passed through.

Compared to passing through wall, sometimes radar can pass around to observe the object behind the wall, which is in fact using the multipath(which is also called multipath effect) of electromagnetic wave. In some particular occasions, it can become a special application of radar, like using multipath effect to detect the existence of foreign matter in the spot beyond our sight. It is for sure that sometimes there will be "Ghost" , namely voice. Radar will detect some objects that don't exist at all (Ghost Object) and it is usually caused by multipath effect.

Sometimes radar can pass around to observe the object behind the wall right: multipath effect can mislead the radar that it has observed something nonexistent; the solution often lies in checking the strength of the backward signal and the phase difference on the signal processing layer to decide whether the reflected signal is from an existed object.

Pattern of the Antenna and Sending Signal

Common electromagnetic waves usually radiate as spherical wave or at least radiate omnidirectionally through one plane, which is excellent for common communication, as it

can assure that the communication is unobstructed in each direction. However, it seems to be insufficient for the specific functions of radar as radar usually needs to be able to emit electromagnetic waves to a certain direction in order to have a definite object in view, and special directed antenna design is thus needed.

When emission frequency is provided, the vectorial property of an antenna (the angle that wave beam diverges) is inversely proportional to the dimension of it. That explains why the radar antenna for deep space antenna needs to be made in such huge size. The design of antenna itself is a considerably profound knowledge. So, our introduction here is just using the little to get the big. It usually contains a protuberant main lope and some side lobes. The width of the main lope decides the width of the wave band of the antenna while the side lobes are treated as voice.

The common pattern of three-dimensional electromagnetic radiation of directed antenna. The basic figure contains a main lobe and some small side lobes around it. The width of the main lope decides the vectorial property of the antenna. Just as the wave band shown in the figure, its beam width is about 60 degree. The beamwidth can be decreased by enlarging the area of the antenna, increasing the amount of the emitter or the emitting frequency. An effective way of decreasing the beam width is combining multiple electromagnetic wave emitter into a radar with narrow wavebeam. This method is also called beamforming.

The Formation of Radar

Normally, radar is made with emitter, receiver, transmitting/receiving antenna, signal processing unit and terminal device. The emitter emits the electromagnetic with FM or AM through an antenna; part of the electromagnetic wave will be reflected back to the emitter when it touches objects, just as the sound will be be reflected when it touches wall. The signal processing unit analyzes the received signal and extracts the useful information through it. The result of the objects distance, angle and the proceeding speed will be displayed on the terminal device in the real time. To adjust the direction of the antenna, the traditional military radar will be fixed with mechanical controlled rotating device, while a new radar can be adjusted through more electronic means.

In order to save material and space, the emitter and the receiver will usually share an antenna with the mean of switching on and off the emitter and receiver alternatively in order to avoid congestion.

Normally, the terminal device is a monitor that can display the position of an object.

Meanwhile, in the application of mini-radar, it's more likely that the physical information extracted by radar will be sent as input signal to electronic devices , like a watch. The originality and spirit of radar lie in the signal processing unit where mathematical and physical analysis and the method that algorithm filter, select the signal of the radar before working out the direction of an object are mainly adopted. On this basis, advanced machines can be used to learn algorithm to capture the signal and identify spatial gesture.

Range and Velocity Measurement

The basic functions of radar at present are still range and velocity measurement. For example, the police usually uses the velocity measuring radar to decide whether the driver has exceeded the speed limit. The basic principles behind range and velocity measurement are not hard to understand. Take range measurement as example, the easiest way is to emit an impulse wave and wait until it is reflected to the receiver. As the electromagnetic wave proceeds speed of light, then merely by measuring the waiting time can we acquire the distance. It's quite easy, isn't it?

When electromagnetic wave meets obstacle, most of the energy will scatter to each corner of the space while a small part of it will be reflected to the receiving antenna. By measuring the time from emitting to receiving wave, the distance between radar and object can be inferred.

Absolutely, the requirement of emitting impulse for the peak power of the emitter is high and the circuit implementation is relatively complex. A relatively common way of acquiring distance information with low power consumption is to modulate the emitting signal. The technical term of this kind of radar is called Frequency Modulation Continuous Wave(FMCW). The operation method is to emit a chirp, the waveform is as fighre10. As the relation between frequency and distance is linear, the object distance can be inferred by checking the difference between reflected wave and transmitting wave. The writer assume that the Project Soli issued by I/O is a kind of mini-radar based of FMCW. The commercial use of FMCW at present is extremely common, mainly because of its low requirement to broadband, low power consumption and its circus design is relatively easy to achieve. Besides, there is also ultra wide-band radar, but we are not going into further introduction.

Another advantage of radar is that it can measure the instantaneous speed of an object and what needs to be mentioned here is the famous "Doppler effect" in the physics field. The main idea of the effect is that the frequency of the reflected wave will change with the change of the proceeding speed of an object. A typical example is about the transmission of

31

sound wave. The whistle of a train coming in high speed from afar become brittle and sharp (namely the frequency become higher) as the speed of the train gets faster while it turns deep when it is going far away as the speed of the train gets slower. Then the speed of an object can be inferred merely with the insight of the frequency change. In fact, the FMCW radar mentioned above can extract the distance and speed of an object in the same time, which is killing two birds with one stone. The concrete detail of algorithm can be seen in the reference.

Doppler effect demonstrates: The frequency of reflected wave changes because of the variety of the speed and direction of the object.

Gesture Recognition

The object is imagined as an abstract spot in the text where the distance and speed measurements are mentioned. For a real object, like a human hand can be considered as the collection of a set of three dimensional points. Applying radar to recognize various gestures in short distance is a relatively new research area. Before this many people have tried to recognize gesture with cameras, but the problem lies in the high cost of the camera as a preferable camera lens is needed to achieve that. Meanwhile, the power consumption is large and it's not suitable to be fixed in a wearable device. Theoretically, mini-radar can be low consumption and low-cost and the camera lens will not protrude from the device. On the aspect of public information, it mainly differentiates different gestures by analyzing the changes of radar signal on the time line. These gestures might be the relaxation and contraction of fingers when one similes or the opening and closing of palms or the front or back position of fingers.

The reflecting wave has already contained many distance of spot and speed signals on your hands. At the meantime, a good method to present these information is called Range-Dopler Map which is RDM in short. In RDM, the horizontal axis represents speed while the vertical axis represents distance. It can be considered as a distributional diagram or probability graph of energy of reflected wave. It's very easy to make RDM with FMCW radar as it only needs to be converted through two-dimensional Fourier. In RDM, the characteristic motion figure of the object is already able to be detected. Base on the RDM and its time series, we can adopt the method of machine learning to recognize specific changes of energy pattern and recognize gestures and moves. Before the release of Project Soli, Nvidia has conducted a very similar research.

Phase Array and Positioning

Apart from easy gesture recognition, radar can also be used for positioning. No matter distance measurement, speed measurement or gesture recognition, none of them can precisely point out the three-dimensional location of an object. Positioning is not hard to achieve. One easy and direct way is to use a directed antenna and a mechanical rotating device to scan every area in the sky.

This method of rotating antenna in mechanical way is clumsy for a mobile product. Besides, it's inconvenient with high power consumption. An intelligent and interesting way to solve this problem is to adjust the synthesis direction through phase array in an electronic way, which is also called beam forming. The main principle is using multiple emitters to control the direction of the synthetic transmitting wave by adjusting the phase of the waveform and constructive and destructive interference between waveforms. Simply speaking, it's making use of time difference. For better understanding, we can imagine the interference fringe between ripples. If antenna can be adjusted randomly, guided by the principle of distance measurement, the automatic positioning of radar can be achieved.

Another commonly used way of positioning is using multiple receivers! As the phases received by the multiple receivers are slightly different, positioning can be carried out by measuring the phase difference of them. From the introduction, we can see that the Google new mini-radar type Project Soli contains two emitters and four receivers, which enables itself to perform palm positioning and tracking and gesture recognition by using methods of beam forming and phase difference at the same time.

Other Amazing Applicaton of Radar

Monster-Revealing Mirror

The method of detecting respiratory rhythm through interval with radar provided by Xethru company is substantially different from the radar of Project Soli. The basis of Xethru radar is ultra wide-band technology while what Soli radar adopts is narrow band technology. Ultra wide-band radar can detect extremely subtle moves by emitting and receiving very short impulse. Normally, the subtleness of movement that radar can perceive is inversely proportional to the board band it use. Xethru uses the frequency band of $3.1 \sim 10.6\text{GHz}$, which brings the perceiving accuracy to a new level. The electromagnetic wave of that frequency band can penetrate wall or cloth easily, and it can even check one's heartbeats through interval. Assuming that in the approaching future, the new radar polygraph will be equipped in police stations and the police can decide whether the suspects are lying by checking their heartbeats through interval; the tech geeks in a blind date can also decide whether the beauties in front of them have a crush on them thus wooing the girls hard...At present, the biggest problem is how to check people's respiratory condition when one is moving. According to the speculation of the writer, the application condition of the Xethru at present might be assumed as a situation that people remain motionless. Of course, even so, it's not bad. At least we can use it like a sorcerer using a monster-revealing mirror and distinguish from between human and demon through interval.

Digging machine

Ground-penetrating radar is a very interesting invention which can take advantage of the property of electromagnetic wave that it can penetrate through and pry into the soil to see whether there is something cannot be divulged. It is an application scene in the movie Jurassic Park. Besides pipelines, fossil under the ground, it is said that someone had discovered gold! If you have a yard, you can have a try and there might be a surprise.

Ground-penetrating radar: Electromagnetic wave will reflect back when it penetrates through pipelines, rock, fossils or gold. Replacing figure is chosen from Jurassic Park and the protagonist is guiding a group to detect dinosaur fossil.

单词释义：

1. 天线	aerial
2. 反射波	reflected wave
3. 瞬时速度	instantaneous velocity
4. 声纳	sonar
5. 雷达阵列	radar array
6. 卫星遥感雷达	satellite remote sensing radar
7. 气象雷达	meteorological radar
8. 深空探测雷达	deep space detection radar
9. 生命体征检测雷达	vital signs detection radar
10. 探底金属雷达	ground-penetrating metal detection radar
11. 穿墙透视雷达	wall-penetrating radar
12. 竞标	competitive tender/binding
13. 频段	frequency band
14. 透视眼	X-ray Vision

15. 多路径效应　　　M-Path Effect

16. 波束宽度　　　　beamwidth

17. 调频　　　　　　FM（frequency modulation）

18. 调幅　　　　　　AM（Amplitude Modulation）

19. 电路实现　　　　circuit implementation

20. 带宽　　　　　　bandwidth

21. 横轴　　　　　　horizontal axis

22. 纵轴　　　　　　vertical axis

23. 波束成形　　　　beam forming

24. 干涉条纹　　　　interference fringe

25. 相位　　　　　　phase

译文解析：

1. 汉译英中词性转换

原文： 能够提前掌握对方的军事动态并且做出预判，从而有效干预，是战场上的制胜法宝之一，所谓知己知彼方能百战不殆。

译文： The capability of getting hold of the military dynamic and make judgement of your opponent in advance to intervene is one of the key to success in the battlefield，and that's what is called "Know yourself and know your enemy, you will win every war."

分析： 该句把情态动词"能够"转换成名词"capability"；中英文一个重要的特点就是，中文强调动作，多用动词，而英文多用名词，这是两者之间的重要差异，所以此处为了避免使译文出现"中式思维"，将中文中的动词翻译成英文中的名词，尽量使译文更加地道。

原文： Xethru 公司提供的利用雷达隔空探测呼吸节律的方法，和 Project Soli 雷达有着本质的不同。

译文： The method of detecting respiratory rhythm through interval with radar provided by Xethru company is substantially different from the radar of Project Soli.

分析： 在该例句中，将译入语的"有着本质的不同"这一成分转换成英文的"is substantially different from…"，原因在于，原文中"有"这一动词的动作执行者为"方法"这一无生命的名词，而英文中一般很少用无生命的名词当主语，所以翻译成英文时，转化为主系表的结构，原文为形容词修饰名词的结构，而转化为英文的主系表结构后，因为没有宾语，就采用副词修饰形容词的结构。

2. 长句中的翻译

原文: 相比于其他隔空检测或者体感技术,例如体感相机、超声波等,雷达有着一些天然优势:首先是稳定性强,无论白天黑夜、暴晒寒风,雷达皆可正常工作;其次是制造起来相对容易且硬件成本低;最后是功能强大,高频雷达测量物体距离通常可以精确到毫米级别;而低频雷达则可以做到"穿墙而过",完全无视遮挡物的存在。

译文: Compared to other technology of detection through interval or motion-sensing technology like motion-sensing camera, supersonic and ect., radar has some inherent advantages: First of all, it has strong stability as it can work normally all day in all weather conditions. Besides, its manufacture process is comparatively easy and the cost of its hardware is low. Lastly, with its advanced function, measurement with HF radar of the distance between objects can be accurate to millimeter while low frequency radar can receive and send signal through walls, which is entirely free from obstacles.

分析: 尽管这个句子比较长,但是汉语中用了几个分号隔开,加上"首先、其次,最后"这些描述顺序的词,使整句话条例清晰,主要使用直译的方法,但是,为了使英文也具有条理性,在翻译的时候也使用了非限定性定语从句,使其更加符合英文的表达习惯。

原文: 在这之前很多人都尝试过使用相机来做手势识别,问题是相机成本较高,需要一个较好的镜头才有可能实现,同时耗电量较大,并不适合放置在可穿戴设备上。

译文: Before this many people have tried to recognize gesture with cameras, but the problem lies in the high cost of the camera as a preferable camera lens is needed to achieve that. Meanwhile, the power consumption is large and it's not suitable to be fixed in a wearable device.

分析: 该例子中,较长的是后半句的分析部分,在翻译成英文时,为了避免句子冗长,将其分成两句。

原文: 它可以认为是一张反射波能量的分布图或概率图。每一个单元的数值都代表了反射波从某个特定距离到达以某个特定速度的运动的物体,所得到的反射波能量。

译文: It can be considered as a distributional diagram or probability graph of energy of reflected wave, in which every value in each unit represents the reflected wave that reaches a certain object that is moving at particular speed from particular distance.

分析: 该例子中,中文用了一个无生命名词做主语,同时还用了主动语句,主要时为了符合中文的表达习惯,但是对于英文而言,较常用的时被动语态,所以在翻译时做了语态的转换;同时,第二句也是对第一句进行补充,所以翻译时用了非限定性定语从句,使句式更加紧凑。

第三章 三维成像

计算的交互与图形时代正在到来，在三维、全息、光场里创造"真实"

对眼聚焦和单眼调焦冲突（Vergence-Accommodation Conflict，简称 VAC）：在传统立体显示的过程中，如果三维成像的虚像位置大于或者等于屏幕的时候，人眼判断距离的两大利器，对眼聚焦和单眼调焦的位置是一致的。但是如果显示虚像的距离比屏幕距离短，那么人的对眼聚焦模式会专注在虚像的位置，而单眼调焦会让眼睛不停试图去聚焦屏幕的位置。这一点会造成极大的不适。这个 VAC 也是困扰了业界多年的著名课题，有很多科学家对这个问题进行了专项的研究。

经过多年立体三维电影的发展，一代代科学家和工程师的探索，最好的解决方案逐渐清晰。1968 年 Ivan Sutherland 发明了世界上第一个虚拟现实头显之后，迎来了一个有趣的学生 Edwin Catmull。这位学生毕业之后，创办了一间专注于计算机生成动画的制作室皮克斯（Pixzar）。他和老师一样坚信计算机生成动画会在动画片以及电影界掀起革命，能以低成本制作比较传统特效更精良更真实的影片效果。后来，Edwin 遇到了自己的投资人和合伙人史蒂夫·乔布斯，再后来皮克斯被迪士尼收购，Edwin 成为迪士尼总裁。他执掌的迪士尼开始和洛杉矶南加州大学，以 Mark Bolas 为代表的几位教授合作，研究新一代基于立体显示的虚拟现实系统。

到了 2010 年，距离 Ivan 教授发明达摩克利斯之剑已经过去了 42 年。电子技术发生了翻天覆地的变化：首先是电子计算机的计算速度和当年相比提高了一亿倍以上；其次图像处理芯片的计算功能进步更加迅猛，计算机图形学的算法、图像显示原理的各方面都获得了许多的重大突破；更重要的是，随着智能手机行业的蓬勃发展，制作小尺度高分辨率的显示屏幕成本越来越低。

2011 年，不到 20 岁的 Palmer Luckey 敲开了 Mark Bolas 教授的办公室门，Palmer Luckey 从小就是个 DIY 爱好者，喜欢捣鼓电子设备。他阅读了 Mark 教授发表的论文，提出一个大胆的设想：如果用两块 iphone4 的屏幕固定在一起，安装合适的光学系统和遮光系统,就可以得到一个高质量的双目显示系统,成本一个只需要 150 美元。Mark Bolas 对这个提议感到非常兴奋,他们立刻和迪士尼的科学家一起动手了,花了几个月的时间,成功地展示了一个高质量低成本的双目立体显示系统。Palmer 非常激动,为了筹集资金开始下一阶段的产品生产,Palmer 在美国著名的众筹网站 Kickstarter 上众筹他的产品想法,最终,筹集到 200 多万美元。然而,Palmer 和 Mark 教授的产品并没有解决之前提到的问题。幸运的是,这款产品引起了 John Carmack 的注意。John Carmack 可不是等闲之辈,他是世界上第一批商用图形处理引擎和算法的发起人,也是虚拟现实行业的老兵。至今那些震惊世界的 2.5D 第一人称视角射击游戏,如毁灭公爵系列、雷神之锤系列、古墓丽影系列的早期类 3D 图形处理引擎都是 John 开发的。他非常清楚双目显示系统存在的问题。Palmer 的产品令他非常激。他发现,通过现有的他所知道的软件、硬件技术和这个产品平台配合,便能解决双目显示的部分问题。通过使用合适的光学放大镜,再用算法输出正确的图像,就可以解决视场角失配问题。也可以通过图像处理轻松解决显示屏幕适配尺寸问题。同时,利用手机中普及的加速计和陀螺仪,可以以低成本、高速精确地测量使用者头部的动作,避免了因使用者头部运动产生的图像畸变问题。2013 年 John 和 Palmer 合力研发的第一代原型机面世,叫作 Oculus DK1。这款产品加入传统光学动捕技术,使其具备位置追踪功能,同时把人通过 motion parallax 来判断距离的功能也融合到产品中。这一款产品成为双目三维显示的划时代产品。同一时间,他们众筹了第二代产品 Oculus DK2。

2014 年,社交网络公司 Facebook 以 20 亿美元收购了 Oculus

双目显示系统是对全息显示一种不严谨的近似,人们一直在追求一种对全息显示的无限近似:这样一种显示系统,可以记录和再现任何光线入射的角度、颜色和强度等所有光信息,就好像透过窗户看外面的风景一样自然。那么这样的显示系统要满足什么样的要求呢? 作为一个窗户一般的显示系统,需要将不同景深的物体进行小孔成像,并且用光学传感器记录下来,再通过一个显示系统同时输出所有图像。这样才能像"窗户"一样忠实地显示所有入射光的信息。其实,这样的系统早就开始有人研究了。早在 1900 年,人们研究照相机成像原理的时候就有这样的苦恼:拍摄的照片永远是一个瞬间在一个景深上的聚焦影像。如果想重新聚焦只能回去再照一次。能不能把所有景深的聚焦信息同时抓拍下来? Gabriel Lippmann 想出了一个办法:使用阵列小透镜对不同深度做成像。在原则上就记录了入射光的所有光场信息,称作光场薄膜。

Lippmann 因为在照相机成像上的卓越贡献获得了 1908 年诺贝尔物理学奖。然而他的贡献远远不止于此,他同时培养出了唯一一个同时获得过诺贝尔物理学奖和诺贝尔化学奖的科学家:居里夫人。若干年后居里夫人在 X 射线的成像和操作方面做出了许多卓著的贡献,这和老师 Lippmann 的影响大有关系。

1936 年,Gershun 发表著名论文(A. Gershun. The Light Field. Journal of Mathematics and Physics. 18:1〔1936〕,51{151.})总结和阐述了光场的定义、理论框架和在相机以及显示方面可能的应用。Gershun 发现,光场相机和显示的原理非常简单。在相机方面,光学传感器相机的每个像素都在一个小孔透镜后面,相当于一个小孔透镜阵列和一个小孔传感器陈列的叠加。在这样的框架下,就可以记录完整的光场信息。在显示方面是一个逆过程。在一系列显示像素前面有一个小孔透镜阵列,把显示的每个像素都通过小孔成像的方式投射出去。根据推算,和传统显示相比,假设要显示 1000×1000 分辨率,也就是 100 万像素的图片,传统显示需要的像素是 1000^2,也就是 100 万像素。但是如果用相应的光场显示系统显示同样分辨率的图片,需要 1000^4,也就是一万亿像素,而且这个透镜阵列也需要 100 万个透镜。今天的技术条件下,5 寸屏幕的显示像素仅仅能达到 500 万像素。未来 10 年内做到 1 亿像素是有可能的,但是一万亿像素,以今天的显示技术来说是无法做到的。

以 Lippmann 为首的光学爱好者发现两条规律:

1. 在像素密集的情况下,系统并不需要许多针孔透镜。也就是说,1000×1000 的显示系统中,只需要 10×10 的透镜阵列也能得到非常好的效果。

2. 无论是拍摄还是电脑制作,都不需要收取所有方向来的光,也能得到非常好的显示效果。

这两条近似法则让光场显示在实用化的道路上有了飞速发展。

在 2013 年的计算机图形学权威年会 Siggraph 上,来自图像显示技术供应商伟达的科学家 Doug Lanman,显示了他根据索尼 HMZ-H1 和其他几款商用头戴显示设备改进是光场显示器。通常来说,头戴显示器因为屏幕离眼睛太近,需要安装大口径放大镜配合屏幕使用才可以让人眼聚焦。而 Doug 的系统通过 10×15 的透镜阵列和显示模组构建的光场显示系统,可以让整个设备极其轻薄,眼睛即使贴着显示器也能看到正确深度的图像。这款光场相机由商用的 OLED 双屏系统,和公司定制的 150 个微透镜阵列组成的近似小孔投影阵列贴片制作而成。有效输出图像约 6000 像素。因为眼近所以完全没有对眼聚焦和单眼调焦冲突即 VAC 问题。左右两条鱼的距离不一样。最上面一排是在理想状况理论计算下得到前后两条鱼对焦后人眼视网膜的图像,中间一排是在这个系统中用计算机模拟后的视网膜的图像。下面一排是样品实测的不同焦距下观察得到的图片。可以看到实际拍摄效果中,图片中央的位置两幅图在不同焦距上聚焦的效果和理论值比较接近。

这是光场显示系统的一大进步。工业界意识到,在现有显示技术的基础上稍作改

动就可以得到不错的光场显示效果了。那么这个技术还有什么局限呢?

1.需要更高分辨率的微显示系统。这个样品使用的显示设备是 720P 分辨率。大概只有 90 多万像素。按照之前给出的理论推导,如果做到 2 亿~3 亿像素,就可能在 1080P 有效分辨率下做不错的光场显示近似系统了。

2.需要更大的微显示芯片。现在的显示芯片的尺寸对角线都在 25 毫米左右。经过计算,这样的视场角非常窄。此文使用的样品视场角在 30~40 度。虽然有的制造商通过倾斜芯片来扩大视场角,但是这会使用户的有效视觉面积变小。最好的方法是增加芯片尺寸。然而增加芯片尺寸的最大困难是坏像素造成的良好率低下。因此这也成为光场相机未来发展最大的制约因素。

3.衍射极限。如果像素之间过于紧密就会接近衍射极限尺寸。经过计算,如果像素有效区域间隔小于 1.3 微米,视网膜就会出现模糊。也就是说,这个系统的角度分辨率极限在 30 像素每度(30ppd)。

时至今日,人类文明走过了一万年,文明从涂鸦到绘画,从纸质呈现到电子显示。随着时间的推移,显示技术给我们带来越来越多的惊喜,展现了出一个愈发真实的世界。光场显示系统在实用化的道路上不断迈出了坚实的脚步,技术宅们正在不断用黑科技颠覆传统,变革世界。

参考译文:

Three-dimensional Imaging

The age of computational interactions and graphics is coming, creating "reality" in three-dimension, holographic and optical fields

Vergence-Accommodation Conflict (VAC): during the process of traditional stereoscopic display, if the position of three-dimensional imaging's virtual image is further than or equal to the screen, the location of vergence and accommodation, which are major tools for people to judging distance, is consistent. However, if the situation is that the distance of the virtual image is shorter than that of screen, like the right picture, people's vergence mode will focus on the position of the virtual image while accommodation will let people's eyes restlessly try to focus on the screen, which will cause considerable discomfort. Many scientists has specifically research the VAC which is a famous subject plaguing the whole industry for many years.

With the development of stereoscopic 3-D films over the years, generations of scientists and engineers ceaselessly explored and gradually found out the best solution. After Ivan Sutherland invented the first virtual reality HMD in the world in 1968, he ushered in an interesting student called Edwin Catmull, who founded Pixzar, a production studio specializing in computer-generated animation since graduation. Both Edwin and his teacher convinced that computer-generated animation would revolutionize the animation and film industry, and would produce more sophisticated and lifelike movie effects than the traditional ones at a lower cost. Afterward, Edwin met Steve Jobs, his investor and partner, and then Pixzar was purchased by Disney and Edwin became the president of the company. Under his leadership, Disney begins to cooperate with several professors represented by Mark Bolas at the University of Southern California in Los Angeles so as to research the new generation of virtual reality systems based on stereo displays.

By 2010, it had been 42 years since Professor Ivan invented the Swords of Damocles. Electronic technology had changed overwhelmingly: firstly, the computational speed of electronic computers were over 100 million times faster than ever before. Secondly, the calculation functions of image processing chips had progressed more rapidly, and all aspects of Computer Graphics algorithms and principle of image display had made significant breakthroughs. What's more, with the booming of smartphone industry, the cost of making display screen with small scale and high resolution became lower and lower. In 2011, the door of Professor Mark Bolas' office was knocked by Palmer Luckey of less than 20 years old, who was a DIY enthusiast since he was a toddler and liked fiddling with electronic equipment. After reading the paper published by Professor Mark, Palmer proposed a bolder idea: if two iphone4 screens were fixed together and appropriate optical system and shading system were installed with them, high-quality binocular display system was available at the cost of only 150 dollars. Mark Bolas was so excited about this proposal that they immediately worked with the scientists of Disney, and successfully, displayed a high-quality binocular display system at a low cost spending a few months later. Palmer was very excited. In order to raise funds to start production in the next stage, he crowdfunded his product ideas on Kickstarter, a well-known American crowdfunding website, and finally raised more than two million US dollars. However, the products of Palmer and Professor Mark did not solve the problems of binocular display mentioned above. Fortunately, this product caught the attention of John Carmack.

John Carmack was not a ordinary person, but the initiator of the first commercial graphics processing engines and algorithms and the pioneer of the virtual reality industry.

So far the early 3D graphics processing engines of those world-shaking 2.5D first-person perspective shooting games, including Destroy the Duke Series, Quake Series, Tomb Raider Series and so on, were developed by John. He was clearly aware of the problems of binocular display system, so Palmer's product excited him a lot. John found that through coordination between existing software, hardware technology and this product platform that he had ever known, some of the problem of binocular display could be settled. By making use of suitable optical magnifiers, and then using algorithm to output correct images, the mismatch of field of view could also be facilely corrected. Besides, the matter of screen's adaptation size could be easily tackled by means of image processing. Meanwhile, taking advantage of the accelerometer and gyroscope accessible to the mobile phone could at a lower costs measure the movement of users' head with high speed and precision, so as to avoid image distortion caused by the movement of the users' head.

The first generation prototype jointly developed by John and Palmer was available in 2013, called Oculus DK1. Equipped with traditional optical dynamic capture technology, the product had the function of location tracking. The function that people could determine the distance through motion parallax was also merged into it at the same time. Therefore, it became the game-changing product for binocular 3D display. At the same time, they crowdfunded the second generation Oculus DK.

Social Network Company Facebook Acquired Oculus for 2 Billion Dollars in 2014

Binocular display system loosely approximated to holographic display, to which people ceaselessly pursued an infinite approximation: such a display system can record and reappear all light information of any incidence of light, for example, angle, color and intensity, which was as natural as we saw outdoor scenery from the window. So what kind of requirements should such a display system meet? As a window-like display system, it needs to perform pinhole imaging towards an object with different depth of fields, document it with an optical sensor, and finally, output all picture simultaneously through a display system. In this way, the system can faithfully show all the information of incident light like a "window". Actually, such a system had long been researched. When people studied the image-forming principle of camera, back in 1900, they were troubled: photographs were always focused images of one moment at a depth of field. If you wanted to refocus, we could only take them again. Whether we can capture focus information of all depth of field at the same time.

Gabriel Lippmann came up with a way: using an array of lenslets to image from different depths. In principle, all light field information of incident light was recorded, namely a light field film. Lippmann was awarded the Nobel Prize in Physics in 1908 for his excellent contribution to imaging of camera. However, his dedication went far beyond it. He also fostered Madame Curie who was the only scientist to win the Nobel Prize in Physics and Chemistry at one time. A few years later, the great madame made outstanding contributions to the imaging and operation of X-ray, which had a lot to do with her teacher Lippmann's sincere teaching.

In 1936, Gershun published one of his most famous paper: (A. Gershun. The Light Field. Journal of Mathematics and Physics. 18: 1〔1936〕, 51{151.}), which summarized and elaborated the definition and theoretical framework of light field and its possible applications to camera and display. Gershun found out that the principles of light field cameras and displays were very simple. In terms of camera, each pixel of optical sensor camera was behind a small aperture lens which is equivalent to the superposition of a small aperture lens array and a small aperture sensor display. Under such a framework, complete light field information could be recorded. As for display, it was an inverse process. An array of aperture lens in front of a series of display pixels cast each pixel of the display through pinhole imaging. It is calculated that the pixel required for conventional display was1000^2, namely 1 million pixels if 1000×1000 resolutions (1 million pixels) were need to display. Compared with the traditional display, however, corresponding light field display system needed 1000^4 (10 million pixels) to display the pictures with the same resolution, and the lens array also needed 1 million lenses. Under the present technology, the display pixels of 5-inch screen could only reach 5 million pixels. It was possible to achieve 100 million pixels in the next 10 years, but 1 trillion pixels were far from our reach with the display technology today.

Optical enthusiasts represented by Lippmann found two laws:

1. In the case of intensive pixels, the system need less pinhole lenses. That is to say, in a 1000×1000 display system, only a 10×10 lens array is required, and a very excellent effect could be obtained.

2. No matter whether it is shooting or computer generation, vivid display effects can be realized without gathering light from all directions.

These two approximate rules rapidly made light field display pragmatic.

At 2013 Siggraph which was a authoritative annual conference on computer graphics, Doug Lanman, a scientist of image display technology supplier Weida, showed his light field display which was improved based on Sony HMZ-H1 and several other commercial head-mounted display devices. Generally speaking, because the screen of head-mounted display was too close for eyes to focalize, it was necessary to install large-diameter magnifying glass to match the screen. Through light field display system structured by 10 × 15 lens arrays and displays module group, however, Doug's system could make the entire device extremely thin so that users could see images of correct depth even though their eyes were close to the display. The light field camera was manufactured from commercial OLED dual-screen system and an approximately small aperture projection array patch which was composed by 150 bespoke micro-lens arrays. Its effective output pixel of image was about 6000. There were also absolutely no Vergence-Accommodation Conflict (VAC) because the distance between eyes and display was too short. The distance between two fishes was distinct. The top row is the images of the human retina after the tow fish were in focus under the theoretical calculation of the ideal condition. The middle row was the pictures of retina through computer simulation in this system. The pictures on the bottom row were observed from authentic samples in different focal lengths. From the actual photographs, it could be seen that the two images have similar focused effects and theoretical value in different focal distance at image center.

It was a huge advancement for the light field display system. Business industry realized that decent light field display effects could be achieved, as long as we improved lightly the existing display technology. So what were the limitations of the technology?

1. Micro-display system with higher resolution was needed. The resolution of the sample display device was 720p, only about over 9000 thousand pixels. According to the previous theoretical derivation, if the pixel could reach 200 million to 300 million, good light field display approximation system would be established at 1080p effective resolution.

2. Larger micro-display chip was necessary. Nowadays, the diagonal dimension of display chip was around 25mm. Through calculation, the field of view would become very narrow. The field of views of samples used in this article was 30 to 40 degree. Although some manufacturers broadened field of view by means of leaning chip, users' effective visual area would be reduced. The best way was to increase the size of chip, but the biggest difficulty is the low rate of good outcomes caused by bad pixels. Therefore, it had become the biggest constraint to the development of light field cameras.

3. Diffraction limit. If the distance among pixels are too dense, it would approach the diffraction limit size. Through calculation, if the effective area interval among pixels was less than 1.3 micrometers, the retina would become blurry. That is to say, the angular

resolution maximum of the system was 30 pixels per degree(30ppd).

To this day, from graffiti to drawing, from paper to electronic display, human civilization has gone through 10 thousand years. With time elapsing, display technology brought us more and more surprises exhibiting an increasingly vivid world. Light field display system has ceaselessly developed with a solid pace in the course of pragmatical use, and tech head goes on subverting the tradition and transforms world with black sciences and technologies.

单词释义：

1. 对眼聚焦　　　　vergence
2. 单眼调焦　　　　accommodation
3. 虚拟显示头显　　virtual reality HMD
4. 分辨率　　　　　resolution
5. 光学系统　　　　optical system
6. 视场角失配　　　field of view mismatch
7. 光学放大镜　　　optical magnifiers
8. 陀螺仪　　　　　gyroscope
9. 小孔成像　　　　pinhole imaging
10. 景深　　　　　　depth of field(DOF)
11. 光场薄膜　　　　light field film\Lippermann film
12. 光学传感器相机　optical sensor camera
13. 像素　　　　　　pixel
14. 针孔透镜　　　　pinhole lens
15. 衍射极限　　　　diffraction limit

译文解析：

1. 合句译法。

把汉语内容关系密切的两个或者多个句子合译为英语一句话就是合句译法。该翻译方法会使译文更加流畅，表述更加清楚。

原文： 他非常清楚双目显示系统存在的问题。Palmer 的产品令他非常激动。

译文： He was very aware of the problems of binocular display system, so it is excited for him to know Palmer's product.

分析： 在科技英语中，英语句子要比汉语句子长，一个英语句子要比一个汉语句

45

子具有更大的容量。因此,我们在汉译英时有必要也有可能把两个以上的汉语句子翻译成一个英语句子。这里译文中将原文中的句号去掉,加入了 so,将两句话译为了一句话。

2. 增译法。

增译是指在不影响原文意思的前提下,在译文中增译一些原文中没有的词汇和表达。增译可以分为两大类:即语法增译和内容增译。

原文: 这款产品加入传统光学动捕技术,使其具备位置追踪功能,同时把人通过 motion parallax 来判断距离的功能也融合到产品中。这一款产品成为双目三维显示的划时代产品。

译文: This product incorporated traditional optical motion capture technology to enable position tracking, and the ability to determine distance through motion parallax is also integrated into the product. Therefore, it became the game-changing product for binocular 3D display.

分析: 汉语重"意合",句子之间的关系一般不必交代得很清楚,自然意在其中,译了反嫌累赘;而英文重"形合",句子之间的逻辑关系一般交代得很清楚。所以,在科技英语的汉译英时有必要借助增补译法,增添原文中暗含而无需明言的词语,把原文意思表达明确。译文中加入了 therefore,将原文中的因果关系体现出来了。

3. 转性译法。

转性译法是指在翻译过程中,根据译语的规范,把原句中某种词类的词转换成另一种词类的词,也就是我们通常所说的词类转换。英语和汉语的词类大部分重合,但是在英语中某个词类的词语可以充当的句子成分相对较少。

原文: 经过多年立体三维电影的发展,一代代科学家和工程师的探索,最好的解决方案逐渐清晰。

译文: With the development of stereoscopic 3-D films, generations of scientists and engineers ceaselessly explored and gradually found out the best solution.

分析: 动词转换成名词的译法,是汉译英是比较普遍的译法。但是也有将名词转换成动词的相反译法。要根据具体情况具体分析。例如,这里"探索"和后面"方案逐渐清晰"有一种递进的关系在,所以分别译为动词"explored"和"found out"。

4. 转态译法。主动语态和被动语态的互译。

原文: 通过使用合适的光学放大镜,再用算法输出正确的图像,就可以解决视场角失配问题。也可以通过图像处理轻松解决显示屏幕适配尺寸问题。

译文: By making use of suitable optical magnifiers, and then using algorithm to output correct images, the field of view mismatch could also be facilely corrected. Besides, the matter of screen's adaptation size could be easily tackled by means of image processing.

分析：被动语态的陈述句意义表达上清晰、简洁和精炼，以至科技英语作品中使用的被动语态达到三分之一到一半多。而汉语中使用被动句的频率远远低于英语，表示被动的手段也十分有限，在科技文本中常用无主句，所以汉译英时将汉语主动句或无主句，转换成英语的被动语态。因此，第一个例子中将主动语态的"固定"和"按照"，分别译为了被动语态的"were fixed"和"were installed"；第二个例子中将主动语态的两个"解决"分别译为了"be facilely corrected"和"be easily tackled"。

5. 调整语序。

利用改变正常语序的方法来增加表达效果，如主谓殊位、主宾颠倒、主居宾位、宾居状位、状语位移、定语后置等。

原文：但是如果到了右图的情况，如果显示虚像的距离比屏幕距离短，那么人的对眼聚焦模式会专注在虚像的位置，而单眼调焦会让眼睛不停试图去聚焦屏幕的位置。这一点会造成极大的不适。

译文：However, if it is the situation that the distance of the virtual image is shorter than that of screen, like the right picture, people's vergence mode will focus on the position of the virtual image and accommodation will let people's eyes restlessly try to focus on the screen, which will cause considerable discomfort.

分析：汉语是比较含蓄的语言，表述习惯上，常常首先交代有关情况和背景，先进行"寒暄"，然后再叙述所发生的事情。但英语是直接的语言，往往先讲结果，再讲条件和背景。在上面两个句子中都将原文中的状语后置，以符合英语的表达习惯。

第四章　磁力魔法

从漂浮滑板到悬浮住宅，技术创造科幻现实

《回到未来》三部曲中的第三部上映于 1989 年感恩节期间。这部电影讲述了男主角马蒂和疯狂科学家布朗博士从 1985 年穿越到 2015 年（也就是这本书出版年份的前一年），而后又穿越回 1955 年，以及这段奇妙的经历如何影响了他们在 1985 年的生活。电影深入浅出地解释了时空因果效应，情节妙趣横生，一时风头无双。即使在二十多年后的今天，我们依然可以在百思买的货架上直接买到《回到未来》三部曲的 DVD（数字激光视盘），可见这部电影在影迷心目中着实经久不衰。

多年后重看这部电影，最为有趣的一件事情是看 20 世纪 80 年代的美国人如何脑补美国在 2015 年的景象。其中不乏充满调侃意味的搞笑，比方说 2015 年的年轻人以把兜掏出来穿衣服为时尚（事实证明我们今天不这么穿），通货膨胀如此之高以至于一杯百事可乐要卖 50 美元，或者使耐克发明了可以自己系鞋带的运动鞋。但是有不少当年的天方夜谭在今天已经成为现实，比方说电影中的可视电话就是我们今天用的 FaceTime。不过，真正让我们这一代科幻迷梦寐以求的却是一种叫做悬浮滑板（hoverboard）的交通工具。

顾名思义，悬浮滑板摒弃传统的小轮支撑，而通过其他反重力手段使滑板悬浮于空中。电影中最为经典的桥段，是几个反派配角和男主角在街头乘悬浮滑板追逐的场景。在第三部中，疯狂科学家布朗博士更是在悬浮滑板上抱得美人归。导演自然不用关心悬浮滑板的物理原理和工程实现，但是这种即使在今天看来都"黑"到骨子里的技术却让广大观众十分着迷，尤其是那些喜欢科幻电影的科学家和工程师总是跃跃欲试想把这项技术从银幕带到现实之中。事实上可随身携带的悬浮（或者说反重力）装置是人类普遍的夙愿，从影视作品中就可见一斑，例如《哆啦 A 梦》中的竹蜻蜓，《阿拉丁》中的飞毯，《007 之雷霆万钧》中的喷气背包。在《回到未来 2》上映之后的二十多年里，不断有公司声称自己开发出了悬浮滑板，可惜都被证明是欺世盗名之举。类似的技术和产品倒是已经存在，例如 Martin Aircraft Co. 推出的个人喷气式飞行器，或者

Jet-Flyer 推出的喷水式飞行器。但是,这两类产品价格昂贵,操作复杂,还需要经过专业训练才能操作,稍有不慎就会受伤甚至机毁人亡,因此离成为大众消费品还有很远的距离。

我们不妨先后退一步,暂时不要求飞得那么高那么快,先着眼于一些简单如悬浮滑板这样的小系统,那么便携式反重力系统的开发和推广就会在一个更为可控的范围之内。物理学中的迈斯纳效应和楞次定律均为反重力系统提供了具有高可行性的解决方案。其中迈斯纳效应是一种量子物理学现象,涉及高温超导;而楞次定律是一种经典力学现象,涉及电场和磁场之间的相互转化。这两种方案呈现出诸多精妙绝伦的物理学原理,而工程实现却又足以简单、链家,所以相信在不太久远的未来,一些人家的后院、学校的操场或者是小区旁边的公园,就会在传统滑板滑道的旁边架起悬浮滑板滑道,以供人们体验飞翔的感觉。我也相信,这些玩家中不仅会有年轻人,更会有像我一样的中年人,因为我们玩的不仅仅是黑科技,也是童年的梦想。

后院里的悬浮装置:Lexus vs. Arx Pax

我学医学的太太只能记住我专业领域的一件事情,就是我总喜欢自诩为万磁王(Magneto)。她每每看到磁学的新闻就会跟我提一下。就在 2015 年 6 月末的时候,她小心翼翼地告诉我悬浮滑板的模型样机已经试验成功,而且是由著名的 Lexus(雷克萨斯,对于我们这一代人来说更乐意称之为凌志)车厂推出,利用的就是磁性。作为一个既热爱磁学又热爱电影的人,十分心痛自己错过了一次创造历史的机会。后来仔细看了 Lexus 的广告,略觉欣慰,因为 Lexus 并不是要推出悬浮滑板,而是用其作为一种噱头来卖车。悬浮滑板示意图中的广告中主人公脚踏悬浮滑板飞跃 Lexus 在 2015 年推出的新款车。请注意滑板的两侧在冒白烟,我将在下一节解释原因。这个广告的设计非常精巧:其一,Lexus 在美国首次亮相并推出了第一款车正好是《回到未来2》上映的那一年,即 1989 年;其二,2015 年距《回到未来1》上映正好是三十周年;其三,当年看《回到未来》那一带的年轻人(也就是最迷悬浮滑板的那一代人)现在处于 35 到 45 岁之间,最有消费能力区购买 Lexus 的新款车。所以 Lexus 就利用这一代消费者的群体记忆制作了这个广告,并且提出了一个口号叫做"利用这一代消费者的群体记忆"制作了这个。这无非是在暗示,即使今天你还买不到一个悬浮滑板,但是你却可以买一辆 Lexus 的车,圆一个儿时的梦。

本着工匠精神,我做了更彻底的搜索,发现自己真的错过了创造历史的机会。一家地处硅谷圣克拉拉市(Santa Clara)叫做 Arx Pax 的公司已经推出了用于个人娱乐的悬浮滑板,也应用了此行原理。这款产品叫做 Hendo,来自于公司创始人 Greg Henderson 的姓氏。请注意这款产品没有在冒白烟,原因我也会在下一节解释。

Henderson 的个人经历颇具传奇色彩。此人在西点军校工程系出身,十年军事生涯,退役后一直在美国顶尖的建筑事务所工作,一直做到合伙人,而后来到硅谷创业。他的公司仍然处于初创阶段,不过 30 人左右的规模,却集中了毕业于美国最顶尖院校的工程师和设计师。在我看来,他们是一群天才的梦想者,也将是新一代娱乐方式的创造者。然而对于 Henderson 来说,悬浮滑板仅仅是开胃菜,"空中楼阁"才是他的终极目标。也就是说个人娱乐产品远远不能满足 Henderson 的野心,从根本上颠覆建筑业才是他真正的着眼点。根据他的网站说,有一天他在遛狗的时候思考了这样一个问题:如果我们可以实现磁悬浮列车,为什么不能建造磁悬浮房屋?

可是为什么要建造磁悬浮房屋呢? 为了抵抗地震和洪水。

我客居美国多年,辗转于东北、东南和中南诸地,最近才搬到了位于西海岸的硅谷。个人认为在这所有的地方之中,硅谷的环境是最差的——常年干旱植被荒芜也倒罢了,更要命的是旧金山地区处在地震带上。君不见好莱坞电影中,纽约往往会灭于外星人入侵(例如《复仇者联盟》),而加州往往毁灭于地震(如《末日崩塌》和《2012》)。虽然电影中描述的那种毁灭性的地震百年难遇,但是各种小震从不间断。美国的住宅以一两层的木质结构居多,抗震性较差,轻则裂缝,重则倒塌,更难以抵抗大地震。另外一项美国居民常见的灾害是洪水,例如 2015 年席卷美国南方诸州的卡特里娜飓风便引发了洪灾。有调查显示,在美国每年有洪水引发的经济损失在 20 亿到 40 亿美元之间。居民的损失往往是双重的:其一为直接经济损失;其二是灾难过后,房屋保险会大幅提高,长远来看,也是一笔巨大的花销。

然而从另外一个方面来看木制房屋,其轻便的特点又可以加以利用。Henderson 认为磁力既然能够拖起磁悬浮列车这样沉重的钢铁结构并且使之高速行驶,那么为什么不能拖起比磁悬浮列车轻得多的住宅呢?一旦住宅和地壳之间存在一个缓冲层,那么地震波就无法直接作用于住宅之上,而是被缓冲层吸收,这样就能确保建筑物的安全。更进一步讲,如果加大磁场(例如另加一个由电流控制的电磁铁),就可以把房屋升的更高一些,洪水就不会溢进屋了。Henderson 认为,磁悬浮是最为高效、简单并且廉价的方法来形成并保持这些缓冲当我读完 Henderson 的这项专利书之后,只觉得这恐怕就是磁学极致"暗黑"的应用了吧。

磁悬浮背后的博弈:迈斯纳效应 vs. 楞次定律

Lexus 和 Arx Pax 虽然都创造出了悬浮滑板,但应用的物理原理并不相同。前者为迈纳斯效应，涉及高温超导，是一种量子力学现象，需要在低温条件下实现；后者基于楞次定律，可以在常温下实现，是一种经典电动力学的现象。简单来说，日本的磁悬浮列车采用超导悬浮技术，利用了迈纳斯效应；而中国浦东机场的磁悬浮列车采用常导磁悬浮，利用了楞次定律。

Lexus 的解决方案：迈斯纳效应

磁性是一个笼统的概念，里面可以细分很多类，包括顺磁性、铁磁性、反铁磁性、亚铁磁性、反亚铁磁性以及抗磁性等。在日常生活中，我们接触到的磁性以顺磁性和铁磁性居多。我们都知道可以用一块磁铁找到掉在地上的针，就是因为磁铁具有铁磁性。磁铁 1 在周围的空间形成了一个磁场的分布；铁制的针感磁，在外加磁场中被磁化产生南北二极，于是成了磁铁 2；顺磁材料此话的方向总与外加磁场方向相同（这也是顺磁性这一术语的由来），导致磁铁和被磁化的顺磁材料相反的两极总相对，于是异性相吸，一根针就可以被磁铁隔空吸引过来。

那么有没有这样一种材料，其磁化的方向于外加磁场的方向相反从而产生斥力？答案是肯定的，而这种性质就被称为抗磁性。抗磁材料不会被磁铁吸引，反而会被推开。有趣的是抗磁材料并不罕见，只不过在日常生活中时常被忽略。例如铜、铅、钻石、水银和铋都是抗磁材料，就连水也是抗磁材料。也就是说，如果我们站在一块磁铁之上，体内的水分子就会和磁铁产生斥力。如果磁铁足够强，就可以让我们悬浮起来。目前在实验室中，科研人员已经可以让一只青蛙浮于空中。可是仅依靠抗磁性很难在现实中实现磁悬浮，原因在于上文所列举的抗磁材料对外加磁场并不那么敏感（即磁化率很低），所以产生的抗磁力往往很微弱。前文所述能够举起青蛙的磁铁需要有 42 特斯拉的强度，而一般工业中用的最强的永磁铁，比如钕铁硼（NdFeB），其磁场强度也不超过 1.5 特斯拉。也就是说用普通的磁铁和抗体材料来实现磁悬浮是不现实的。

那么如何提高材料的抗磁性呢？人们发现超导材料具有巨大的负磁化率，Lexus 就采用了高温超导体作为解决方案。超导体简单来说就是电阻为零的导体。电流可以在超导体中无限循环。超导材料往往只能在低温下运作，所谓高温超导是针对绝对零度而言，而不是基于日常生活中我们对温度的感知。已知的高温超导的操作温度至少要低于 -135℃。假设小球为超导材料暴露于外加磁场中；Tc 为超导材料的临界温度。当材料的温度高于 Tc 时（例如在室温），小球不显示超导性质，外加磁场穿透小球但是小球没有任何电磁感应；然而一旦材料的温度低于 Tc，小球就会显示超导特性，并产生感应电流。考虑到小球处于超导态，电流可以无限循环，而且感应电流有会诱发另一个磁场；这个被诱发的磁场总和外加磁场方向相反，因此小球受到的磁力和外加磁场方向相反，表现为斥力。此时的物理图景就好像外加磁场会可以绕过小球。

这种现象被称为迈斯纳效应,而我们称此时的小球具有超导抗磁性。

悬浮滑板在冒白烟就是因为在滑板中安装了高温超导材料——石墨,并由液氮冷却使其温度低于临界温度,液氮不断蒸发,从两侧泄露,所以看起来滑板在冒烟。而地面下又铺设了一层磁铁,于是处于超导态的石墨和磁铁之间产生巨大的抗磁力以至于可以撑起一个人的重量。如果读者有兴趣,可以自己动手做一个简单的实验,只需一块热解石墨、几块磁铁和一些液氮。前者大概 100 元人民币就可以买到,液氮可以在大学实验室或者是液氮冰激淋店得到。实验时先把磁铁静置于石墨之上,使其冷却到超导临界温度以下,以诱发迈斯纳现象。此贴和石墨之间缠上抗磁力,就会出现磁悬浮现象。

Arx Pax 的解决方案:楞次定律

高中物理学是这样描述楞次定律的:感应电流的效果总是反抗引起感应电流的原因。说得通俗一些,导体感应外加磁场的变化产生感应电流,电路又可以诱发磁场,而被诱发的磁场方向又总和外加磁场相反。也就是说,外加磁场和诱发磁场相当于两块磁铁总是同极相对,于是产生斥力,而利用这种斥力也可以实现磁悬浮。如图 7 所示为楞次定律示意图。当一块磁铁(北极向右,南极向左)向右移动靠近螺旋线圈时,螺线圈感应到周围磁场的加强产生感应电流;根据右手定则,感应电流诱发磁场。此时等价于磁铁在靠近另一块北极面左而南极面右的磁铁,并且两块磁铁总是同极相对,于是永磁铁和螺线圈之间产生斥力。

Hendo 天才地运用了楞次定律。Hendo 需要在导电但不感磁的特殊地面上运行,这就排除了不导电的水泥或者是木头,也排除了导电却又感磁的铁质材料。理想的材料是有良好地导电性又不感磁的铜。在 Hendo 的背面有一套传动装置,简单来说就是一个中心马达带动若干个转子,每个转子安装在磁铁构成的定子之上。马达开转后,转子切割磁场产生感应电流从而产生反应磁场;感应磁场是时变磁场,和导体地面发生感应诱发第二个感应电流,而该感应电流又诱发第二个感应磁场。在此过程中那个,两个感应磁场总是同极相对,从而产生斥力。需要读者留意的是,定子提供的磁场是静磁场,与导电但不导磁的地面不能直接产生感应,所以只要马达不启动就不会有任何斥力。唯有转速高于一定阈值时,斥力才足够撑起操作者的重量实现悬浮。

读者还可以做一个简单的验证实验。这个实验需要一个约 30 厘米长的空心铜管,一块磁铁和另外一块没有磁性的金属小块,如铝块;注意空心半径略大于小块即可。先竖直放置铜管,再把磁铁和铝块分别放入铜管使其自由滑落。虽然铜这种材料不会吸附磁铁,但磁铁下落的速度明显慢于铝块的下落速度。这就是楞次定律的应用:铜管感到周围的磁场变化,产生感应电流,感应电流又产生感应磁场从而对磁铁产生斥力,所以磁铁下降的速度就被减慢了。

"黑"磁性

除了磁悬浮之外,此行还有这其他五花八门的应用。这一小节将介绍一些有趣的磁学应用和研究领域。

磁性细胞分选技术

细胞分选对生物研究、生物医学工程和临床医学都是不可或缺的步骤。以骨髓移植手术为例,骨髓捐献者所提供的样品包含多种细胞,不能直接用于移植,需要先把骨髓细胞隔离出来进行纯化和培养,否则会发生危险的排异现象。于是细胞分选便成了骨髓移植中异常关键的一步。人类细胞非常小,直径在 5 ~ 10 微米,传统的操作和工具难以进行分选,此时就要引入纳米技术。总体来说荧光分选和磁性分选是最主流的手段,而且都需要具有生物兼容性的纳米粒子。在美国,磁细胞分选因为能够保证分选腔不受过去样品的影响,是最为广泛应用于临床的分选手段。具体来讲,很多细胞在其表面会具有一些特定的分子,即分化簇。以从血液样本中分离淋巴细胞为例。某些淋巴细胞的分化簇为 CD12,于是我们可以方便地在顺磁纳米粒子表面植入 CD12 的配体,再将这些纳米粒子和细胞样品混合。因为一种分化簇只和特定的配体结合(类似于抗体和抗原),所以血液样品中那些具有 CD12 的淋巴细胞会被磁性纳米粒子吸附;接下来仅需一块磁铁牵引住这些被磁性标记的淋巴细胞,倾倒其他细胞,于是特定的淋巴细胞就从血液样品中分离出来了。

磁性细胞分选的前沿研究是如何分理出具诊断价值的细胞,例如随血液循环的肿瘤细胞。这种技术讲究在肿瘤恶化之前能够直接从血液样本中分离出癌变细胞以实现癌症的预诊——要知道,若是能够在癌症早期就做出诊断并加以治疗,患者的存活率和生活质量都能够大幅度提高。

磁性结构组装

上文中所说的抗磁材料具有颇为"叛逆"的特性,可惜在常温下磁化率太小难有用武之地。保留磁性微结构组装的研究另辟蹊径,在常温的情况下使顺磁和抗磁两种粒子共存,并且通过改变外加条件使两种粒子相互作用从而产生各种奇异的微观结构。用磁流体(有大量半径在 50 纳米左右的铁磁性纳米粒子构成)、顺磁(铁的氧化物)微粒子和聚苯乙烯(通俗讲就是塑料)微粒子就可以实现一系列微观晶格结构的组装。具体来说,在水中加入一些磁流体以提高介质的磁化率;于是在这样的介质中,原本不感磁的聚苯乙烯粒子因为磁化率低于周围的介质而展现出抗磁性,而顺磁粒子则继续展示顺磁性,于是二者产生很多有趣的作用。这类技术的一个潜在应用使 3D 微纳米结构的成型。具有特定晶格结构的微纳米材料可用于操控光波、声波和热传递。例如对电磁波、光波和声波隐蔽的材料往往需要极为特殊的晶格结构。如果有一天我们能够任意地控制晶格结构的形成,我们就可以任意地创造出各种自然中不存在的材料,譬如《哈利·波特》中的隐身衣。

磁冰箱

电冰箱利用制冷剂的液化和汽化来制冷,而磁冰箱则利用热磁材料的磁化和去磁化制冷。二者的原理都基于熵变。通俗来讲,熵变表征了构成物质的原子或分子的无序状态。假象一块材料由很多磁旋子组成(所谓磁旋子就像指南针一样,拥有南北两极可以自由旋转),如果这些磁旋子的指向完全随机,熵值就很高,从整体来看物质就不具备磁性。我们平时接触的水、空气和桌子都处于这种状态。与此相反,如果磁旋子的指向都相同,系统的熵值很低,从整体看物质就具备磁性,例如磁铁。有趣的是,热力学认为只要温度足够低,任何物质,包括木头、水、空气,甚至是人体,都具有自发的磁性。这是因为原子的无序的热运动会随温度的降低而减少,相应的熵值也会变低,从而使得磁旋子统一指向。

磁冰箱不需要压缩机,却需要一块热磁材料和一块电磁铁。从 a 到 b,电磁铁开启产生磁场来磁化热磁材料,热磁材料被磁化时熵值降低并且放热。由 b 到 c,热磁材料产生的热能被释放到周围的空气之中,温度降低并趋于稳定。从 c 到 d,将热磁材料靠近冰箱并关闭电磁铁,此时热磁材料熵值升高并且吸热,于是冰箱中的热量被转移到热磁材料之中从而实现制冷。通过循环 a 到 b 这种制冷办法甚至能够把温度降低至 $0.3°K$,逼近绝对零度($-273.15℃$)。

与电冰箱相比,磁冰箱只消耗三分之一的电力就能达到相同的制冷效果,可以节省大量的能量,而且它也不需要制冷剂,所以备受环保人士推崇。在 2015 年的 CES 之上,中国的海尔、美国的 ACA 和德国的 BASF 都发布了作为家用电器的磁冰箱,这种产品能否走入千家万户取代传统的电冰箱,我们拭目以待。

磁单极子

细心的读者会发现,刚刚提到的所有磁铁都有南北二极,没有单独存在的南极或者使北极。经典物理学认为磁的本质是电,然而电磁二者有一个明显的区别,即带电体可以以单极子的形式存在,例如电子只带负电而质子只带正电,然而题词总是二极共存,即以偶极子的形式存在。目前还没有确凿的证据显示磁单极子(即某种磁荷仅有南极或仅有北极)存在。如果我们把一块磁铁从中间截开,那么两块小磁铁就会马上产生引得南极或北极以保证自己时偶极子。所以从根本上讲,磁铁的产生不是因为同极磁荷的聚集,而是因为大量磁偶极子的有序排列。

磁单极子对于完善物理学模型亦有重要意义,也是物理学家孜孜以寻的一种存在。举一个简单的例子,百年来物理学家和数学界对麦克斯韦方程组都不能完全释怀,就是因为缺少了一个磁单极子,这组方程的对称性就减了一分,美感也减了一分,这也算是一种书呆子气十足的美学追求吧。然而冷酷的事实是,在一直的物理研究中,大至外太空,小至原子核,温度高如热核反应,温度低如绝对零度,无数才华横溢的物理学家为此呕心沥血,仍然没有任何确凿证据证明磁单极子的存在。

有一次我问教量子力学的老师:量子力学是否还是肯定磁单极子的存在。这位老师岔开话题,给我讲了他的老师 Blas Cabrera 教授的一则往事。Cabrera 教授是斯坦福大学(硅谷的心脏)物理系教授。早在 1982 年,他就声称自己在实验中发现了磁单极子,并且在物理界最权威的期刊《物理评论快报》发表了结果,不过也老实地指出这次实验仅捕捉到了一个磁单极子。事后 Cabrera 本人和其他实验小组投入了大量的人力物力去重复这个实验以证明磁单极子的存在,可惜都无功而返。时至今日,磁单极子依然是物理学中一个悬案。而 Cabrera 的这次发现,因为孤证不立,并没有给物理学带来一场新的革命,而更多的成了物理学家和如我这样的物理学票友的一种谈资。

事实上磁悬浮滑板涉及的物理原理早已为人们所知:楞次定律早在 1834 年就已经被提出;超导现象发现于 1911 年,而迈斯纳效应发现于 1933 年。过去的磁悬浮技术的焦点在运输业,而硅谷则要把这些已经熟知的物理原理推到一个更"黑",更为颠覆,但是同时又更接近大众的领域。以大众消费者为终极目标是硅谷科技公司最重要的一个特征,这也解释了为什么硅谷人总喜欢把"用户体验"四字箴言挂在嘴上。就拿我的工作来举例,我要花相当可观的时间和公司的艺术家做斗争。因为这些艺术家根据用户体验划定产品设计的框架,于是从美感上固然没得挑,但是其研发难度也称几何倍数加大。于是工程师们就绞尽脑汁围绕这些框架做设计,真是白首穷经,艰苦卓绝。

然而这就是硅谷精神之所在,即不断发掘和满足消费者的需要,不论这种需要多么"逆天"。于是我们有了允许你通过旋转拇指选择歌曲的音乐播放器,旨在给你 10 的 10 次方的结果的搜索引擎,可以在驾驶座以上人性打游戏的无人驾驶汽车,给你的生活和职业带来极大方便的社交网络,有贴心管家一般照顾你生活起居的恒温仪,甚至是带来飞翔快感的悬浮滑板和抵抗自然灾害的悬浮房屋等等黑科技。

硅谷的成功之道从来都不仅仅是技术,理想主义甚至是天马行空的白日梦永远是

保持公司活力不可或缺的元素。正如《回到未来》中的一句台词所说的那样："如果我的计算准确的话,你将会看到令人震惊的结果。"

参考译文:

Magnetic Magic

From Hoverboard to Floating Apartment, Technology Creates Magical Reality

Back to the Future II was released on 1989s Thanksgiving Day. The film tells a story of its leading actor Marty and eccentric scientist Dr. Emmett Brown travel from 1985 to 2015, which is one year before the book published, and flash back to 1955. Also, it shows us how that amazing experience changes their lives. With its amusing plots, the film explains the causal effect of the time in a simple way, attracting the eyes of the world. Though 20 years after that, Back to the Future trilogy DVD, Digital Videodisc, is still easily found on BestBuy's shelves, which indicates the popularity of the film among the audience.

The most interesting thing is the people living in 1980s image the life in 2015, when I re-watched the film after many years. Some plots might as well express ridiculous and funny meanings: the youth in 2015 see wearing clothes with pockets off as fashion, actually, we do not wear like that today; inflation is exceedingly high that a cup of Pepsi cost 50 dollars; or Nike invites self-lacing shoes. But many impossibilities in the film have come true today, for example, the video phone in the film is just today's FaceTime. However, what fiction fans of our generation dream of is a transportation facility called hoverboard.

Hoverboard, as it names, is suspended in the air by other antigravity means instead of traditional small wheels. The most impressive scene in the film is villains chasing the hero by hoverboard on the street. In the Back to the Future III, the eccentric scientist Dr. Emmett Brown even falls in love on the board.

Of course, directors don't need to consider the physical principles and engineering realization of hoverboard, but such a extremely Black magic, even in the eyes of people today, fascinates the audience, especially for those scientists and engineers who love fiction. They have been seeking ways to make the magic true. In fact, inventing portable hovering, antigravity devices in other words, has been our dream for a long time. Just like what we can see in the films such as, the Bamboo-copter in Doraemon, flying carpet in Aladdin, and jetpack in Thunderball. More than 20 years after Back to the Future II was released, companies continuously announced the invention of hoverboard, but only for their reputation, and proofed to be ostensible. On the other hand, similar technology and products have been developed. For example, personal jet aircraft by Martin Aircraft Co., or water jet aircraft by Jet-Flyer. But those two are expensive and hard to manipulate, therefore, both the driver and machine might be badly hurt without professional training. Therefore, they still have a long way to go to be publicly available.

We might as well step back, focus on tiny systems like hoverboard instead of the requirement of height, and speed so that the development and spreading of portable antigravity system would be within our control. Meissner effect and Lenz's law in physics have the potential for the development of the system. Meissner effect is a quantum physics phenomenon involving high temperature superconductivity, and Lenz's law is a classical mechanical phenomenon consisting of the mutual transformation between power and magnetic fields. Those two concepts present many exquisite physical principles while the engineering realization is simple and cheap enough. Thus I believe, in the near future, in people's backyards, on schools playgrounds, or parks near the communities, slide ways for hoverboard will be built next to that of the traditions for us to experience "flying". Also, I'm sure that the players would not only be the youth, but also be middle-aged ones like me, because we are for the dream of childhood more than black magic.

Lexus vs. Arx Pax

My wife, major in medical science, can only remember one thing about my field: I always praise myself as Magneto. Every time reading news on magnetic, she would mention it to me. On the end of June 2015, she cautiously told me that the prototype of hoverboard using magnetism had been successfully tested, and was launched by famous car producer Lexus (more widely known to our generation as LING ZHI). As a fan of magnetic and film, I felt quite pity missing a chance to make history. But after watching the ads of Lexus, I was relieved. Lexus' intention was not to develop hoverboard, but to regard it as a gimmick to sell cars. The hero of advertisement flying over Lexus 2015 new car by hoverboard. Please pay attention to the smoke on the two sides, I will explain that in detail in the next section.

The design of advertisement is very ingenious. First, Lexus debut and launching in America in 1989 was just the same year Back to the Future II was released. Second, the

year 2015 was the thirty years anniversary of Back to the Future I . Third, audiences of Back to the Future, who love hoverboard most, are now between the age of 35 to 45, and they have a great potential buying the new car of Lexus. So by using the collective memory of that generation of consumers, Lexus made that advertisement, and put forward a slogan called "Amazing in Motion". It was no more than a clear hint: though you cannot purchase a hoverboard now, you could have a Lexus to fulfill your dream.

In the spirit of craftsman, I did a further exploration and found I really missed the opportunity to create history. Located in Santa Clara, silicon village, a company called Arx Pax has introduced hoverboard for personal recreation. The product, Hundo, is named after the last name of the founder of company, Greg Henderson. Attention that there's no smoke in this product, and I will explain it in the next section, too.

Henderson has a wonderful life experience. Graduated from the Engineering Department of West Point Military Academy, he had been in the army for ten years. After his retirement, Henderson had been working in the top architectural firms, and was promoted to be a partner at last. And then, he came to the silicon village for entrepreneurship. Although in the initial stage, with no more than 30 staff, he centralized engineers and designers graduated from top universities of the States. In my view, they represent talented dreamers and creators of new entertainment.

For Henderson, however, hoverboard is nothing but only an appetizer, while floating apartment is his ultimate dream. In other words, personal recreation products cannot meet Henderson's ambition, and his mind and spirit lie in overhauling the traditional construction industry. According to Henderson's website, one day when he was walking his dog, a question occurred to his mind: since we are able to make magnetic levitation train, why can't we make magnetic levitation house?

But What's the Meaning of Building a Magnetic House? For the Resistance of Earthquake and Flood.

Living in the States for many years, I lingered around northeast, southeast, and central-west parts of America, and recently, I have just moved to the silicon village in the west coast. Personally speaking, silicon village has the worst environment among all of those places. Despite of its drought and sparsely vegetation covering, even worse, the San Francisco area is in the earthquake zone. Hollywood films are usually seen the New York City destroyed by the invasion of aliens (such as The Adventurers), California, however, are often ruined by earthquake (such as San Andreas and 2012). Although devastating earthquakes in the film are rarely witnessed in reality, small ones never stop. In America, most of buildings are built in wooden structure with one to two floors, and their ability to resist shock is quite weak, crack or collapse, which makes it even hard to bear severe quake. Another usually-seen disaster in America is flood. For example, Hurricane Katrina, which swept through the southern states of the United States in 2015. Surveys suggest that economic loss trigger by flood in America per year accounts 2 to 4 billion U. S dollars. Generally speaking, residents have to meet not only direct economic loss, but also insurance rise after the disaster, which is, as well, a big deal in the long term.

On the other hand, because of the lightness of wooden buildings, we might as well make use of them. Henderson believes that since magnetism is able to hold such heavy iron maglev trains, and keep them in such a high speed, why can't it push up buildings much lighter than trains? Once the buffer layer between house and crust was made, the seismic wave cannot directly act on the house, but absorbed by the layer, so that, in this way, we keep the buildings safe. Furthermore, if settling a larger magnetic field (e. g. an additional current-controlled electromagnet), we could elevate buildings higher, and flood would not spill into houses. Henderson thinks Henderson believes that magnetic levitation is the most efficient, simple and inexpensive way to form and maintain these buffers.

When I finished reading Henderson's patent, the only thought of mine was it must be the coolest application of magnetic.

Game Behind Maglev: Meissner Effect vs. Lenz's Law

Both Lexus and Arx Pax, though, create hoverboard, they follow different physical principles. The former is with Meissner effect involving high temperature superconductivity, which needs to be realized in low-temperature condition, while the latter with Lenz's law, a classical electrodynamics phenomenon, which can be achieved in normal temperature. Simply speaking, Japan's maglev train uses superconducting levitation technology and uses Minas effect, while China's Pudong Airport's maglev train uses constant conducting levitation and Lenz's law.

Lexus Solution: Meissner Effect

Magnetism is a general concept, which can be subdivided into many categories, including paramagnetism, ferromagnetism, antiferromagnetism, ferromagnetism, antiferromagnetism and diamagnetism. In our everyday life, paramagnetism and ferromagnetism are more often seen. We all know using a magnet to find a needle because of the magnetism. It explains the physical process that paramagnetic material is magnetized: magnet 1 forms a magnetic field distribution in the surrounding space, as shown by the arrow in the figure; iron-needle is induced by magnetism, which is magnetized in the external magnetic field to produce the north and South poles, thus becomes magnet 2; paramagnetic material always has the same direction as the external magnetic field (where the term comes), which causes the opposite poles of the magnet and the magnetized paramagnetic material to always be opposite, so that the opposite poles attract, and a needle can be attracted by the magnet. So is there existing such a material whose magnetization is opposite from external, generating repulsion? Absolutely, it is called diamagnetism. It indicates the magnetization of it. Diamagnetic material will instead repulse magnet.

Interestingly, diamagnetic materials are not uncommon, but usually ignored in our daily life. For example, copper, lead, diamonds, mercury and bismuth are antimagnetic materials, and even water is antimagnetic materials, too. That is to say, if we stand on a magnet, the water inside of us will generate repulsion against magnet. Once the magnet is strong enough, we can hover in the air. At present in the lab, researchers have enable a frog hovering.

However, it is extremely hard to achieve magnetic levitation in reality only by antimagnetism. The reason lies in the weak sensitivity of antimagnetic materials we mentioned above against external magnetic field, or low magnetic susceptibility, so it can merely generate weak antimagnetic force. The magnet that hover the frog we mentioned above needs 42 teslas, whereas the strongest permanent magnet used in general industry such as NdFeB, with a magnetic field intensity no more than 1.5 teslas. In other word, it is impossible to achieve magnetic levitation by normal magnet and antimagnetic materials.

How to improve the antimagnetism of material? People found that superconducting materials boats enormous negative susceptibility, thus Lexus adopted high superconductor as solution. Simply speaking, superconductor is the conductor whose resistance is zero. Electrical current circulates indefinitely in the superconductor. It usually works only in low temperature environment. The so called high temperature is based on absolute zero instead of what we know in our daily life. Supposing the ball is in the exposure of an external magnetic field; Tc means the critical temperature of superconductor. When the temperature of material is high than Tc, e. g. normal temperature, ball doesn't shows superconductivity. At this juncture, external magnetic field goes though the ball but no electromagnetic induction occurrence. However, once the temperature decreases below Tc, the ball indicates superconductivity and generates induced currents. Considering the state of superconducting of the ball, currents can circulate infinitely, moreover, induced currents will generate another magnetic field. The generated field is always opposite to the external field, repulsing. External field seems to bypass the ball. This phenomenon is called Meissner effect. The ball boasts superconductivity at the moment.

Hoverboard's white smoke is due to the installment of superconductor, graphite, which is, then, cooled by liquid nitrogen, and released from the both sides. Therefore, it seems as if smoking. Additionally, the underneath magnets generate a magnificent antimagnetic force between superconductive graphite and magnets, which is able to bear the weight of a person.

If you, readers, are interested in it, you can do a simple experiment yourself. You only need to prepare a piece of pyrolytic graphite, some magnets and liquid nitrogen. The former two objects can be gotten within the expense 100RMB, whereas liquid nitrogen can be gained in universities' labs or liquid nitrogen ice cream bar. During the experiment, we should first put magnets on the graphite still, and pour liquid nitrogen onto them so as to low them under superconductor critical temperature to induce Meissner Effect. The phenomenon antimagnetic force between magnets and graphite.

The Solution of Arx Pax: Lenz's Law

High school physics describes Lenz's law in this way: the effect of induced current always resists the cause of induced current. More generally speaking, the conductor feels the change of the external magnetic field to produce the induced current, and the circuit can induce the magnetic field, and the induced magnetic field direction is opposite to the total external magnetic field. That is to say, the external magnetic field and induced magnetic field are equivalent to two magnets which are always opposite to each other at the same pole, thus repulsion is generated, and magnetic levitation can be realized by using this repulsion. It shows a schematic diagram of Lenz's law. When a magnet (north pole to right, south pole to left) moves to the right near the coil, the coil induces an induced current by the enhancement of the surrounding magnetic field; according to the right-hand rule, the induced current induces a magnetic field. This is equivalent to a magnet near the left of another north pole and the right of the south sole, and the two magnets are always opposite to each other at the same pole, so the repulsion force between the permanent magnet and the screw coil occurs.

Hendo used Lenz's law with genius. Hendo needs to operate on a special ground that

is conductive but not magnetically sensitive, which excludes non-conductive cement or wood, as well as conductive but magnetically sensitive ferrous materials. The ideal material is copper with good conductivity and no magnetism. On the back of Hendo, there is a set of transmission device. Simply speaking, a central motor drives several rotors, each of which is mounted on a stator made of a magnet. When the motor is turned on, the rotor cutting magnetic field generates induction current, which generates reaction magnetic field. The induction magnetic field is a time-varying magnetic field, which induces the second induction current with the conductor ground, and the induction current induces the second induction magnetic field. In this process, the two induced magnetic fields are always opposite to each other, which generates repulsion. It should be noted that the magnetic field provided by the stator is static magnetic field, which can not directly induce with the conductive but non-magnetic ground, so as long as the motor does not start, there will be no repulsion. Only when the speed is higher than a certain threshold, the repulsion force is enough to support the operator's weight to achieve suspension.

Readers can also do a simple verification experiment. This experiment requires a hollow copper tube about 30 centimeters long, a magnet and another non-magnetic piece of metal, such as aluminum; note that the hollow radius is slightly larger than the small piece. First, the copper pipe is placed vertically, and then the magnet and the aluminum block are put into the copper pipe to make it slide freely. Although copper does not absorb magnets, the falling speed of magnets is significantly slower than that of aluminum blocks. This is the application of Lenz's law: the copper tube senses the change of the magnetic field around it and generates the induced current, which in turn generates the induced magnetic field and repulses the magnet, so the speed of the magnet's descent is slowed down.

Black magnetism

In addition to magnetic levitation, magnetism has many other applications. This section will introduce some interesting applications and research fields of magnetism.

Magnetic Cell Sorting Technology

Cell sorting is an indispensable step in biological research, biomedical engineering and clinical medicine. In the case of bone marrow transplantation, the samples provided by bone marrow donors contain many kinds of cells, which cannot be directly used for transplantation. Bone marrow cells need to be isolated for purification and cultivate, otherwise dangerous rejection will occur. So cell sorting has become a critical step in bone marrow transplantation. Human cells are very small, with diameters ranging from 5 to 10 microns. Traditional operations and tools are difficult to separate. At this time, nanotechnology will be introduced. Generally speaking, Fluorescence-Activated Cell Sorting (FACS) and Magnetically-Activated Cell Sorting (MACS) are the most mainstream means, and they all need biologically compatible nano particles. In the United States, magnetic cell sorting is the most widely used clinical sorting method because it can ensure that the sorting chamber is not affected by past samples. Specifically, many cells have specific molecules on their surface, namely Cluster of Differentiation, CD. Taking the separation of lymphocytes from blood samples as an example. Some lymphocyte clusters are CD12, so we can easily implant CD12 ligand on the surface of paramagnetic nano particles, and then mix these nano particles with cell samples. Because a CD binds only to specific ligand (similar to antibodies and antigens), the lymphocyte with CD12 in the blood sample is absorbed by magnetic nano particles; then only a magnet is needed to pull the magnetic labeled lymphocyte and pour other cells, so the specific lymphocyte is separated from the blood sample.

Frontier research in magnetic cell sorting is how to isolate diagnostic cells, such as circulating tumor cells in the blood. This technique focuses on the ability to isolate cancerous cells from blood samples directly before the cancer deteriorates in order to achieve cancer diagnosis-you know, if you can diagnose and treat cancer early, the survival rate and quality of life of patients can be greatly improved.

The Assembly of Magnetic Structure

The diamagnetic materials mentioned above are quite rebellious. Unfortunately, the susceptibility at room temperature is too small to be useful. Reserved magnetic micro structural assembly is a new way to make paramagnetic and antimagnetic particles coexist at room temperature. By changing the external conditions, the two particles interact to produce a variety of strange microstructures. A series of Colloidal Crystal assembly can be achieved by magnetic fluid (composed of a large number of ferromagnetic nanoparticles with radius of about 50 nanometers), paramagnetic (iron oxide) particles and polystyrene (plastics in general) particles. Specifically, some magnetic fluids are added to water to increase the susceptibility of the medium; therefore, in such a medium, the original non-magnetic polystyrene particles exhibit diamagnetism because the susceptibility is lower than that of the surrounding medium, while paramagnetic particles continue to exhibit paramagnetic, so

they have a lot of interesting effects. One potential application of such technologies is the formation of 3D micro and nanostructures. Micro nanomaterials with specific lattice structures can be used to manipulate light, sound and heat transfer. For example, materials concealed from electromagnetic, optical and acoustic waves often require very special lattice structures. If one day we can arbitrarily control the formation of lattice structure, we can arbitrarily create materials that do not exist in nature, such as the invisibility cloak in Harry Potter.

Magnetic Refrigerator

Refrigerators use liquefaction and vaporization of refrigerants to refrigerate, while magnetic refrigerators use magnetization and demagnetization of thermo magnetic materials. Both of them are based on the principle of entropy change. Generally speaking, the entropy change represents the disorder of atoms or molecules that make up matter. Imagining a material consisting of many magnetic rotors (so-called magnetic rotors, like compasses, have two poles that can rotate freely). If the direction of these magnetic rotors is completely random, the entropy value is very high, and the material as a whole does not have magnetism. The water, air and tables we normally touch are in this state. On the contrary, if the orientation of the magnetic rotors is the same and the entropy value of the system is very low, the matter as a whole has magnetism, such as magnets. Interestingly, thermodynamics considers that as long as the temperature is low enough, any substance, including wood, water, air and even the human body, has spontaneous magnetism. This is because the disordered thermal motion of atoms decreases with the decrease of temperature, and the corresponding entropy value decreases, which make the magnetic rotors point uniformly.

The refrigerator does not need a compressor, but needs a thermo magnetic material and an electromagnet. From a to b, the magnet opens to produce a magnetic field to magnetize the thermo magnetic material. When the thermo magnetic material is magnetized, the entropy decreases and the heat is released. From b to c, the thermal energy generated by thermo magnetic materials is released into the surrounding air, and the temperature decreases and tends to be stable. From c to d, the thermo magnetic material is close to the refrigerator and the electromagnet is closed. At this time, the entropy value of the thermo magnetic material increases and the heat is absorbed, so the heat in the refrigerator is transferred to the thermo magnetic material to achieve refrigeration. The refrigeration method of circulating a to b can even reduce the temperature to 0.3 degree k, approaching absolute zero (-273. 15 degrees C).

Compared with refrigerators, magnetic refrigerators consume only one third of the electricity to achieve the same refrigeration effect, save a lot of energy, and does not need refrigerants, so it is highly praised by environmentalists. On CES in 2015, Haier, China, ACA, the United States and BASF, Germany have all released magnetic refrigerators as household appliances. Whether this product can go into thousands of households to replace traditional refrigerators remains to be seen.

Magnetic Monopole

Some readers might find that all the magnets just mentioned have North and South Poles, and separate North or South doesn't exist. Classical physics holds that the essence of magnetism is electricity, but there is an obvious difference between magnetism and electricity, that is, charged bodies can exist in the form of monopoles. For example, electrons are only negatively charged while protons are only positively charged. However, the inscription always coexists in the form of dipoles. There is no conclusive evidence for the existence of magnetic monopoles (i. e. a magnetic charge that exists only in the North or only in the South). If we cut a magnet in the middle, two small magnets will immediately produce a dipole that attracts either the North or South to ensure its own balance. So fundamentally, magnets are produced not because of the accumulation of magnetic charges at the same pole, but of the orderly arrangement of a large number of magnetic dipoles.

Magnetic monopoles are also of great significance to the improvement of physical models, and they are also a kind of existence that physicists are diligently seeking. For a simple example, physicists and mathematicians have been unable to completely agree with Maxwell's equations for centuries, because without a magnetic monopole, the symmetry of these equations is reduced by one point, and the aesthetic sense is also reduced by one point, which might as well be a kind of bookworm's aesthetic pursuit. However, the cruel fact is that in all the physical studies, from outer space to atomic nuclei, the temperature is as high as thermonuclear reaction, and the temperature is as low as absolute zero. Countless talented physicists have worked hard for this, but there is still no conclusive evidence to prove the existence of magnetic monopoles.

Once I asked the teacher of quantum mechanics whether quantum mechanics still affirmed the existence of magnetic monopoles. The teacher turned aside and told me a story about his teacher, Professor Blas Cabrera. Professor Cabrera is a professor of physics at

Stanford University (the heart of Silicon Valley). Early in 1982, he claimed to have found a magnetic monopole in his experiment and published the results in the most authoritative journal Physical Review Letters, but honestly pointed out that only one magnetic monopole was caught in the experiment. After that, Cabrera himself and other experimental groups devoted a lot of manpower and material resources to repeat the experiment so as to prove the existence of magnetic monopoles, but failed. Today, magnetic monopoles are still an unsolved problem in physics. Cabrera's discovery, however, did not bring a new revolution in physics because of its lack of evidence, became a conversation topic for physicists and physics fans like me.

In fact, the physical principles involved in magnetic hoverboard have long been known: Lenz's law was proposed early in 1834; superconductivity was discovered in 1911, and the Meissner effect was discovered in 1933. In the past, the focus of magnetic levitation technology was mainly on transportation, while Silicon Valley was pushing these well-known physical principles into a more black and subversive field, but at the same time closer to the public. Taking consumers as the ultimate goal is one of the most important characteristics of Silicon Valley Technologies, which also explains why Silicon Valley people always like to talk about the motto of "user experience". Taking my work as an example, I will spend considerable time fighting with the artists in the company. Because these artists delimit the framework of product design according to user experience, they have no defeats in aesthetics, but their research and development difficulty is also geometrically multiplied. So engineers racked their brains to design around these frameworks. It's really hard and painstaking.

However, this is the spirit of Silicon Valley, that is, to constantly explore and meet the needs of consumers, no matter how "impossible" such needs. So we have music players that allow you to choose songs by turning your thumb, search engines that aim to give you 10th power results, driverless cars that can play games with people above the driver's seat, social networks that bring great convenience to your life and career, thermostats that caring housekeepers generally take care of your life and living, and even hoverboard that bring you flight pleasure, and black technology such as panels and floating houses to resist natural disasters.

Silicon Valley's road to success has never been just technology. Idealism and even fantasy are indispensable elements to keep the company alive. As a line in Back to the Future says, "If my calculations are accurate, you will see shocking results."

单词释义：

1. 楞次定律   Lenz's Law
2. 磁单极子   Magnetic monopole
3. 悬浮滑板   hoverboard
4. 迈斯纳效应   Meissner Effect
5. 磁化率   magnetic susceptibility
6. 热解石墨   pyrolytic   graphite
7. 分化簇   cluster of differentiation
8. 国际消费类电子产品展览会   International Consumer Electronics Show(ICES)
9. 熵变：   Entropy change

译文解析：

1. 英汉语言形合意合之差异

原文：具体来说，在水中加入一些磁流体以提高介质的磁化率；于是在这样的介质中，原本不感磁的聚苯乙烯粒子因为磁化率低于周围的介质而展现出抗磁性，而顺磁粒子则继续展示顺磁性，于是二者产生很多有趣的作用。

译文：Specifically, some magnetic fluid is added to the water to increase the magnetic susceptibility of the medium where the originally non-magnetic polystyrene particles exhibit diamagnetic resistance because of a lower magnetic susceptibility than the surrounding, and paramagnetic particles continue to be paramagnetic. Therefore, interactions between the two make many interesting effects.

分析：在汉语行文中，"具体来说，在水中加入一些磁流体以提高介质的磁化率"与"原本不感磁的聚苯乙烯粒子因为磁化率低于周围的介质而展现出抗磁性"两部分之间是通过"于是在这样的介质中"相连接的，符合汉语意合的语言特征，且句子多逗号、多短句。在翻译为英文时，需注意英文意合的句式特点，用 where 连接定语从句，从句与句关系层面体现出前后文的顺承关系。

2. 归化与异化

原文：电影深入浅出地解释了时空因果效应，情节妙趣横生，一时风头无双。

译文：The movie, then, caught the world's eyes by its simple expression of casual effects of time and place with vivid plot.

分析：原文中，"深入浅出"、"妙趣横生"、"一时无双"等词语，均为十分具有中国特色的语言，在翻译为中文时，不应刻板的一一对应翻译，可考虑采取归化的翻译策略,用简洁轻快的英语代替生动明了的汉语,进而达到相同的语用效果。

3. 长句翻译

原文: 因为一种分化簇只和特定的配体结合(类似于抗体和抗原),所以血液样本中那些具有 CD12 的淋巴细胞就会被磁性纳米粒子吸附;接下来仅需一块磁铁牵引住这些被磁性标记的淋巴细胞,倾倒其他细胞,于是特定的淋巴细胞就从血液样本中分离出来了。

译文: Because a CD binds only to specific ligand (similar to antibodies and antigens), the lymphocyte with CD12 in the blood sample is absorbed by magnetic nano particles. Then only a magnet is needed to pull the magnetic labeled lymphocyte and pour other cells, so the specific lymphocyte is separated from the blood sample.

分析: 长难句分析主要需要厘清句子逻辑关系,在本句中,分号之前的半句是由"因为"、"所以"引导的原因状语从句,翻译时有两种选择,可以用 because 引导或是用 so 引导,自由度较高。汉语的行文特点使然,全句中多次使用标点符号,句子冗长,因此在汉译英的过程中,我们可以适当地对原句进行切分。在整个句子中间部分出现了一个分号,我们不妨在此处进行切分,将原本的一句汉语转变为两句英语,这样不仅可以使行文简洁,也更利于读者理解原文。

第五章　暗物质探索

宇宙的组成

　　随着时间的推移以及人类科技的高速发展,人类的活动半径不再被限于地球或者太阳系,而是逐渐拓展到整个宇宙空间。想要进行星系(galaxy)级别的航行,宇宙飞船的推动力是必须要解决的问题,学会使用如电影《变形金刚4》出现的暗物质引擎(Dark Matter Engine)进行星际穿越,或许会成为22世纪人类星际航行的必备的技术。

　　利用暗物质作为动能的飞行器之所以成为星际穿越的首选,是因为暗物质和反暗物质的湮灭产生高能粒子过程,可以将暗物质粒子的所有质量转化为能量,这将比核裂变和核聚变的能量要强大数万倍甚至更多。而且暗物质散布于宇宙空间,在宇宙飞船飞行的过程中,暗物质可以不断被收集并被用于飞船的加速或者维持飞船系统的正常运行上。像暗物质引擎这样的超级装备,目前虽然还只停留在是科学幻想中,但仍然引起了人们的好奇和追问——它真的有可能实现吗?

　　要了解暗物质,需要先从宇宙说起。宇宙是什么组成的?你可能会想到星球、尘埃以及广阔无垠的空间。从微观角度上看,你或许会想到分子、原子、质子,等等。但是这些只是人类对于宇宙组成所理解的一小部分,我们可以用三个部分来概括宇宙的组成部分:正常的物质,暗物质,还有暗能量。若按照宇宙所有能量的配比划分,正常物质只占有宇宙所有能量的4%,暗物质占22%,而余下74%左右的宇宙能量则都为暗能量!

　　虽然暗能量在整体宇宙能量中所占的比重巨大,但却不知道它由什么组成,未知大于已知。自然引起了众多理论物理学家的极大兴趣。由此也应运而生了相当数量的理论模型来解释暗能量的性质以及由来,这是一个单独的学科方向。有相当一部分学者相信,暗能量只是广义相对论中的宇宙学常数(出现在爱因斯坦场方程中的一个常数),只不过人们还不理解为什么宇宙学常数是我们现在所观测到的取值。如果是这样的话,在某种意义上,人们已经理解了暗能量的本质。相比暗能量,以物质形态存在的暗物质,则给我们提供了非常多的线索。如果想要了解暗物质,我们先来看看什

么是正常的物质。

正常物质，便是我们通常意义上的物质，包括中子、质子、电子、中微子以及其他粒子。化学家研究分子，原子物理学家研究原子，核物理学家研究中子和质子的世界，而一部分高能物理学家则在研究组成中子和质子的更微小的粒子比如夸克和胶子，W/Z规范粒子以及希格斯玻色子等等。人们对于这些物质结构已经有了非常深刻的理解。但这些都不足以满足科学家们的好奇心，神秘的暗物质便成了热门的研究方向。暗物质到底具有什么样的属性？至今仍然有太多的未知，它是一种新的、不发光、不在人们现在所理解范畴内的物质。目前，暗物质之间以及暗物质与正常物质之间有着什么样的联系，这些都还是未知，我们仅仅知道的是暗物质参与引力相互作用，这是暗物质存在的唯一一个确凿的证据。

暗物质存在的证据

既然我们对暗物质的了解如此之少，那么我们为什么能够确定暗物质的存在呢？这还要归功于天文学家。根据牛顿万有引力定律，我们知道任何两个有质量的物体之间都存在着引力作用，物体的质量越大，相互之间的引力就越强；物体之间的距离拉近，引力也会增强。由于万有引力的存在，银河系中的恒星均围绕着银河系的中心旋转。根据天文学观测，人们可以计算银河系中不同位置恒星的运行速度，物理学家们便可以此推测出银河系中的质量分布。你可能会好奇天文学家是如何测量遥远的恒星的运行速度的？这是通过恒星发出光线的红移（光波的多普勒效应）进行测量的。

多普勒效应是指波源和观察者有相对运动时，观察者接收到波的频率与 J 波源发出的频率并不相同的现象。光的多普勒效应是指，当星体远离地球运动时，它的运动速度越快，在地球上的观测者看来它所发出的光线能量就越低。相反的，如果它朝向地球运动，那么运动速度越快，在地球上观测到光线的能量就越高。由此，可以通过银河系中恒星的分布而粗略估计银河的质量分布。

证据一：行星运动速度分布曲线

当天文学家在对银河系中行星运动的速度分布曲线，以及对银河系总质量的估计进行比较时，他们惊奇地发现行星围绕银河系中心旋转的速度比他们预估的要高出许多，这说明有很多质量来源于一种新的、不发光、不在人们现在所理解范畴内的物质！而这一部分没有被观测到的质量即被天文物理家称为暗物质，这概念最初是在 1932年由简亨德里克·奥尔特（Jan Hendrlk Oort）基于银河系恒星运行的轨道速度推测出的。

类似的，就像行星组成银河系一样，星系也可以组成星系团。这些宇宙的大尺度结构都是从宇宙形成之初分布非常均匀的气体逐渐由万有引力聚拢坍缩形成的，而每一个星系中心都可能存在超大质量黑洞。星系也同样通过万有引力围绕星系团的中心旋转。正如前面所讲到的恒星与银河系那样,星系的速度也与星系团的中心质量发生了很大偏差。也就是说,暗物质也广泛存在于星系团中。这里需要强调的是,暗物质与银河系中心超大质量黑洞并没有直接的联系。黑洞可以近似理解为银河系中心的一个质点。

证据二:宇宙大尺度结构的形成

你也许会好奇有没有能证明暗物质存在的其他更直接的证据?近年来的天文观测及理论研究确实为我们找到了许多证据,其中一个显著的证据就是宇宙大尺度结构的形成。宇宙在最开始形成的时候经历过一次大爆炸的过程。这时整个宇宙的温度非常之高,甚至高过太阳内部的温度,宇宙中充满了气体(主要是氢气和氦气),这些气体均匀地弥散在整个宇宙。随着宇宙的膨胀,宇宙中的气体逐渐冷却。由于气体分布的微小不均匀性,开始在万有引力的作用下逐渐聚拢成团。气体成团后,便逐渐形成了星系团,而在星系团内部,便形成了一个个星系,我们所处的银河系便是其中一个。当宇宙逐渐从高温状态冷却下来时,物质分布的微小的不均匀性导致了星系的形成。简单地说,物质分布较多的区域会在引力的帮助下从附近的区域吸引更多的物质,而不断累积则形成了宇宙的大尺度结构,如星系团和星系。

在通过一系列精确的数值模拟计算后,人们发现暗物质存在与否,很大程度地影响了宇宙大尺度结构的形成。由于暗物质占据宇宙能量配比的很大一部分,并且它也像正常物质一样参与引力作用,如果我们现在的宇宙没有暗物质或暗能量,而仅仅存在正常的物质,那么宇宙大尺度结构将与观测结果截然不同。这就是物理学家普遍认为暗物质存在的第二个重要证据。精确的数值模拟还可以估算出暗物质的许多性质,其中最重要的一个性质便是暗物质是"冷"的! 所谓的暗物质的"冷",主要描述的是暗物质在宇宙大尺度结构形成时的速度:如果此时暗物质的速度接近光速并做相对论运动,便称为"热";反之若其速度远远小于光速,则被称为"冷"。

中微子是一种近似没有质量,与其他粒子几乎没有相互作用的特殊粒子。在暗物质模型提出之初,中微子曾是极好的暗物质的选项。然而人们发现,由于中微子的质量极小,在宇宙的演化过程中很难产生运动较慢的中微子。假如"热"的中微子是所谓的暗物质,那么在宇宙大尺度结构形成之时,由于中微子的运动速度接近光速,高速运动的中微子不能轻易地聚拢成团,这会改变宇宙的物质分布,这与我们实验观测到的数据不符。另一方面,如果暗物质通过引力相互作用,而被束缚在银河系中,那么暗物质的速度不应该超过光速的千分之一。如果速度过快,暗物质则会直接飞出银河系!

证据三:引力透镜实验

以上的两个证据都是基于暗物质是均匀分布的稳定的引力源的假设而得到的。

有些物理学家提出,实验数据与理论不符并不是由于暗物质的存在,而是对于宇宙大尺度结构,牛顿万有引力不再适用,而有可能需要另外的理论来描述。基于这样的动机,人们开始修改万有引力定律。更具体地说,我们知道牛顿万有引力是与距离的平方成反比。这个定律只在太阳系尺度以下的距离得到了验证,并不知道在银河系尺度下这一定律是否仍然成立。当物理学家发现在银河系尺度下如果不加入暗物质的话,牛顿万有引力定律不能做出正确的预测时,便提出修改牛顿万有引力定律的可能性。然而我们将要说到的第三个证据,将在很大程度上把修改引力模型这种可能性排除在外。

由于引力场的存在,组成光线的光子不再沿直线运动,其运行轨迹变得"弯曲"。若在宇宙中某一个空间存在暗物质团而没有正常的物质,那么这一区间内的暗物质也可以通过其引力相互作用将光线弯曲,这便是著名的引力透镜效应。在引力透镜的实验观测中,人们发现某些区间几乎没有正常物质的存在,然而光线在这些区域仍然有很大的弯曲,这就说明这些区间存在正常物质以外的其他物质。这便是引力透镜实验提供的暗物质存在的一个非常重要的证据。也正由于没有正常物质的存在,也即不存在万有引力定律中可以提供引力源的物质,也就使得修改万有引力定律的尝试不再是合理的研究方向。

证据四:子弹星系团

最后,我们再来说说最为重要的一个证据——子弹星系团研究。子弹星系团,是两个星系团由于引力相互靠近并发生撞击而组成的一个处于非平衡态的星系团。由于正常的物质之间存在很强的相互作用,当两个星系团撞在一起时,由正常物质组成的星系团部分将会纠缠在一起。而另一方面,如前面所说,暗物质不与正常的物质相互作用,所以当撞击发生时,没有其他相互作用使得暗物质部分减速。

用天文望远镜进行观测,可以看到这样一个非常有趣的现象:正常物质由于相互作用几乎停在星系团中心,而属于原本星系团中的暗物质由于不受任何阻碍,顺利地穿过了对方,从而形成了正常物质与暗物质组成部分互相分离不再重叠的神奇结构。通过引力透镜效应的观测,人们发现,子弹星系团中存在两个相互不重叠的引力中心,然而这两个引力中心附近并没有可以发光的正常物质。这便说明,星系团中大部分的质量来源于人们尚未了解的暗物质,这也成为暗物质存在的最直接的证据。

研究暗物质的实验方法

暗物质与正常物质之间的作用

虽然我们对于暗物质的性质有各种推测,但是暗物质与正常物质之间真的没有任何除引力以外的相互作用吗?暗物质之间到底又有什么关系?这些都是近年来高能物理学家研究的热门话题。可是如何研究暗物质与正常物质的相互关系呢？

一、直接观测法

观测宇宙背景中的暗物质与正常粒子相互碰撞的实验，如果暗物质存在于宇宙背景中，而又与正常物质存在相互关系，那么理论上暗物质是可以与正常粒子相互碰撞。人们知道，即使暗物质与正常粒子存在相互作用，那也是非常微弱的。因此直接观测对于精度的要求非常高，要对实验中可能遇到的背景噪音进行非常有效的控制。

我们会遇到什么样的背景噪音呢？比如，地球表面会接受大量的宇宙射线，如果这些射线进入探测器中并与探测中的粒子发生相互作用，便很有可能被误认为是暗物质的信号，所以这类实验都要在与外界尽可能隔绝的封闭空间进行。例如将探测器埋藏于地下几百至几千米深的废弃矿井中，从大气层来的宇宙射线，就可以很好地被这几百至几千米深的土壤所阻挡，这便形成一个利于暗物质直接探测的环境。或许你会怀疑这样的环境是否能满足实验的精度？事实上目前世界上有过若干暗物质直接探测的实验（比如在美国 LZ 和 CDMS 实验，意大利的 XENON 实验，以及中国的 PandaX 实验等），这些实验的精度均已达到极高的水平。有多高呢？由于暗物质与正常物质极微弱的相互作用，暗物质能轻易地穿过几百米深的地壳，但是即使进入探测器，暗物质与正常物质发生相互作用的几率还是非常之小，不过即使一年中只有一两个暗物质与探测器中的正常物质发生作用，我们也可以清晰地判断出哪些是暗物质与正常粒子的相互作用。在不久的将来，这些实验便会告诉我们暗物质是否具有特定的与正常物质之间的相互关系。

二、间接探测法

如果当宇宙中既存在暗物质又存在暗物质的反粒子，那么暗物质就能与它的反粒子发生碰撞而相互湮灭，即正反粒子相遇后通过质量—能量转换关系将粒子的质量转化成为比如光子这样的较轻粒子的能量的过程。又如果它们湮灭后产生的粒子为正常物质的话，那么我们便可以通过高能的、由正常物质组成的宇宙射线来观察在宇宙中是否存在暗物质与暗物质的反粒子的湮灭发生。当然正常的天体物理活动也能够产生高能的宇宙射线，所以这类间接观测就需要我们对宇宙的背景射线有很好的了解，以分辨射线产生的源头。

由于暗物质湮灭时产生的宇宙射线多种多样，所以我们会同时研究几种暗物质间接观测的信号通道。比如说，若暗物质湮灭产生高能光子，那这些高能光子便可能被伽马射线望远镜卫星所观测到。由于光子的传播几乎不受宇宙中其他物质的干扰，人们便能清晰地判断高能光子来自的方向，这便能更好地帮助我们判断到底高能光子是来自于正常的天体物理活动还是来自暗物质的湮灭。另外，暗物质的湮灭还能产生反物质，质量相同但具有相反量子数的粒子，所有粒子都有自己的反粒子，有些粒子的反粒子就是它们自己，比如光子；有些则是另外的粒子，例如正电子、反质子或反中子。

由于宇宙主要由正物质构成,反物质粒子在宇宙射线中相对少见,所以来自正常天体物理过程(正常的物质粒子参与的天体过程)中的反粒子一般会比正粒子少很多,于是宇宙射线中的反粒子便成为背景较为"干净"的、探测暗物质湮灭的主要通道。当然,这些反粒子也有不完美的地方。比如由于反电子带有电荷,当它们在宇宙中,尤其是在银河系中穿行时,它们的轨迹会由于星际磁场的存在而被弯曲,这便使人们失去了这些反电子来源方向的信息。

总的来说,暗物质的间接观测可以通过许多信号通道来完成,这些信号通道之间可能可以相互验证或相互证伪。若在几个不同的信号通道中,均发现与正常的天体物理过程所预测的宇宙射线不符,这些不符便极有可能来自于暗物质与它的反粒子的相互湮灭。当然并不是只有暗物质的湮灭能够产生高能宇宙射线。类似的,暗物质也很有可能自己衰变而产生这种高能宇宙射线。暗物质并不一定是一种稳定的粒子,它们的寿命要远远长于宇宙现在的年龄。如果暗物质粒子可以衰变,而它的衰变产物又恰恰是我们可以观测到的正常粒子的话,人们同样可以通过这种间接观测来探测暗物质的衰变,目前还没有观测到这样的衰变产物,相关的研究仍在进行中。

参考译文:

Dark Matter Exploring

The Composition of the Universe

Over time, with the high-speed development of human science and technology, the range of human activities are no longer restricted by the earth or Solar System, but are gradually expanded to the whole universal space. Wanting to navigate at the galaxy level, impetus of the spacecraft is a problem that must be solved. Learning to use the dark matter engine, which appears in the movie Transformers4, for interstellar crossing may become a necessary technology for human interstellar navigation in the 22nd century.

The aircraft using dark matter as engine become the first choice of interstellar crossing is because the process that appear and annihilation of dark matter and anti-dark matter produce high energy particle can convert all the mass of dark matter particle into energy, which is ten thousand times or even more powerful than the energy produced by Nuclear fission and Nuclear fusion. Moreover, the dark matters are scattered in space, while in the process of the spacecraft flying the dark matter can be collected and used to accelerate the spacecraft or maintain the normal operation of the spacecraft system. Super equipment such as Dark Matter Engine, although a scientific fancy at present, still arouses people's curiosity and question about if it is possible to come true.

To understand dark matter, it is necessary first to talk about the universe. What is the universe composed of? Perhaps you may think of the planets, dirt and expanded space. From a micro perspective, you are likely to think about molecules, atoms, and protons etc. However, these are only a small part of people's understanding of the universal composition. we can use three parts—normal matters, dark matters and dark energy—to conclude the universal composition. If they are divided according to the ratio of all universal energy, the normal matters only account of 4%, the dark matters 2% and the rest of about 74% are dark energy.

Although the dark energy is largely accounted in the whole universal energy, we have no idea about what it's composed of. What we unknow is more than known. It naturally has aroused great interest among many theoretical physicists. Therefore, there appear large amounts of theoretical models to explain the property and origin of the dark matter, which is a separate discipline direction. And quite a lot of scholars believe that dark energy is only a cosmological constant in the general relativity, which is a constant that appears in the Einstein field equation, but people do not understand why the cosmological constant is the number we've watched. If so, in a sense, people have understood the essence of dark energy. Compared to dark energy, the dark matter that existence in the shape of matter provide us with many clues. If we want to know about dark matter, we should have a look at what is the normal matter.

Normal matter is what we usually mean by matter, including neutron, proton, electron, neutrino and other particles. Chemists study molecule, atom physicists study atom, nuclear physicists study the world of neutron and proton, while a part of high energy physicists study the smaller particle that consist of neutrons and protons, such as quarks and gluons, W/Z standardized particle and Higgs bosons and etc. It has already had a profound understanding in the structure of these matters. But these are not enough to cater the curiosity of scientists, therefore the mystery dark matter became a popular research direction.

So far, there are still too many unknown properties of dark matter. It is a new kind of matter that is not luminous and is beyond the scope of people's understanding. At present, what kind of connection that between dark matter and between normal matter have is still unknown, what we know is that dark matter participates in gravitational interaction, which

is the only solid provident that certify dark matter's existence.

Evidence of the Existence of Dark Matters

Since we know so little about dark matters, but why can we determine the existence of dark matters? This can be attributed to astronomers. According to Newton's law of universal gravitation, we know that there is a gravitational action between any two objects with mass. The larger the mass of an object, the stronger the gravitation between them, the closer between objects the gravity strengthen as well. Due to the existence of gravitation, stars in the Galaxy all spin around the center of Galaxy. On the basis of Astronomy speculation, people can calculate the moving speed of stars in different positions in galaxy, thus physicists can speculate the mass disputation in the Milky Way Galaxy according to this. You may be curious about how astronomers to measure the moving speed of far stars, which is through red shift, Doppler Effect of light wave, of the light lines discharged by stars.

Doppler Effect refers to the phenomena that when wave source and the observer are moving relatively, the wave frequency received by the spectator differ from the frequency discharged by J wave source. Doppler Effect of light wave is that when stars moving far away from earth, the faster they moving, the higher energy of light observed in the earth. In that, the mass distribution of Milky Way Galaxy can be roughly estimated through stars distribution in the Milky Way.

Evidence one: Moving speed distribution curve of planets

On comparing velocity distribution curve of planetary motion in the Milky Way, and the estimation of Milky Way's gross mass, astronomers surprisingly found that the speed that planets spin around Milky Way Galaxy center was much higher than their estimation, which means there is much mass coming from a matter that is new, dark and beyond present understanding. While this part of mass that hasn't been observed is called dark matters, which is originally speculated by Jan Hendrlk Oort in 1932 based on the orbital speed of the Milky Way Galaxy planets.

Similarly, like planets forming the Milky Way, galaxies can also form clusters of galaxies. These large-scale structures of universes are formed by the gravitational gathering of gases that are very uniform from the beginning of the formation of the universe, and supermassive black holes may exist in the center of each galaxy. The galaxy also rotates around the center of the galaxy cluster by gravitation. As with the stars and the Milky Way, the speed of the galaxy is also very different from the center mass of the galaxy cluster. That is to say, dark matter is also widely present in clusters of galaxies. It should be emphasized here that dark matter is not directly related to the supermassive black hole in the center of the Milky Way. A black hole can be approximated understanding as a particle in the center of the Milky Way.

Evidence two: Formation of large-scale structures of the universe

You may be wondering if there is any other more direct evidence that can prove the existence of dark matter. Astronomical observations and theoretical studies in recent years have indeed found a lot of evidence for us. One of the notable evidence is the formation of large-scale structures of the universe. The universe experienced a big explosion at the very beginning. At that time, the temperature of the entire universe is very high even higher than the temperature inside the sun and the universe is filled with gases (mainly hydrogen and helium), which are evenly dispersed throughout the universe. As the universe expands, the gas in the universe gradually cools. Due to the small inhomogeneity of the gas distribution, it began to gather together in a gravitational force. After the gas was agglomerated gradually formed a cluster of galaxies, and galaxies were formed within them. The Milky Way is one of them. When the universe gradually comes down from the high temperature, the tiny inhomogeneity of the distribution of matter leads to the formation of galaxies. Simply, the area with more material distribution with the help of gravity will attract more matter from nearby areas, and accumulating to form such large-scale structures of the celestial as clusters of galaxies and galaxies.

After a series of accurate numerical simulations, it was found that if the dark matter exist or not will greatly affect the formation of the large-scale structure of the universe. Since dark matter occupies a large part of the universe energy ratio, and it also participates in gravitation like normal matter. If our current universe has no dark matter or dark energy, but only normal matter, then the large-scale structure of the universe will absolutely be different from the observations results. This is the second important evidence that physicists generally believe that dark matter exists.

Accurate numerical simulation can also estimate many properties of dark matter. One of the most important properties is that dark matter is "cold"! The so-called "cold" of dark matter mainly describes the speed that when the formation of dark matter in the large-scale structure of universe. If the speed of dark matter is close to the light speed and does relativistic motion, which is called "hot", otherwise if it is much slower than the speed of light, which is called "cold".

Neutrino is a special particle that has no mass and has little interaction with other particles. At the beginning of the dark matter model, neutrinos were an excellent option for dark matter. However, it has been found that due to the extremely small mass of neutrinos, it is difficult to produce slow-moving neutrinos during the evolution of the universe. If the "hot" neutrino is the so-called dark matter, then when the large-scale structure of the universe was formed, because the movement speed of the neutrino is close to the light speed, the high-speed moving neutrino cannot easily gather into a group, which will change the material distribution of the universe, which is inconsistent with the data observed in our experiments. On the other hand, if dark matter is bound in the Milky Way through the interaction of gravitational force, the speed of dark matter should not exceed one thousandth of the light speed. If the speed is too fast, the dark matter will fly directly out of the Milky Way!

Evidence three: Gravitational lens experiment

The above two evidence is based on the assumption that dark matter is an even distributed gravitational source. Some physicists have suggested that the experimental data does not conform to the theory not because of the existence of dark matter, but for the large-scale structure of the universe, Newton's universal gravitation is no longer applicable, and may require additional theories to describe. Based on this motivation, people began to modify the law of universal gravitation.

More specifically, we know that Newton's universal gravitation is inversely proportional to the square of the distance. This law has only been verified at distances below the solar system scale, and it is unknown whether this law is still true at the Milky Way scale. When physicists discovered that Newton's law of universal gravitation could not make a correct prediction without adding dark matter at the Milky Way scale, so that it proposed the possibility of modifying Newton's law of universal gravitation. However, the third evidence we will be talking about will largely exclude the possibility of modifying the gravity model.

Due to the existence of the gravitational field, the photons that make up the light no longer move along straight line, and their trajectory becomes "bent". If there is a dark matter mass in a space in the universe while without normal matter, then the dark matter within this interval can bend the light through its gravitational interaction as well. This is the famous gravitational Lens effect.

In the experimental observation of the gravitational lens, it was found that there were almost no normal substances in some intervals, but the light still had a large curvature in these areas, which indicates that there are other substances other than normal substances in these intervals. This is very important evidence for the existence of dark matter provided by gravitational lens experiments. It is also because there is no normal material, that is, there is no material that can provide the source of gravity in the law of universal gravitation, and the attempt to modify the law of universal gravitation is no longer a reasonable research direction.

Evidence four: Bullet Cluster

Finally, let us talk about one of the most important evidence that the bullet cluster study. The bullet cluster is a non-equilibrium cluster of galaxies formed by the gravitational coming close to each other and colliding. Due to the strong interaction between the normal substances, when two cluster of galaxies collide the parts of the cluster of galaxies composed of normal substances will be entangled with each other. On the other hand, as mentioned above dark matter does not interact with normal matter, so when an impact occurs there is no other interactions that decelerate the dark matter.

Observing with an astronomical telescope can see such a very interesting phenomenon that normal matter is almost stopped at the center of the cluster of galaxies because of the interaction, while the dark matter belonging to the original cluster of galaxies passes through each other without any hindrance. A miraculous structure in which normal and dark matter components are separated from each other and do not overlap is formed. Through the observation of the gravitational lens effect, it is found that there are two gravitational centers that do not overlap each other in the bullet cluster. However, there are no normal substances that can emit light near the two gravitation centers. This shows that most of the mass in the galaxy cluster comes from dark matter that people have not yet understood, which is also the most direct evidence of the existence of dark matter.

Experimental Method of Studying Dark Matter

The interaction between dark matter and normal matter

Although we have all kinds of speculation about the properties of dark matter, are there really no other interactions between dark matter and normal matter except gravity? What is the relationship between dark matter? Above these are the hot topics of high energy physicists over recent years. But how to study the relationship between dark matter and normal matter?

Direct Observation

In the experiment of observing the collision between dark matter and normal particles in the background of the universe, if dark matter exists in the background of the universe

and has mutual relationship with normal matter, then, theoretically, dark matter can collide with normal particles. It is known that even if there is interaction between dark matter and normal particles, it is very weak. Therefore, direct observation requires high accuration, and it is necessary to control the background noise that may be encountered in the experiment very effectively.

What kind of background noise would we encounter? For example, there are a large number of cosmic rays be received by the earth's surface. If these rays enter the detector and interact with the particles in the detector, they are likely to be mistaken as signals of dark matter, so such experiments should be conducted in a closed space as isolated as possible from the outside world. For example, if the detector is buried in in an abandoned mine that lies hundreds to thousands of meters below, the cosmic rays from the atmosphere can be well blocked by its soil with the depth of hundreds to thousands of meters, so as to form an environment conducive to the direct detection of dark matter. Maybe you wonder whether such an environment is accurate enough to the experiment? In fact, there were several experiments of direct detection of dark matter in the world (such as LZ and CDMS experiments in the United States, XENON experiments in Italy, and PandaX experiments in China). The accuracy of these experiments has reached a very high level.

What Level Is It?

Because of the very weak interaction between dark matter and normal matter, the former can easily pass through hundreds of meters deep crust. But even if it enters the detector, the dark matter and normal matter is little possible to be interacted with each other. However, even if only one or two dark matters interact with normal matter in the detector in a year, we can clearly determine which are the interactions between dark matter and normal particles. In the near future, these experiments will tell us whether dark matter has a specific integration with normal matter.

Indirect Observation

If there are both dark matter and dark matter antiparticles in the universe, then dark matter can collide with its antiparticles and annihilate together. That is, the process of transforming the mass of particles into the energy of lighter particles such as photons through the mass—energy conversion after the encountering of positive and negative particles. In another case if the particles produced by their annihilation are normal matter, we can observe whether there is the annihilation of dark matter and antiparticle of dark matter in the universe through high-intensive and cosmic rays composed by normal matter. Of course, normal astrophysical activities can also produce high-intensive cosmic rays, so this indirect observation requires us to have a good understanding of the background rays of the universe, so as to identify the source of ray generation.

Because there are all kinds of cosmic rays produced by dark matter annihilation, we will study several signal channels indirectly observed by dark matter at the same time. For example, if dark matter annihilates produces high-energy photons, these high-energy photons may be observed by gamma ray telescope satellites. Since the propagation of photons is almost free from the interference of other matters in the universe, people can clearly determine the direction of high-energy photons, which can better help us to determine whether high-energy photons come from normal astrophysical activities or from the annihilation of dark matter. In addition, dark matter annihilation can also produce antimatter, which is the particles that have the same mass but opposite quantum number. All particles have their own antiparticles, some of which are themselves, such as photons, while some are other particles, such as positrons, antiprotons and antineutrons. As the universe is mainly composed of positive matter and antimatter particles are relatively rare in cosmic rays, so the number of antimatter particles from normal Astrophysical Processes, celestial processes involving normal matter particles, is generally much less than that of positive particles, so the antimatter particles in cosmic rays become the main channel for detecting dark matter annihilation with a cleaner background. Of course, there are also imperfections in these antiparticles. For example, due to electric charge of antiparticles, when they are in the universe, especially travelling the galaxy, their tracks will be bent due to the existence of the interstellar magnetic field, which makes people lose the information of the source direction of these antielectrons.

In general, the indirect observation of dark matter can be done through many signal channels, which may be mutually verified or falsified. If it is found in several different signal channels that they are not consistent with the cosmic rays predicted by normal Astrophysical Processes, these inconsistencies are likely to come from the mutual annihilation of dark matter and its antiparticle. Of course, it is not only the annihilation of dark matter that can produce high-energy cosmic rays. Similarly, dark matter is likely to decay on its own to produce such high-energy cosmic rays. Dark matter is not necessarily a stable particle; but may live much longer than the universe. If dark matter particles can decay, and its decay products are exactly the normal particles we can observe, people can also detect the decay of dark matter through this indirect observation. At present, no such decay products have been observed, and relevant research is still in progress.

单词释义：

1.	《变形金刚4》	Transformers 4
2.	暗物质引擎	Dark Matter Engine
3.	星际穿越	interstellar crossing
4.	星际航行	interstellar navigation
5.	湮灭	annihilation
6.	高能粒子	high energy particle
7.	核裂变	nuclear fission
8.	核聚变	nuclear fusion
9.	暗能量	dark energy
10.	中微子	neutrino
11.	宇宙学常数	cosmological constant
12.	坦场方程	Einstein field equation
13.	胶子	gluons
14.	W/Z 规范粒子	W/Z standardized particle
15.	希格斯玻色子	Higgs bosons
16.	天文学家	astronomers
17.	牛顿万有引力定律	Newton's law of universal gravitation
18.	光波的多普勒效应	Doppler Effect of light wave
19.	J 波源	J wave source
20.	银河系	Milky Way Galaxy
21.	星系团	clusters of galaxies
22.	大尺度结构	large-scale structures
23.	超大质量	supermassive
24.	引力透镜实验	Gravitational lens experiment
25.	引力场	gravitational field
26.	光子	photons
27.	暗物质团	dark matter mass
28.	子弹星系团	Bullet Cluster
29.	宇宙射线	cosmic rays
30.	反粒子	antiparticles
31.	伽马射线望远镜卫星	gamma ray telescope satellites
32.	正电子	positrons

33.	反质子	antiprotons
34.	反中子	antineutrons
35.	星际磁场	interstellar magnetic field
36.	高能宇宙射线	high-energy cosmic rays

译文解析：

1. 词性转译法。

汉英常用词语词性略有不同，如汉语造句时动词用得多，名词介词用得少；英语恰好相反，动词用得少，名词介词用得多，当然其他词性之间也有转换。

原文：因此直接观测对于精度的要求非常高，要对实验中可能遇到的背景噪音进行非常有效的控制。

译文：Therefore, direct observation requires high accurate, and it is necessary to effectively control the background noise that may be encountered in the experiment.

分析：名词"精度"译为形容词"accurate"，形容词"有效的"译为副词"effectively"。

2. 语态转译法。

英语中被动语态使用比较多，汉语被动语态一定要译成英语被动语态；有些句子，汉语不用被动语态，英语也要使用被动语态，因此在汉英翻译中要注意语态的转换。

原文：这些宇宙的大尺度结构都是从宇宙形成之初分布非常均匀的气体逐渐由万有引力聚拢坍缩形成的，而每一个星系中心都可能存在超大质量黑洞。

译文：These large-scale structures of universes are formed by the gravitational gathering of gases that are very uniform from the beginning of the formation of the universe. , and supermassive black holes may exist in the center of each galaxy.

分析：在这里明显有汉语被动标记语"由 形成的"，而汉语被动语态一定要译成英语被动语态，因此选择译成被动语态"成的"，而汉语被动语态一定要。

3. 分译法。

当原文有多个分句，逻辑不够清晰时，翻译时就应采用分译法，将原文逻辑重新归纳处理。

原文：想要进行星系（galaxy）级别的航行，宇宙飞船的推动力是必须要解决的问题，学会使用如电影《变形金刚4》出现的暗物质引擎（Dark Matter Engine）进行星际穿越，或许会成为22世纪人类星际航行的必备的技术。

译文：Wanting to navigate at the galaxy level, impetus of the spacecraft is a problem that must be solved. Learning to use the dark matter engine, which appears in the movie Transformers 4, for interstellar crossing may become a necessary technology for human interstellar navigation in the 22nd century.

分析：前面两个分句和后面的几个分句逻辑关系不是很明显，完全可以成为两个独立的句子，这样结构会更清晰些，利于读者理解。

4. 逻辑转译法。

英语形合，汉语意合，英语注重外部结构，因此要使用增译法把原文内含的逻辑结构译出来，必要的连接词是不可以像中文一样省略的。

原文：而且暗物质散布于宇宙空间，在宇宙飞船飞行的过程中，暗物质可以不断被收集并被用于飞船的加速或者维持飞船系统的正常运行上。

译文：Moreover, the dark matters are scattered in space, while in the process of the spacecraft flying the dark matter can be collected and used to accelerate the spacecraft or maintain the normal operation of the spacecraft system.

分析：前面两个分句主语各不相同，可以看出句子之间有转折关系，因此增加连词"面两个分句主来表达汉语中蕴含的转折关系，而不是像汉语一样省略。

5. 无主句翻译策略。

汉语中无主语句和省略主语的句子很常见。而英语中，主语和谓语相互依附而存在，有谓语就应该有主语，主语不可缺少。

原文：要了解暗物质，需要先从宇宙说起。

译文：To understand dark matter, it is necessary first to talk about the universe.

分析：这是一个科技英语文本中常见的无主语句，前一句译为目的状语，而后一句由于主语较长，为了平衡句式在主语的位置放一个形式主语"it"起到了强调作用。

原文：自然引起了众多理论物理学家的极大兴趣。

译文：It's naturally has aroused great interesting among many theoretical physicists.

分析：这是一个无主语句，翻译时增加了 it 做主语，但是并不是形式主语，而是指代前文内容，由于英文一定要有主语，这样处理能使句子变得完整。

6. 省略法。

由于汉英两种语言的差异，汉译英时常常要省略英语中所没有的词，如量词和语气词，或因表达习惯不同而造成的省译。

原文：如果是这样的话，在某种意义上，人们已经理解了暗能量的本质。

译文：If so, in a sense, people have understood the essence of dark energy.

分析： 这里"如果是这样的话"直接译成了"如"，简单易懂，完全符合原文想要表达的意思，但又没有过多的赘述，符合科技英语中简洁的用词原则。

7. 合译法。

为了符合目的语的表达习惯，有时会使用合译法，即把两个或两个以上简单句或复合句在译文中用一个句子表达出来。有时也会采用分译法，即把一个简单的句子分成两个或两个以上，使逻辑更加清晰。

原文： 暗物质到底具有什么样的属性？至今仍然有太多的未知，它是一种新的、不发光、不在人们现在所理解范畴内的物质。

译文： So far, there are still too many unknown properties of dark matter. It is a new kind of matter that is not luminous and is beyond the scope of people's understanding.

分析： 这句话同时使用了分译与合译，将前面两个分句结合在一起，而把后面一个句子拆开，因为前面两个分句的意思比较连贯可以合译，逻辑更明显些，而后面是描述其特征的句子，可以作为一个完整的句子。

8. 代词翻译的处理。

在英语中经常使用"这"字句，汉语语法称"这"前面的文字为外位成分，"这"字有点像英语中的形式主语。其翻译方法有很多种，具体使用可以根据其句式特点来决定。

原文： 我们仅仅知道的是暗物质参与引力相互作用，这是暗物质存在的唯一一个确凿的证据。

译文： What we know is that dark matter participates in gravitational interaction, which is the only solid provident that certify dark matter's existence.

分析： 在这句的翻译中，把"这"字句译成了一个复合句，即把其汉语外位成分"我们仅仅知道的是暗物质参与引力相互作用"译为主句，引导非限制性定语从句，说明主语。

9. 倒装句的翻译。

汉语偏正句一般"偏"句在前，"正"句在后。若把"正"句提到偏句之前，就成了倒装句式。汉语中倒装句使用比较少，而英语中倒装使用较多。

原文： 根据牛顿万有引力定律，我们知道任何两个有质量的物体之间都存在着引力作用，物体的质量越大，相互之间的引力就越强；物体之间的距离拉近，引力也会增强。

译文： The larger the mass of an object, the stronger the gravitation between them, the closer between objects the gravity strengthen as well.

分析：这一句是明显的比较结构,译成英文时采用倒装结构,符合英语的使用规律,在修辞上,为了强调,为了平衡结构,英语常用倒装句。

10.名词化的处理。

科技英语中多使用名词化结构,增强科技英语的客观性。

原文：若在几个不同的信号通道中,均发现与正常的天体物理过程所预测的宇宙射线不符,这些不符便极有可能来自于暗物质与它的反粒子的相互湮灭。

译文：If it is found in several different signal channels that they are not consistent with the cosmic rays predicted by normal Astrophysical Processes, these inconsistencies are likely to come from the mutual annihilation of dark matter and its antiparticle.

分析：原文中"与正常的天体物理过程所预测的宇宙射线"译成了 ophysical Processes, these inconsisten 使用名词化结构,增强科技英语的客观性。

11.无主句的处理。

科技英语翻译中,广泛使用非谓语动词,动词不定式、分词、动名词。

原文：多普勒效应是指波源和观察者有相对运动时,观察者接收到波的频率与 J 波源发出的频率并不相同的现象。

译文：Doppler Effect refers to the phenomena that when wave source and the observer are moving relatively, the wave frequency received by the spectator is differ from the frequency discharged by J wave source.

分析：Doppler Effec 作后置定语修饰"后置定语修饰 Effecy",过去分词作定语,典型的非谓语结构,符合科技文本的特点。

12.否定转移法。

汉语和英语两种语言表达否定的方式大多数情况下是相同的或基本相同的,但是少数情况下不同,甚至差别很大,所以要注意差异,在翻译时做出必要的正反处理,使译本符合中文的表达习惯。

原文：简单地说,物质分布较多的区域会在引力的帮助下从附近的区域吸引更多的物质,而不断累积则形成了宇宙的大尺度结构,如星系团和星系。

译文：Simply, the area with more material distribution with the help of gravity will attract more matter from nearby areas, and accumulating to form such large-scale structures of the celestial as clusters of galaxies and galaxies.

分析：在原句中"不断积累"在英语中的名词化肯定结构,因为在英语中习惯使用肯定形式的词义含蓄否定句来表达否定含义。

13. 增译法。

增译法是为满足目的语读者需求,增加补充翻译内容的一种方法。由于英汉两种语言具有不同的思维方式、语言习惯和表达方式,在翻译时,为了使译文更符合目标语的语境和语用习惯,译文需要增添一些词、短句或句子,以便更准确地表达原文的意义。

原文: 由于气体分布的微小不均匀性,开始在万有引力的作用下逐渐聚拢成团。

译文: Due to the small inhomogeneity of the gas distribution, it began to gather together in a gravitational force.

分析: 在做主语,采用增译一方面能有效保证译文语法结构完整;另一方面能确保译文意思明确。

14. 合译法。

为了符合目的语的表达习惯,有时会使用合译法,即把两个或两个以上简单句或复合句在译文中用一个句子表达出来,使逻辑更加清晰。

原文: 气体成团后,便逐渐形成了星系团,而在星系团内部,便形成了一个个星系,我们所处的银河系便是其中一个。

译文: After the gas was agglomerated gradually formed a cluster of galaxies, and galaxies were formed within them. The Milky Way is one of them.

分析: 原文是两个句子,但是通过分析发现,译成英语的逻辑层次略有不同,前面四个分句描写形成星系的过程,最后一个分句是总结性话语,可以单独成句,因此把这个句子拆分开来。

15. 名词化结构。

科技英语中,广泛使用名词化结构,即用名词短语表示动作,用名词短语表示调价、原因、目的、时间等状语从句。

原文: 所谓的暗物质的"冷",主要描述的是暗物质在宇宙大尺度结构形成时的速度。

译文: The so-called "cold" of dark matter mainly describes the speed that when the formation of dark matter in the large-scale structure of universe.

分析: 这里的"形成"译成了 cold,符合科技英语善用名词化结构的特点,使文章客观性增强。

第六章　天空互联网:连接未来世界

　　在"黑科技"前面十多个篇章里,技术创新正在缔造的纷繁世界已经浮现眼前,而未来,这一切将如何连接起来? 什么? 宽带、IPv6? 问题不在这些层面,尽管宽带正在走向超宽带。无论是下一代互联网、下一代信息技术设施意义上的NGI,还是下一代网络,"未来网络"的形态和架构,不是我们眼前看到的互联网,也不是过去一个阶段讨论较多的NGN。在ITU国际电信联盟、IETF国际互联网工程任务、IEEE电气和电子工程师协会,以及OneM2M、3GPP、OMA、ISO/IEC JTC1等组织,以及ZigBee、Z-Wave、AllSeen等方向的各种联盟那里,有些许答案,但技术走得更远,且不是在类似"从http/1向http/2演进"这样的层面。未来已经开始发生。会发现未来网络在一定程度上是多路力量齐头并进、多维创新协同作用的结果。能量密度、连接密度、数据密度,材料尺度、感知尺度、网络尺度,计算速度、移动速度、融合速度,这9个度在影响网络演进的速度和形态,不同领域和层面的基础科学、应用科技、产业群落在9个维度的突破创新,使得未来的网络——智慧网络若隐若现。未来的智慧网络,从放眼天空、仰望星空开始。空间互联网、天空互联网,也就是正在到来的"天网",是未来智慧网络当中最基础的"连接"。这注定是一个极富探索性同时也充满争议的话题。

太阳能,无人机,未来的"天网"平台?

　　2016年6月29日,Facebook太阳能无人机Aquila在亚利桑那州完成首次试飞。原计划飞行30分钟,由于进展顺利,最终飞行了96分钟,成功收集了与模式、飞机架构有关的飞行数据。在以无人机作为网络平台的方向,Facebook成功地迈出一大步。两年前收购英国太阳能无人机研发企业Ascenta之后成立的Project Aquila项目,给Facebook的internet.org计划成功带来"巨大里程碑"。Internet.org是Facebook CEO扎克伯格着力甚多的未来项目,目标是通过网络连接世界上的每个人。尤其是尚未接入网络的40亿人——贫困、偏远、网络状况比较差甚至没有网络覆盖的地区的人们。Intermet.org多方努力,与手机厂商和运营商合作,Free Basics项目免费为民众提供300多项简化的互联网服务。名为项目免费为民众提供300多项简化的互的部门也

88

因 Internet. org 项目而成立,专门负责寻找激光、无人机等网络通讯新方法,包括将人工智能与上网服务结合起来,并最大程度的推进上网技术的开源、开放与共享。Internet. org 的目标是将互联网连接数量增加 10 倍,将上网价格降到目前 1/10 水平。

这时候有人说了,Facebook 式的激光通信要变成可以规模化应用的商业项目需要10 年。可是那又怎样,Facebook 还是出发了。成功首飞的 Aquila 有和波音 737 一样长的 43 米的翼展,机身重量约 454 公斤,相当于载人客机的百分之一;由氦气球提升至气候环境稳定的平流层,Aquila 白天飞行于 27432 米左右,避开飞机航线高度,吸收和贮存太阳能;晚上飞行于 18288 米以节约电能。Facebook 的目标是 Aquila 太阳能一次可以在高空自持飞行 90 天。90 天?听起来这个数字挺骇人,甚至耸人听闻,但未来这是可行的。这里又是一个能量密度材料尺度等涉及 9 度理论的问题。Facebook有意未来在全球各地的天空部署 1000 架甚至 10000 架这样的无人机,每架无人机在直径 60 英里范围内来回转圈。这么多无人机在天上转来转去不是为了刷存在感,而是提供普遍的互联网接入服务。这个时候激光通信技术就派上用场了,"意未来在全球各地的天空部署 1000 的负责人表示,他们的激光通信数据传输速率能达到10Gbps,近乎地面光纤水平,是标准激光信号的约 10 倍。通信过程是,地面母基站与无人机之间进行激光与电磁通讯,无人机也可将信号通过激光发送到其他无人机进行中继。机群将激光光束向下发送到地面子基站的收发器,以收发器为圆心,网络信号可以覆盖半径 30 英里的地区以便上网。系统会将信号转化为 Wi-Fi 或者 4G、5G 网络。

这个时候问题来了,Project Aquila 说白了还是需要地面子基站的,无人机只是发挥了类似传统电信网络的骨干网的作用,这就还不如传统卫星通讯服务商 Iridium 的老思路了,后者至少全程都是在空中,以每个人都可以使用卫星手机为目标。Iridium的卫星在太空,Aquila 无人机在大气层内距地面 20 公里左右的平流层,电离层之下,理论上 Aquila 是可以直接做空中基站,省去地面基站环节,但是这样一来,重量仅仅454 公斤的 Aquila 怎么能够受得了重量远在自身之上的收发设备,又怎么能够仅靠太阳能飞行 90 天不掉下来,还能够进行能耗巨大的天地通讯,地面上网设备比如手机的天线之类的一整套东西也要跟着制定相关标准了。到这里,又是一个能量密度、连接密度、材料尺度等涉及 9 度理论的问题,但并非不可逾越。

Free Basics 已经为地球不同角落的上千万民众提供了互联网接入服务,但是如同在印度、埃及等国家招到一些政府和民间组织的强烈反对一样,Project Aquila 在全球各地面临的阻力不会比空气的飞行阻力小。在印度,Free Basics 甚至被禁止。那些或援引既有法律,或以网络中立原则或税收问题阻止,或振振有词或冠冕堂皇的反对理由,也许没有一项是站得住脚的,但是触及传统利益,就会招致传统力量的反对,更何况这一次无人机要飞临的是传统主权国家的边界。相比之下,频谱资源不是最大的那个问题。

在太阳能无人机天空互联技术的探索者中,还有波音公司、AT&T、英特尔、空客公司、ZephyrS/T 等大大小小的企业和团队。Google 也没有落后,在美国新墨西哥州的美国太空港逾 15000 平方英尺的机坪上,Google 正在进行着一项名为 Project Sky Bender 的新计划,同样是采用无人机作为网络平台,不同的是 Google 采用未来高速无线通讯技术之一的毫米波通讯技术,号称比 4G LTE 传输速度快 40 倍,甚至可能成为 5G 网络通讯骨干,Project Sky Bender 因此颇有空中 5G 网络平台的意思。

气球的可靠性虽然有待观察,但也已经被作为空中网络平台的探索方向之一。Google 在进行的 Project Loon 和 Project Sky Bender 同属一个项目,它将高空聚乙烯氦气球送入平流层作为基站,形成网络覆盖,将互联网带到地球各个角落。Project Loon 的气球高度约为商务飞机的 2 倍,已进行的试验能够在高空停留 180 天以上。WiFi 飞艇 Google 也有在尝试。中国的北京航空航天大学正在实验临近空间飞艇,以此实现无线网络覆盖。中国深圳的光启公司,就是投资入股新西兰研发个人飞行器的 Martin 公司的那家民企,他们号称"云端号"WiFi 飞艇已经进行初步测试,不过其技术细节还需要推敲。总体看来,气球和飞艇在未来天空互联网中的位置,属于处在补充、次要地位的网络平台。

智能宽带卫星网络,比无人机更遥远,但更贴近"天网"未来? 无论叫 Sky-Fi 还是天空互联网、空间互联网,并不重要,重要的是未来的网络信号必然首先来自天空、星空。卫星比无人机更遥远,但是从技术成本效率来看反倒更贴近未来。技术在 6 个方面的快速进化,是"天空互联网"越来越逼真的关键:发射成本大幅度降低、轨道近地化、波段高频化、卫星智能化和小型化、天线与终端小型化与低功耗、天地一体组网技术等。卫星制造成本、发射成本不断降低,带宽、可支持用户量不断提高,智能宽带卫星网络日趋可行。这是一条不断接近性价比临界点的路,尽管眼前和电信固网、移动网络相比,性价比还不够高。

Space X、Google、Facebook 是这场未来网络游戏的大玩家。Space X 一开始计划发射 700 多颗低成本的低轨道卫星,为地面提供上网服务。不过根据 Space X 在 2016 年向联邦通信委员会提交的最新报告,这项向全球提供卫星宽带网络服务的计划,将发射 4425 颗卫星。迄今为止着重点依然是发射服务,以及不断提高自己的火箭回收重复利用技术,为将航天发射成本降低到新的临界值而努力,甚至在此基础上宣布了雄心勃勃的火星计划。Space X 同时也为 Facebook 和法国卫星运营商 Eutelsat 合作的宽带卫星上网项目提供卫星发射服务。遗憾的是,2016 年 9 月第一颗卫星就被 Space X 失败的猎鹰火箭发射送到火焰里去了。以色列公司制造的这颗 5 吨重、造价 2 亿美元的 Amos-6 卫星化为灰烬,原本它要为撒哈拉沙漠以南部分非洲地区提供互联网服务,对 imenet.org 来说这是个不小的挫折。这场未来网络空间竞赛游戏里也有创业公司的身影,一家以色列公司干脆把自己公司的名称命名为 SkyFi,且对外宣布计划向太空发射 60 颗微型卫星。卫星上天看起来悬,SkyFi 的微型卫星天线却反倒有些自己的独到技术,引来多个买家与其接触。

One Web 比 Space X 低调得多,但是股东背景一样来头不小,站在后面的是 Virgin Galactic、Qualcomm、Honeywell Aerospace 等。One Web 的信号处理芯片就来自高通,后者利用其终端与基站之间的切换技术,帮助建立卫星通信网络,解决诸多卫星在掠过一个个地面基站过程中的交接、切换问题。和 Space X 一样,One Web 要用小型低轨道卫星网络覆盖地球,计划发射 648 个小型卫星到近地轨道,终端接入速率约为 50Mb/s,每颗卫星的制造费用在 35 万美元左右,项目总成本约 20 亿美元。One Web 为航空公司、灾难救援组织、个人家庭客户、偏远山区的学校和村落提供服务。不过,尽管是近地轨道,OneWeb 的天线和功耗技术似乎一般,设备小型化程度还是不够高。地面基站的设备尺寸依然不小,虽然可以用太阳能电池板供电,但体积未来还是需要缩小。

美国 MDIF 公司 2014 年曾经发布的 Outnet 外联网计划是个插曲,MDIF 向近地轨道发射数百颗卫星以支持全球免费 WiFi 的实现,虽然貌似动人,但这个 Outnet 的技术思路显然有问题,终端用户只能单向接收经过挑选的网络内容,不能互动,仅仅只是单向广播,不合时宜。

OneWeb 的频谱资源通过 O3b 获得,而 O3b 是这个领域另一个重要角色,可以称之为中轨道玩家。起步不晚、开局不错,不过现阶段有些问题。O3b 的成立之意,在于解决地球上 other 3 billion——也就是另外 30 亿人没能上网的问题。Google、SES、汇丰银行等不同行业巨头是其重要投资人。相比原来 35000 多公里高度地球同步轨道通讯卫星存在的时延问题,处于 8000 公里中轨道的 O3b 卫星网络时延低于 150 毫秒,且中继带宽达到常规光纤水平,这意味着网络品质可以规模化商用了。O3b 在 2013 和 2014 年通过阿丽亚娜火箭已经分别发射两个批次的 Ka 波段卫星,实现 8 颗在轨。O3b 卫星网络计划达到 12 至 16 颗卫星,利用这些成本比过往地球同步轨道卫星低廉得多的卫星,覆盖非洲、中东、亚洲、拉丁美洲等区域,提供最快可达 10Gbps 的速度和总容量 84Gbps 的网络服务给非洲、中东、亚洲、拉丁美洲等区域的发展中国家。此前,尽管已经有号称 140Gbps 全球最高容量的宽带通信卫星 ViaSat1 在轨,同为 Ka 波段,但 ViaSat 公司是高轨道玩家,地球同步轨道以及 10 倍于中轨道 O3b 卫星的总质量,使其成本极为高昂。ViaSat1 的地面系统包括卫星用户终端——Ka 波段蝶形天线和卫星调制解调器,网关卫星地面站及网络操作中心,对企业、家庭用户的服务能力相对较强。相比之下,O3b 的天网定位于骨干网络而不是最终用户接入,O3b 采取一颗卫星下降到地平线之后由另一颗卫星接力的网络策略,使得制造、发射成本大幅降低,O3b 自己的数据是有希望让非洲等地区的上网成本降低 95% 以上。太平洋岛国、非洲、美洲等区域的 40 多家 3G 和 LTE 移动运营商、互联网接入服务商已经成为客户。

当然,位于加利福尼亚州的 Viasat 公司也不会满足于现状,随后的 Viasat2 卫星带宽会是 Viasat 1 的 2 倍,容量 2.5 倍,为 250 万用户提供服务,宽带互联网服务下载速度从 Viasat 1 的 12/15Mbps 提升到 25Mbps。重组之后的 Google 在天空互联网方向越来越没了感觉,先是从 O3b 退股,后来又取消了 10 亿美元打造 180 颗高性能绕地卫星网络计划。但是,在关乎未来的重大方向上,Google 不会一去不回。在卫星图像领域 Google 已经有多起投资,间接拥有多颗在轨卫星。三星尽管没有行动,但在口头上也表示自己也是一家世界级的、关注全球网络问题的大公司。三星称,未来要发射 4600 颗微型卫星,为用户提供低成本的互联网接入。

中国企业和相关机构尽管技术实力、所处发展阶段不太一样,但是在高中低不同轨道的发展方向与前述项目大致相同。中国航天科技集团在进行高通量宽带卫星项目,2016 年发射了第一颗地球同步轨道移动通信卫星天通一号 01 星,为船舶、飞机、车辆等大型移动用户以及手持终端提供通信、短报文、语音和数据传输服务兼备。中国卫通也在实施 Ka 频段宽带卫星计划。十几年前,前北电网络公司(Nortel Networks)CEO 欧文斯曾提出和华为合作做低轨道卫星,类似今天 Facebook 和 Google 的方案,相关讨论未能继续。不过在低轨道方面,2014 年清华大学与信威集团联合研制的首颗灵巧通信试验卫星完成发射并进行在轨测试。卫星重量约 130 公斤,运行高度约 800 公里,通信覆盖直径约 2400 公里。测试验证了星载智能天线、星上处理与交换、天地一体化组网、小卫星一体化集成设计等多项技术,实现手持卫星终端通话、手持卫星终端与手机通话、互联网数据传输等业务。为了实现未来天空互联网布局,信威集团甚至通过其子公司卢森堡空天通信公司向以色列 Space-Com 发出收购邀约,而 Space-Com 就是被 SpaceX 的猎鹰火箭送到火焰里的那颗 2 亿多美元的宽带卫星 Amos-6 的制造者。香港上市企业中国趋势控股有限公司与美国休斯飞机公司(Hughes Aircraft)合作,计划通过采用最新的大容量 Ka 波段宽带卫星资源,在亚太地区打造免费卫星移动互联网,用户可使用指定终端实现卫星上网、拨打卫星电话、收看卫星电视。

未来每个人都可以通过移动手持设备上"天网"吗?

或者说,未来人人都会通过天空互联网上网,天空互联网会成为未来连接的基础网络吗?眼下,包括一部分电信通信甚至卫星通信从业者在内,恐怕许多人会这么说:手持设备怎么可能卫星上网?打打卫星电话还行,上网尤其是宽带上网还是算了吧,发射功率太小,天线体积太大,上行速率难以提高,用卫星网络作为骨干网为移动运营商或者接入服务提供商的地面基站提供网络中继服务还行,直接向个人用户提供大规模互联网接入服务,难!摩托罗拉耗资数十亿美元的铱星计划不就这么破产的吗?1996 年开始发射,1998 年开始提供服务,到 1999 年 3 月破产的时候,铱星手机在全球才发展了 5.5 万多用户；而此时世界各地的电信运营商已经把更便宜、更便携、通讯性能更好的手机送到千百万用户手上了。和铱星计划同时代的 Globalstar 也死得很难看，Teledesic 尽管有比尔盖茨甚至沙特王室出手资助，却连项目成型的那天都没有等到。

但是技术驱动的创新进化，就是这样一个不断前仆后继、生生死死、死而复生的过程。私募基金后来接盘铱星计划，蛰伏数年后甚至成为 8 亿美金市值的美股上市公司铱星通讯（IridiumCommunications），尽管这只是铱星计划当时庞大投资的一个零头。2014 年，铱星通讯推出全球覆盖卫星 WiFi 热点服务 IridiumGO，而这个时候，已经是 OneWeb、SpaceX、Facebook 甚至 Sky-Fi 们的真正的天空互联网开始风起云涌的日子了。无论怎么应景和努力，铱星通讯都只是明日黄花，因为从技术层次和通信体制的角度看，铱星通讯都首先是一个卫星电话网络，而新生力量们所要实现的，是真正以数据通讯为基础的天空互联网，而不是电话或 GPS 网络。这个阶段，火箭重复利用等技术使得卫星发射成本大幅度降低，而近地轨道卫星组网不仅能够有效解决时延问题，信号质量也远比地球同步轨道好，因此也有助于更小的天线工作；波段高频化尤其是 Ka 波段的大规模深度开发利用成为现实，为海量用户提供宽带服务的技术障碍已经扫除；卫星的智能化和小型化、天线与终端小型化与低功耗以及天地一体组网技术，这些也都成为不仅看得见也能够落得实的技术趋势。天空互联网领域，已经不仅是休斯、劳拉、波音等传统卫星制造商和卫星通讯运营商的天空，IT 企业、互联网巨头、新创企业、新生力量们已经当仁不让。

站在技术角度，微型天线技术、小型手持设备、卫星手机、手机卫星上网方面的产品动向尤其值得注意。未来最鼓舞人心的变化，也会发生在这个部分。看过移动卫星通信服务提供商 Inmarsat 的卫星热点设备 ISatHub 就知道，地球同步轨道卫星的地面设备已经可以小型化到半个笔记本电脑大小，这还是天线和 Modem 等不同部分共同加起来的体积。它很容易让你想起 20 年前电脑拨号上网阶段的 Modem，具有同样的体积。在一些国家，X 波段、Ku 波段和 Ka 波段移动卫星通信兼备的设备已经实现车载、背负甚至单人手持，而过去没有一口如同大锅的卫星天线和工作站级别的沉重设备，是无法想象的。至于中、低轨道尤其近地轨道卫星的地面终端设备，普遍可以手持。从 Inmarsat 到铱星、全球星、亚星电话、Spot 等卫星电话，最突出的是粗壮的通讯天线，主流卫星电话的体积已经远小于最初的 GSM 移动蜂窝电话。而下一步，随着天线、电池技术的进一步提升，以及卫星的规模化、高带宽网络服务能力的提升，有希望逐步创造与普通智能手机相近的卫星上网体验。

总部位于迪拜的卫星运营商 Thuraya 的智能手机卫星适配器是个有趣的方向。即使没有卫星电话，用户的 iPhone 或者 Android 手机只要套上 SatSleeve 适配器，在应用商店免费下载安装 SatSleeve App，智能手机即可与适配器有效连接，然后用户就可以在卫星网络模式下拨打电话、收发短信和电子邮件，使用一些社交、即时通信软件也没问题。Thuraya 的 SatSleeve 适配器,样子和厚一点的手机保护壳看上去没有太大不同,除了粗壮的天线,其他方面是一眼看不出来它竟然能让智能手机秒变卫星电话的。如此看来,卫星上网距离每个手机用户有多远?

超级 WiFi,微基站,Mesh 互联,未来地网与天网形成自联网?

理想而言,所有设备都可以通过天空互联网连接起来,但是天地一体、多网混合、应需组网,将是未来最广泛的应用形态,各种不同特性的网络在不同场景发挥各自所长。近距离通讯过去是 Zigbee、ZWave、AllSeen、Bluetooth 们的专长,但是 WiFi 正在快速切入,中近距离是 WiFi、超级 WiFi 和移动运营商的基站的空间,骨干网在天空有卫星在地面有光纤,有些场所的固网接入依然是光纤,激光等在骨干网和接入网之间发挥中继作用。在局部,越来越强大的 WiFi 正在部分取代原来必须由移动通信基站发挥的作用,超低功耗 WiFi 则在充分替代 Bluetooth。这里的超级 WiFi 是指信号距离远、穿透力强、超高带宽、多路的 WiFi 网络。而 IoT 物联网、车联网,既是融合传感,网络环境必然是卫星、基站、WiFi 和 Zigbee 的融合应用。

天网部分最值得关注的是规模化面向最终用户的低轨道智能宽带卫星网络,地网部分最值得关注的当然是 WiFi 以及 WiFi 互联。更高更快更强,WiFi 的发展并非同一维度的渐进,而是有望在全新维度创造全新的网络环境,以至于很多时候人们将会遗忘电信。这是场正在地球表面的空气中进行的无线革命。无所谓基站,每一个热点都是一个微基站,无数强有力的 WiFi 热点彼此互联且与低轨道卫星网络实时互联,包括手机在内的每一部稍具能力的智能设备也都是一个 WiFi 热点、中继点、微基站,这就是未来最具效率且分布最为广泛的网络环境,其他特性的网络作为局部补充,未来的网络、网络的未来已经若隐若现。

6 个方面的技术突破正在驱动 WiFi 创造未来的网络:传输距离、穿透能力、超低功耗、带宽容量、多用户、不是 Mesh 的 Mesh 互联。速度方面,WiFi 的 5GHz 频带吞吐量预计可达 10Gbps,60GHz 可以达到 20Gbps,理论上端到端点对点突破 100Gbps 不是问题。不过印象最深刻的技术突破是,华盛顿大学研究人员利用电磁后向反射技术,研制出的全新超低功耗 WiFi 技术,也被称之为无源 WiFi,发射功率仅为 10-50 微瓦,是传统 WiFi 路由器的万分之一;更重要的 是,可以进行 WiFi 充电。而 WiFi 充电,是前景广阔的无线充电领域非常有趣的方向之一,WiFi 充电技术的研究,在美、日、以、中、欧等世界主要创新经济体都已经不乏其人。

6 种技术当中最具生态影响力是:不是 Mesh 的 Mesh,将驱动网络、设备之间通过协议实现应需互联,移动自组网,称之为自联网并不为过。Mesh 无线网格网络由 ad hoc 网络发展而来,可以与其他区网络协同通信。自组织、无中心、无边界、动态扩展、任意设备均可互联、每个设备都可中继是 6 个特点。

20 年前有这么一句话，"全世界 PC 连接起来，internet 一定会实现"；今天，我们要说的是，"全地球 thing 连接起来，自联网一定会实现"，这里说的不是 IoT 物联网，而是由设备和设备、设备和人自由连接起来的自组织网络。第一个阶段的 OTT（Over the Top）是在电信网络之上虚拟业务，也是数据业务对话音业务的碾压，而第二个阶段的 OTT 则意味着用户可以脱离甚至完全抛却电信网，用户彼此之间自己连接起来。

驱动自联网成为可能的四种力量：首先是基于 ID 的开放协议、算法撮合等智能耦合；其次是 WiFi 技术的演进正在极大程度上解决端到端的通讯距离、带宽以及多用户多通道能力的提升问题；第三个动力，设备密度和大量非电信网络使得有效的网络连接获得必要的不是基站的微基站密度和网络补充；而第四个也是最关键的一个动力则是，天空互联网就是自联网最强有力的那个"转接"网络，用户随时可以经由这里连接到别处，且路径最短。

传统通讯产业和电信业者深知，电信网络资源在很大程度上耗费在了大量的路由转接上，一个用户到达另一个用户，一个终端连接另一个服务，往往要经过大量的转接过程，造成拥堵，这也是妨碍带宽提高的重要原因。而任何两个点经过一个转接点就能接通，减少网内转发量和转发次数，必然有助于降低成本，提高用户实际能够体验到的带宽。

Mesh 在电信业者眼里不仅新鲜，而且乏善可陈，但自联网的网络原理恰恰基于不是 Mesh 的 Mesh。一说到 Mesh 有人容易首先想到那个令人提心吊胆的 Firechat，MeshMe 等。FireChat 通讯应用基于 MESH 思想的自组网，依靠蓝牙或 WiFi 信号在附近的用户之间传输消息，只要有安装 FireChat 的设备充当节点，FireChat 的网络就不存在地域限制。但我们所说的自联网不是 Firechat，自联网是密度、尺度、路径最为优化的那个网络之上的网络，是基于软件、数据、传感的多方协议体系。每个智能设备都是一个热点、基站甚至路由，自联网的网络形态首先是 P2P。是不是又要有人说，哦，P2P？太 out 了，可是，真的吗？

还记得无尺度网络吗，大量数据用户、服务集中在极少数重大节点上，网络巨头们的这个状况并非没有它的对立面，自联网在一定程度上有助于消解无尺度网络。无论是经济学人杂志担心的互联网巨头的狼性问题，还是科幻电影所呈现的 Matrix 母体对人的集中控制，都会有另外一种力量与之对冲，尽管不一定能够产生和谐与平衡。集中与分布同在，自联与节点同在。

人类登陆火星，飞行器探索宇宙，星际互联网是未来网络的终极边界吗？

美国在外太空建立星际互联网的长期构架 Inter Planetary Internet（IPN）当中，DTN

（Disruption-Tolerant Networking）是重要试验内容。互联网之父温顿·瑟夫早在最初架构互联网之时，就已经有星际网络的概念构想雏形。温顿·瑟夫后来也成为实现行星和航天器间远距离可靠数据传输的新太空协议科学家团队成员。

短期而言，地球与飞向火星等外太空的飞行器失去联系、数据传输速率极低、数据丢失、时延较大等问题，是星际互联网诞生的问题根由。苏联发射的火星探测器绝大多数以失联收场。数据从火星传到地球需要 6 至 20 分钟时间，而地球与冥王星之间的通讯时延高达数小时。长期而言，星际互联网将把各种相关轨道飞行器探测器、登陆车、航天发射装置、宇航员通信装置、卫星等发射接收和通讯中继设备连接起来的互联网络，甚至分布在太阳系的所有装置互联起来形成一个巨大的接收器。IPN 的体系结构设计很多方面都参考了 Internet 的体系结构，在其中可以看到卫星、激光、网关、中继、存储、分布、转发等熟悉的字眼内容。基站与航天器之间也可以用激光来通讯，地球与月球之间的激光通讯已经成功测试，速率达到 600Mbps。

星际互联网，距离普通人似乎像外星和地球的距离一样遥远。但是，在有生之年，激动人心的时刻还是会到来，改变世界的黑科技会一个接一个被人类创造出来。2020年，以生命探测为主要目的的下一代火星车将飞往火星，奥巴马声称 2030 年将送人类第一次到达火星，而 SpaceX 公布的计划如果一切就绪前往火星的载人发射时间窗口在 2022 年就开始到来，但是 SpaceX 的 AllenMusk 话音未落，波音 CEO 就表示波音用于火星载人飞行的太空发射系统 SLS 计划于 2019 年首飞，首个踏上火星的人将坐波音火箭。一切并不遥远。

通讯如何先行？目前在火星表面活动的火星车"好奇号"，与地球之间采取其他方式通讯。"好奇号"与先期发射到火星轨道的火星卫星通讯，然后卫星与地球进行接力通讯。"好奇号"与卫星之间在波长很短的 X 频段以 UHF（超高频）每天进行最多 8 分钟时间的通讯，速率 2MB 及 256KB 不等，也就是窄带互联网的速率水平。那么，如果是量子通讯呢，2020、2030 年的时候，量子通讯网络是否堪用？

参考译文：

Sky Internet：Connecting the Future World

The diverse world created by technological innovation has appeared before us in the past ten-odd chapters of Black Technology, while how will all these be connected in the future? How? By broadband or IPv6? The point is not on these levels, despite that broadband is moving towards ultra-one. The form and framework of the "future network", whether it is the next generation Internet, the NGI in the sense of the next generation info-tech facilities, or the next generation network (NGN), are not the Internet we see now, nor the NGN that has been discussed a lot in the latest stage. There are some solutions one can find from ITU International Telecommunication Union, IETF Internet Engineering Task, IEEE Association of Electrical and Electronic Engineers, organizations such as OneM2M, 3GPP, OMA, ISO/IEC JTC1 as well as ZigBee, Z-Wave, AllSeen and other kinds of alliances alike. Nevertheless, tech goes further, and is not at the level of "evolving from http/1 to http/2". Things that happen in future have appeared now.

It will be found that the future network is to some extent the result of multi-forces advancing side by side and multi-dimensional innovation coordinating with each other. Such nine dimensions as energy, connection and data density, material, perception and network scale as well as speed in computing, moving and convergence affect the speed and form of network evolution. Breakthrough and innovation of basic science, applied science and technology and industrial communities in varied fields and levels within nine dimensions facilitate the appearance of future network—intelligent network. The future intelligent network starts from looking up at the starlit sky. Space Internet or Sky Internet, namely the coming "Sky Net", is the most fundamental "connection" among future intelligent networks, which is bound to be an exploratory and controversial topic.

Solar Energy, UAV, Skynet Platform in the Future

On June 29, 2016, Facebook solar-powered UAV Aquila completed its first test flight in Arizona. As it went well, the flight lasted 96 minutes instead of 30 minutes as planned and successfully collected relevant data about flight mode and aircraft structure. In the field of using UAVs as a network platform, Facebook has taken a giant step. The Project Aquila, launched two years ago after the acquisition of Ascenta, a British solar-powered UAV research and development enterprises, has brought a huge milestone to the success of Facebook's Internet. org project, which is a future project Facebook's CEO Zuckerberg mainly focused on with a goal to connect everyone in the globe through the Internet, especially the 4 billion people who have not yet connected to the network—people in poor, remote, poorly-connected areas or even areas without network coverage.

Intermet. org works with mobile phone manufacturers and operators to provide over 300 simplified Internet services free of charge to the public by Free Basics project. The "Connectivity Lab" department, established as a result of the Intermet. org project, is specifically responsible for searching new methods of network communication such as laser and UAV, including combining artificial intelligence with Internet services, and advancing

to the maximum extent the open source, openness and sharing of Internet technology. The goal of it is to increase the number of Internet connections by 10 times and reduce the price of Internet access to 1/10 of the current level.

At this time, some people said that it will take 10 years for Facebook-style laser communications to become a commercial project that can be applied on a large scale. Despite that, Facebook still took a move. Aquila that succeeded in its maiden flight has a wingspan of 43 meters as long as Boeing 737, and its fuselage weighs about 454 kilograms equivalent to one percent of manned aircraft. It is lifted by helium balloon to stratosphere with stable climatic environment. Aquila flies around 27432 meters in the daytime to avoid the airline altitude, absorbing and storing solar energy, and flies 18288 meters in the evening to save electricity. What Facebook intends is that solar-powered Aquila can fly for 90 days at a time at high altitude. Ninety days sound incredible, even sensational, but it's possible/viable in the future. Here is another problem involving the 9-degree theory, such as the energy density and material scale.

Facebook intends to deploy 1,000 or even 10,000 such UAVs in the future in the skies around the world, each of which circles within 60 miles in diameter. Instead of drawing people's attention, these UAVs moving around in the sky are designed to provide universal Internet access services. At this time, laser communication will come in handy. Leaders of "Connectivity Lab" said that their laser communication data transmission rate can reach 10 Gbps, near the level of ground optical fiber and 10 times of the standard laser signal. Its communication process is laser and electromagnetic communication between the ground master station and UAVs. UAV can also send signals to other UAVs through laser for relay. The fleet sends the laser beam down to the transceiver of the ground sub-base station. With the transceiver as the center, the network signal can cover an area with a radius of 30 miles for Internet access. The system will convert signals into Wi-Fi, 4G or 5G networks.

Here comes a problem. Frankly speaking, Project Aquila still needs ground sub-base stations. UAVs only play a role of backbone network similar to traditional telecommunications network, which is inferior to the old-fashioned idea of Iridium, a traditional satellite communication service provider, for at least the latter stays in the air all the time with a goal of letting everyone have access to satellite mobile phones. Iridium's satellite is in space, while Aquila UAV is in the stratosphere about 20 kilometers above the ground. Under the ionosphere, theoretically, Aquila can be used as an air base station directly without the link of ground master station. But in this way, how can Aquila, weighing only 454 kilograms, withstand the transceiver equipment with a/whose weight far above itself, and how can it fly only by solar energy for 90 days without falling down as well as carry out energy-intensive space-to-Earth communication? Ground-based Internet devices such as a whole set of things like the antenna of mobile phones should follow the development of relevant standards as well. Here again, it is a matter of the theory of 9 degrees, but it is not insurmountable.

Although Free Basics has provided Internet access services to tens of millions of people in different corners of the globe, nevertheless, the worldwide resistance before Project Aquila is no less than air drag in flight, just like the strong opposition of governments and civil organizations towards India, Egypt and such countries. Free Basics is even banned in India. Those objections are either cited from existing laws, or from Internet neutral principle or tax issue, or merely said presumptuously and pretentiously, none of which is tenable. When it comes to traditional interests, it will be against by traditional forces, let alone that the UAV now is going to cross the borders of traditional sovereign states. By contrast, spectrum resources are not the biggest problem.

Among the explorers of solar-powered UAV sky interconnection technology are Boeing, AT&T, Intel, Airbus, Zephyr S/T and other large and small enterprises and teams. Google is not lagging behind. On the 15,000 square foot apron of the American Space Port in New Mexico, Google is launching a new project called Project Sky Bender, which also applies UAVs as network platform. The difference is that Google uses millimeter-wave communication, one of the high-speed wireless communication technologies in the future, which claims to be 40 times faster than 4G LTE transmission speed and may even become the backbone of 5G network communication. Consequently, Project Sky Bender has the meaning of 5G network platform in the air.

Although the reliability of balloon remains to be observed, it has been used as one of the exploration directions of air network platform. Project Loon and Project Sky Bender, which Google is working on, are part of the same project. They deliver high-altitude polyethylene helium balloons to the stratosphere as base stations, forming network coverage and bringing the Internet to all corners of the globe. Project Loon's balloon is about twice the height of a commercial aircraft and has been tested to stay at high altitudes for more than 180 days.

Google is also trying WiFi airship. Beijing University of Aeronautics and Astronautics, a university in China, is experimenting with near-space airships to achieve wireless network coverage. Guangqi, the private enterprise in Shenzhen, China, invests in Martin Company,

which focuses on New Zealand's research and development of personal aircraft. They claim that the WiFi airship "Cloud" has undergone preliminary testing, but the technical details still need to be deliberated. Overall, the position of balloons and airships in the future Sky Internet belongs to the network platform in supplementary and secondary position.

Intelligent broadband satellite networks are more distant to achieve than UAVs, but closer to the future of Skynet? It doesn't matter whether it's called Sky-Fi, Sky Internet or Space Internet. What's important is that the future network signals must come from the sky and stars first. Satellites are farther away than UAVs, but they are closer to the future in terms of technology cost efficiency. The rapid evolution of technology in six aspects, i. e., the substantially reduction of launch cost, near-Earth orbit, high frequency band, intelligent and miniaturization of satellites, miniaturization and low power consumption of antennas and terminals, and the network technology integrating space and earth, is the key for "Sky Internet" to be realized. The decreasing cost of satellite manufacturing and launching and the increasing bandwidth and the number of users facilitate the intelligent broadband satellite network becoming more and more feasible. Although the cost-performance ratio is not high enough compared with fixed-line telecommunication network and mobile network, nevertheless, this is a way that keeps approaching the cost-performance critical point.

Space X, Google and Facebook are big players in this future online game. Space X initially planned to launch more than 700 inexpensive low-orbit satellites to provide Internet access on the ground. However, according to Space X's latest report to the Federal Communications Commission in 2016, the plan to provide satellite broadband network services globally will launch 4,425 satellites. So far, the focus remains on launching, as well as continuously improving their rocket recycling and reuse techniques. To reduce the cost of space launching to a new threshold, they even declare ambitious Mars mission on this basis. Space X also provides satellite launching services for the broadband satellite Internet project, which Facebook is working with Eutelsat, a French satellite operator. Unfortunately, the first satellite was launched into flames by Space X's failed Falcon rocket in September 2016. The 5-ton Amos-6 satellite, built by the Israeli company at a cost of \$200 million, was intended to provide Internet services to parts of sub-Saharan Africa, while now it reduced to ashes, which was a great setback for imenet. org.

Startups can also be found in this future cyberspace competition. An Israeli company simply calls itself SkyFi and announces plans to launch 60 microsatellites into space. Although the plan seems to be difficult to achieve, SkyFi has attracted many buyers for its own unique technology in microsatellite antenna.

One Web is much lower-key than Space X, but the background of its shareholders is also uncommon. Behind it are Virgin Galactic, Qualcomm, Honeywell Aerospace, etc. One Web's signal processing chip comes from Qualcomm, and the latter uses the switching technology between its terminal and base station to help build satellite communication network and solve the shift and switching problems of many satellites in the process of passing over ground base station. Like Space X, One Web intends to cover the earth with small low-orbit satellite network. It plans to launch 648 small satellites into low earth orbit (LEO) with terminal access rate of about 50Mb/s. The manufacturing cost of each satellite is about 350,000 US dollars, and the total cost of the project is about $ 2 billion dollars. One Web provides services for airlines, disaster relief organizations, personal family customers, schools and villages in remote mountainous areas. However, despite into LEO, One Web's antenna and power technologies seem to be modest, and devices are not small enough. The size of the equipment in ground base station is still too large and the volume needs to be reduced in the future although it can be powered by solar panels.

MDIF's Outnet Extranet Project released in 2014 is an episode, which launches hundreds of satellites into LEO to support the implementation of free WiFi worldwide. This seemingly appealing project is obviously questionable in its technical thinking. End users can only receive selected network content unilaterally, and cannot interact with each other. Only one-way broadcasting is out of date.

OneWeb's spectrum resources are available through O3b, another important player in this field, which can be called a mid-orbit player. Although it starts early with favorable beginning, nevertheless, there are some problems at the present stage. O3b was founded to solve the problem of other 3 billion on the planet - another 3 billion people who did not have access to the Internet. Google, SES, HSBC and other industry giants are its important investors. Compared with the time delay of the previous geosynchronous orbit communication satellite of more than 35000 kilometers, the time delay of the O3b satellite network in mid-orbit of 8000 kilometers is less than 150 milliseconds, and the relay bandwidth reaches the level of conventional optical fibers, which means that the network quality can be commercially used on a large scale. O3b has separately launched two batches of Ka-band satellites through the Ariana rocket in 2013 and 2014, and eight satellites are in orbit. With an intention to launch 12 to 16 satellites much cheaper than previous geosynchronous orbit satellites, the O3b satellite network plans to cover Africa, the Middle East, Asia and Latin America, and provides network services with a maximum speed of 10

Gbps and a total capacity of 84 Gbps to the developing countries in these regions. Before that, ViaSat1, a broadband communications satellite with the so-called world's largest capacity of 140 Gbps, was in orbit, and also belongs to the Ka band. Whereas ViaSat was a high-orbit player and is extremely expensive due to its geosynchronous orbit and 10 times the total mass of the med-orbit O3b satellite. ViaSat1's ground system includes satellite user terminals—Ka-band butterfly antenna and satellite modem, gateway satellite ground station and network operation center, which has relatively strong service capability for enterprises and household users. In contrast, O3b's Skynet is located in the backbone network rather than the end-user access. O3b adopts the network strategy of one satellite dropping to the horizon and relaying by another satellite, which greatly reduces the cost of manufacturing and launching. O3b's own data is expected to reduce the cost of Internet access in Africa and other regions by more than 95%. Over 40 3G and LTE mobile operators and Internet access service providers in Pacific island countries, Africa and the Americas have become its customers.

Of course, Viasat, a California-based company, will not be satisfied with the status quo. The next Viasat 2 satellite will have twice the bandwidth and 2.5 times the capacity of Viasat 1, serving 2.5 million users. The download speed of broadband Internet services will be improved from 12/15 Mbps of Viasat 1 to 25 Mbps.

After the restructuring, Google became less and less aware of /lost its nerves the direction of the Sky Internet, first withdrawing from O3b, and then canceling the $ 1 billion plan to build 180 high-performance Earth-orbiting satellites network. Despite that, Google will return someday in the major direction concerning the future, for it has invested in satellite imagery, indirectly owning several satellites in orbit. Although Samsung hasn't take action, it also verbally states that it's a world-class company concerning global network issues. Samsung said it would launch 4,600 microsatellites in the future to provide users with low-cost Internet access service.

Despite that Chinese enterprises and related institutions differ in the technological strength and developmental stage, the developing direction of different tracks in high, middle and low levels is roughly the same as the aforementioned projects. China Aerospace Science and Technology Group is carrying out a high-throughput broadband satellite project. In 2016, the first geosynchronous orbit mobile communication satellite Tiantong 01 was launched, which provides communication, short message, voice and data transmission services for large mobile users such as boats and ships, aircraft, vehicles and handheld terminals. Chinese Satcom are also implementing the Ka-band broadband satellite program.

More than a decade ago, Owens, the former CEO of Nortel Networks, proposed to cooperate with Huawei to build low-orbit satellites, similar to today's plans between Facebook and Google, while the discussions failed to continue. In terms of low orbit, however, the first smart communication test satellite jointly developed by Tsinghua University and Xinwei Group was launched and tested in orbit in 2014. The satellite weighs about 130 kilograms, operates at an altitude of about 800 kilometers and covers a diameter of about 2400 kilometers. The test validates many techniques, such as spaceborne intelligent antenna, space processing and switching, space-earth integrated networking, small satellite integrated design and so on, realizing the services of handheld satellite terminal communication, handheld satellite terminal and mobile phone communication, Internet data transmission, etc. In order to achieve the future layout of Sky Internet, Xinwei Group even offered an acquisition invitation to Israel Space-Com through its subsidiary Luxembourg Aerospace Communications Corporation. Space-Com is the manufacturer of Amos-6, the $ 200 million broadband satellite that was sent to the flames by SpaceX's Falcon Rocket.

Hong Kong listed company China Trends Holdings Limited, in cooperation with Hughes Aircraft, plans to creat a free satellite mobile Internet in the Asia-Pacific region by using the latest large-capacity Ka-band broadband satellite resources. Users can use designated terminals to get online, dial satellite phones and watch satellite televisions.

Will Everyone Be Able to Use Skynet on Mobile Handheld Devices in the Future?

In other words, if everyone will get online through the Sky Internet in the future, will it become the basic network for future connections? At present, perhaps many people, including some telecom and even satellite communication practitioners, will say: How can handheld devices access the Internet by satellite? It's possible to make satellite phone calls, but not with Internet access, especially broadband Internet access, on account of the small transmitting power, large antenna and difficult improvement in upstream speed. It's still good to use satellite network as the backbone network to provide network relay services for mobile operators or ground base stations of access service providers, while directly providing large-scale Internet access services to individual users seems difficult. Isn't that why the Iridium Star project, which Motorola spends billions of dollars, went bankrupt? Launched in 1996, it started providing services in 1998, and Iridium phones developed only 55,000 users worldwide by the time of March 1999; while at the same time telecom operators

around the world have delivered cheaper, more portable and better communications phones to millions of users. Globalstar, a contemporary of the Iridium Star project, also suffered a great failure. Despite of having Bill Gates and even the Saudi Royal family as its supporters, Teledesic did not even wait for the day when the project took shape.

But innovation and evolution driven by technology are such a continuous process of succession. Private-equity funds, offeree of Iridium Star, became the Iridium Communications of US listed companies with a market capitalization of $800 million after so many years' dormancy, even though it was only a fraction of Iridium's huge investment at the time. Iridium Communications in 2014 launched Iridium GO, the global coverage satellite WiFi hotspot service, while the sky Internet of OneWeb, SpaceX, Facebook and even Sky-Fi have already spread like a storm. Iridium satellite communication is only a thing of the past no matter how it works hard to keep abreast of the changing times. Because Iridium satellite communication is first and foremost a satellite telephone network from the technical level and communication system point of view, while what the new bloods wanted is the sky Internet based on data communication, not the telephone or GPS network. At this stage, the technologies like rocket reuse make the cost of satellite launching greatly reduced, and the networking of near-earth orbit satellites can not only effectively solve the problem of time delay, but also improve the signal quality far more than that of geosynchronous orbits, so it is also helpful for smaller antenna work; band high frequency, especially the large-scale depth development and utilization of Ka band, has become reality, which has cleared away the technical barriers to providing broadband services for mass users; intelligent and miniaturization of satellites, miniaturization and low power consumption of antennas and terminals, as well as the network technology integrating space and earth, have become a visible and realistic technological trends. Sky Internet is not only belonging to Hughes, Laura, Boeing and other traditional satellite manufacturers and satellite communication operators. IT companies, Internet giants, start-ups and other new forces are ready to take over them.

From a technical point of view, the product trends about micro antenna technology, small handheld devices, satellite mobile phones and mobile satellite Internet access are particularly noteworthy. The most inspiring changes in the future will also take place in this part. Anyone who have seen ISatHub, the satellite hotspot device of Inmarsat, a mobile satellite communications service provider, will realize that the ground equipment of geosynchronous orbit satellite can be miniaturized to half the size of a laptop computer, which is the combined volume of antenna and Modem anyway. It's easy to recall the Modem in the computer dial-up stage 20 years ago, sharing the same size. In some countries, X-, Ku-, and Ka-band mobile satellite communications equipment has been implemented in vehicles, backpacks and even single-handed, which would have been inconceivable in the past when there was no pot-like satellite antenna and heavy workstation-level equipment. As for the ground terminal equipment of medium and low orbit satellites, especially near-earth ones, it can be generally handheld. From Inmarsat to satellite telephones such as Iridium, Global, Asian and Spot, the most prominent feature is the thick communication antenna. The volume of the mainstream satellite telephones is much smaller than the original GSM mobile cellular phones.

In the next step, along with the further development of antenna and battery technology, as well as the application of satellite's large-scale and high-bandwidth network service capability, it's hopeful to gradually produce Satellite Internet experience similar to ordinary smartphones.

The smartphone satellite adapter of Thuraya, a satellite operator based in Dubai, is an interesting direction. Even without a satellite phone, users' iPhone or Android phone can effectively connect to the adapter by downloading and installing SatSleeve App free of charge in the application store after using SatSleeve adapter. Then the user can make phone calls, send or receive short messages and e-mail in the satellite network mode, as well as use some social and instant messaging software. Thuraya's SatSleeve adapter looks like a thicker mobile phone covers. Apart from a robust antenna, it can't be seen at first glance that it can change a smartphone into a satellite phone by second. So, how far is the satellite internet access from each mobile phone user?

Super WiFi, Micro Base Station, Mesh Interconnection, Future Ground Network and Skynet Form Self-Networking?

Ideally, all devices can be connected through the Sky Internet, while in the future, the combination of space and earth, multi-network and networking on demand will be the most widely used form. Networks with different characteristics will exert their respective strengths in different fields. Near field communication used to be the expertise of Zigbee, ZWave, AllSeen and Bluetooth. As WiFi develops rapidly, mid and near field ones are occupied by the base station of WiFi, super WiFi and mobile operators. The backbone network has satellites in space and optical fibers on the ground. In some places, fixed network access is still optical fibers. Things like lasers are playing a relay role between the backbone and access network. In the local range, the increasingly powerful WiFi is partly replacing the role that mobile communication base stations must play, while WiFi with ultra-low power consumption is fully replacing Bluetooth. The super WiFi here refers to the WiFi network with long signal distance, strong penetration, ultra-high bandwidth and multi-channels. IoT (Internet of Things) and Vehicle Networking are not only fusion sensors, but also integration applications of satellite, base station, WiFi and Zigbee.

The most noteworthy part of Skynet is the low-orbit smart broadband satellite network for end users on a large scale, while that of Ground Network is of course WiFi and WiFi interconnection. Higher, faster and stronger, WiFi is not progressing gradually in one dimension, but is expected to create a new network environment in a new dimension, so that people will often forget about telecommunications. This is an on-going wireless revolution in the air of the earth's surface. No matter the base station, every hotspot is a micro-base station. Countless powerful WiFi hotspots are interconnected with each other and with the LEO satellite network in real time. Every smart device, including mobile phones, is also a WiFi hotspot, relay point and micro-base station. This is the most efficient and widely distributed network environment in the future. In addition, networks with other characteristics will supplement partially. The future of the network has been looming.

Six technological breakthroughs are driving WiFi to create future networks, namely, transmission distance, penetration capability, ultra-low power consumption, bandwidth capacity, multi-user, Mesh interconnection with no Mesh characteristics. In terms of speed, WiFi's 5 GHz bandwidth throughput capacity is expected to reach 10 Gbps, 60 GHz or even 20 Gbps. Theoretically, 100 Gbps from end to end, point to point is possible. But the most impressive breakthrough is the new WiFi with ultra-low power consumption by researchers at the University of Washington using electromagnetic backward reflection technique, also known as passive WiFi, which has a transmitting power of only 10-50 microwatts, one-tenth of that of traditional WiFi routers; more importantly, WiFi charging can be carried out. WiFi charging is one of the most interesting directions in this field with broad prospects. The research of WiFi charging has been well established in the major innovative economies, like the United States, Japan, Israel, China and Europe.

Among the six technologies, Mesh interconnection with no Mesh characteristics is the most ecologically influential. It is deserved to be called self-networking for on-demand internet and mobile ad hoc networks are achieved through protocols driving networks and devices. Mesh wireless grid network is developed from ad hoc network and can communicate with other regional networks in collaboration. Self-organization, no center, no boundary, dynamic expansion, interconnection among devices, relaying are its six characteristics.

Twenty years ago, there was a saying that "If all PC around the world were connected, the Internet will come true"; today, we want to express that "If all thing around the world were connected, the self-networking won't be far." What we mentioned here is not the IoT (Internet of Things), but the self-organizing network which is connected unrestrainedly among equipment and people. The first stage of OTT (Over the Top) is the virtual service on the telecommunication network, namely, the winning of data service over voice service. The second stage of OTT means that users can connect with each other, instead of through the telecommunication network.

There are four forces make it possible to drive the self-networking. First is intelligent coupling/linkage based on the open protocol and algorithm matching of ID. The next one is that the evolution of WiFi is solving the problems of end-to-end communication distance and how to improve bandwidth and multi-user multi-channel capability. The third is that device density and a large number of non-telecommunication networks facilitate effective network connection to obtain necessary base station density and network supplement. And the last but the most critical driving force is that Sky Internet is the most powerful "switching" network of self-networking, where users can connect to other places at any time and with the shortest path.

Traditional telecommunications industry and telecom practitioners are well aware that telecommunication network resources are largely consumed on vast routing transfers. An important reason hindering bandwidth improvement is the congestion caused by many transfer from one user arriving at another user and one terminal connecting another service. Reducing the amount of forwarding and the number of it through the practice of any two points being connected through a switch point will inevitably help to reduce costs and improve the bandwidth that users can actually experience.

Mesh is new yet common for telecom industry, while the network principle of self-networking is based on Mesh interconnection with no Mesh characteristics. Speaking of Mesh, one can easily associate it with Firechat, MeshMe, etc. FireChat communication application is based on ad hoc network featuring MESH thought, relying on Bluetooth or WiFi signals to transmit messages between nearby users. As long as equipment installed with FireChat are applied as nodes, its network does not have geographical restrictions. But what we call self-networking is not Firechat. Rather, it is a network surpassing the one with the most optimized density, scale and path. It is a multi-party protocol system based on software, data and sensor. Every intelligent device is a hotspot, base station and even routing. The network form of self-networking is first P2P. Would someone say, oh, P2P?

Too out, but, is that true?

Do remember scale-free networks, where a large number of data users and services are concentrated on a very small number of major nodes. These network giants have their opposite too, and self-networking helps to eliminate scale-free networks to a certain extent. Whether it's the wolf problem in the Internet giant feared by the Economist magazine or the centralized control of human beings by Matrix in science fiction movies, there is another force to counter it, although it may not necessarily produce harmony and balance. Centralization and distribution coexist, so does self-connection and nodes.

As human land on the Mars and aircrafts are used to explore the universe, will the Inter Planetary Internet be the ultimate end of future networks? DTN (Disruption-Tolerant Networking) is an important part of the long-term framework of the Inter Planetary Internet (IPN) established by the United States in outer space. Winton Cerf, the father of the Internet, had a prototype of the concept of IPN as early as he thought over the Internet. He later became a member of a scientific team on a new space protocol to deliver reliable data over long distances between planets and spacecraft.

In the short term, the root causes of the birth of the IPN are problems including the losing contact between spacecraft flying to Mars and other outer space and that on Earth, the extremely low data transmission rate, the loss of data and the large time delay. Most of the Mars detectors launched by the former Soviet Union ended up in disconnection. It takes 6 to 20 minutes for data to transmit from Mars to Earth, while time delay in communication between the Earth and Pluto reaches to several hours. In the long run, the IPN will form a huge receiver by connecting all kinds of related orbiter detectors, landing vehicles, space launchers, astronauts' communication devices, satellites and other launching and receiving and communication relay devices, and even all devices distributed in the solar system. The design of IPN architecture refers to many aspects of the Internet one, in which you can find satellite, laser, gateway, relay, storage, distribution, forwarding and other familiar terms. Laser can also be used to communicate between the base station and the spacecraft and laser communication between the earth and the moon has been successfully tested at a rate of 600 Mbps.

The IPN seems as far away as the distance between extraterrestrial and Earth to ordinary people. But in our lifetime, exciting moments will come as well, and black technology that will change the world will be created one after another by human beings. In 2020, the next generation of Mars Rovers with life exploration as its main purpose will fly to Mars. Barack Obama claims to send humans to Mars for the first time in 2030. The human launch service announced by Space X will begin in 2022 if everything is well ready. As voice from Space X's Allen Musk is still in the air, however, the CEO of Boeing said Boeing's space launch system SLS for manned Mars flights is scheduled to have a maiden-fly in 2019. Thus it can be seen that it will be soon for the first person to set foot on Mars by a Boeing rocket.

How does communication go first? Curiosity, a Mars rover currently operating on the surface of Mars, communicates with Earth in other ways. Curiosity communicates with the satellite that was launched into Mars orbit in advance, which then relay with the Earth. Curiosity communicates with the satellite in a very short X-band with UHF (UHF) for up to 8 minutes a day at rates of 2 MB to 256 KB, i.e. the rate level of the narrowband Internet. So, if it is quantum communication, will its network be available in 2020 and 2030?

单词释义：

1.	卫星适配器	satellite adapter
2.	在轨卫星	satellites in orbit
3.	卫星调制解调器	satellite modem
4.	地球同步轨道通讯卫星	geosynchronous orbit communication satellite
5.	频谱资源	spectrum resources
6.	地面母基站/子基站	ground master/ sub-base station
7.	星载智能天线	spaceborne intelligent antenna
8.	毫米波通讯技术	millimeter-wave communication
9.	无源 WiFi	passive WiFi
10.	激光通信技术	laser communication technology
11.	骨干网	backbone network
12.	平流层/电离层	stratosphere/ ionosphere
13.	天地通讯	space-to-Earth communication
14.	高空聚乙烯氦气球	high-altitude polyethylene helium balloons
15.	中继带宽	relay bandwidth
16.	移动蜂窝电话	mobile cellular phones
17.	电磁后向反射技术	electromagnetic backward reflection technique
18.	星际互联网	the Inter Planetary Internet (IPN)
19.	窄带互联网	narrowband Internet
20.	量子通讯	quantum communication
21.	空中基站	air base station

22. 收发设备	transceiver equipment
23. 飞行阻力	air drag in flight
24. 网络中立原则	Internet neutral principle
25. 高速无线通讯技术	high-speed wireless communication technology
26. 智能宽带卫星网络	intelligent broadband satellite networks
27. 低轨道卫星	low-orbit satellites
28. 网关卫星地面站	gateway satellite ground station
29. 发出收购邀约	offer an acquisition invitation
30. 微型天线技术	micro antenna technology
31. 卫星热点设备	satellite hotspot device
32. 移动蜂窝电话	mobile cellular phone
33. 应需组网	networking on demand
34. 近距离通讯	near field communication
35. 无线革命	wireless revolution
36. 中继点	relay point
37. 宽带容量	bandwidth capacity
38. 电磁后向反射技术	electromagnetic backward reflection technique
39. 移动自组网	mobile ad hoc networks
40. 智能耦合	intelligent coupling/linkage
41. 算法撮合	algorithm matching
42. 端到端的通讯距离	end-to-end communication distance
43. 路由转接	routing transfer
44. 无尺度网络	scale-free network
45. 火星探测器	Mars detector
46. 轨道飞行器	orbiter detector
47. 航天发射装置	space launcher
48. 火星载人飞行	manned Mars flight
49. 接力通讯	relay communication

译文解析：

1. 汉译英中词性转换

原文：技术在 6 个方面的快速进化，是"天空互联网"越来越逼真的关键：发射成本大幅度降低、轨道近地化、波段高频化、卫星智能化和小型化、天线与终端小型化与低功耗、天地一体组网技术等。

译文：The rapid evolution of technology in six aspects，i. e.，the substantially reduction of launch cost，near-Earth orbit，high frequency band，intelligent and miniaturization of satellites，miniaturization and low power consumption of antennas and terminals，and the network technology integrating space and earth，is the key for "Sky Internet" to be more vivid.

分析：原文运用了大量的名词性短语，译文为了符合英语的表达习惯而使用了名词转化为动词的翻译技巧。

2. 长句的翻译

原文：无论是下一代互联网、下一代信息技术设施意义上的 NGI，还是下一代网络，"未来网络"的形态和架构，不是我们眼前看到的互联网，也不是过去一个阶段讨论较多的 NGN。

译文：The form and framework of the "future network"，whether it is the next generation Internet，the NGI in the sense of the next generation info-tech facilities，or the next generation network（NGN），are not the Internet we see now，nor the NGN that has been discussed a lot in the latest stage.

分析：这句话的翻译首先要厘清逻辑，确定主语的位置，然后将各个分句整合串联，并且还要注意关联词的多样性，不可一个词重复使用。

原文：能量密度、连接密度、数据密度，材料尺度、感知尺度、网络尺度，计算速度、移动速度、融合速度，这 9 个度在影响网络演进的速度和形态，不同领域和层面的基础科学、应用科技、产业群落在 9 个维度的突破创新，使得未来的网络——智慧网络若隐若现。

译文：Such nine dimensions as energy，connection and data density，material，perception and network scale as well as speed in computing，moving and convergence affect the speed and form of network evolution. Breakthrough and innovation of basic science，applied science and technology and industrial communities in varied fields and levels within nine dimensions facilitate the appearance of future network—intelligent network.

分析：首先本句话有两个意群，转换成英文要注意断句。前半句中文原文使用了九个四字并列结构，属于举例说明，译文应体现这一点，且尽量使语言像中文一样有美感。后半句只需抓住句子的主语部分，按顺序翻译即可。"智慧网络若隐若现"这句我使用意译法进行处理。

原文：这个时候问题来了，Project Aquila 说白了还是需要地面子基站的，无人机只是发挥了类似传统电信网络的骨干网的作用，这就还不如传统卫星通讯服务商 Iridium

111

的老思路了,后者至少全程都是在空中,以每个人都可以使用卫星手机为目标啊。

译文: Here comes a problem. Frankly speaking, Project Aquila still needs ground sub-base stations. UAVs only play a role of backbone network similar to traditional telecommunications network, which is inferior to the old-fashioned idea of Iridium, a traditional satellite communication service provider, for at least the latter stays in the air all the time with a goal of letting everyone have access to satellite mobile phones.

分析: 本句偏口语化,语言通俗易懂,翻译的时候可以适当体现口语化的特质。不过句子仍需理清结构,切分句子意群,不可一逗到底。

3. 特色词句的翻译

原文: 那些或援引既有法律,或以网络中立原则或税收问题阻止,或振振有词或冠冕堂皇的反对理由,也许没有一项是站得住脚的。

译文: Those objections are either cited from existing laws, or from Internet neutral principle or tax issue, or merely said presumptuously and pretentiously, none of which is tenable.

原文: 他们号称"云端号"WiFi飞艇已经进行初步测试,不过其技术细节还需要推敲。

译文: They claim that the WiFi airship—"Cloud" has undergone preliminary testing, but the technical details still need to be deliberated.

原文: 而这个时候,已经是 OneWeb、SpaceX 、Facebook 甚至 Sky-Fi 们的真正的天空互联网开始风起云涌的日子了。

译文: While the sky Internet of OneWeb, SpaceX, Facebook and even Sky-Fi have already spread like a storm.

原文: 和铱星计划同时代的 Globalstar 也死得很难看,Teledesic 尽管有比尔盖茨甚至沙特王室出手资助,却连项目成型的那天都没有等到。

译文: Globalstar, a contemporary of the Iridium Star project, also suffered a great failure. Despite of having Bill Gates and even the Saudi Royal family as its supporters, Teledesic did not even wait for the day when the project took shape.

原文: 这么多无人机在天上转来转去不是为了刷存在感,而是提供普遍的互联网接入服务。

译文: Instead of drawing people's attention, these UAVs moving around in the sky are designed to provide universal Internet access services.

第七章　智能微尘——终极的信息化利器

2060 年,你坐在 Tesla Model Z 里,电车正悬浮在空中,以亚音速自动前往目的地 A。A 地机器人暴动,破坏基础设施,出现不少人类伤亡。你是救援人员之一,心急如焚。你转动手中的苹果戒指 5,心想:"我要看看当地的伤员分布图。"戒指感应到了你的思想,随之在面前的空气中浮现出了 3D 虚拟现实界面——是 A 地的鸟瞰街景,其中的红点标注着危急伤员的位置。你在空气中滑动手指,界面立刻放大到了一个最严重伤员的位置,详细地显示出他的一系列生理指标、受伤的部位和预估的救援剩余时间。你心想:"我需要一个计划。""话音"未落,眼前的地图自动更新显示出一个线路图,指出了最佳的救援行动路径。但你不知道,在几十年前信息化不足的时代,搜救工作从来不是这么高效。

MH370 的搜救正在紧张地进行,失联飞机的黑匣子的电量还剩下最后一天。如果电量耗尽,黑匣子就将成为一个永远的谜团沉没在深不见底的印度洋。在浩瀚的印度洋找一架飞机,就如同在足球场上找一根绣花针一般艰难。救援队所拥有的探测器,每次只能收集非常有限的信息和数据。这决定了如果想以地毯式扫描印度洋海底来追寻飞机的去向,必定会是一个漫长而艰苦的过程。即便能够通过足够长的时间来建构海底的形貌图,这些数据也无法做到实时。况且洋流也会导致目标黑匣子的移动,即便通过线式扫描收起整个海洋地貌,也还是有可能错过探测目标。

波及大半个中国的 8.0 级地震,使得城市沦为废墟,交通中断,道路损毁,救灾工作异常艰难。搜救的第一要务,是尽快定位废墟中可能存活的伤员,这样才能和时间赛跑,及时展开搜救。在这个过程中,及时有效的信息同样重要。如果没有明确的目标,搜救工作就是事倍功半。传统的非信息化人力搜救方式,很难实现精确定位,往往有很多伤员不能在第一时间被及时发现,坐失黄金救援时机。

近 10 年的等待,50 亿公里的飞行,新视野号探测器如期飞抵冥王星上空,首次拍下了这颗位于太阳系最深处行星,为世人揭开了它的神秘面纱。从遥远的太阳系深空到地球表面,传回一张高清照片要 4.5 小时之久,数据传输速率仅为 1Kbps。要想获得这颗遥远星体的表面形貌和地质特性的更多信息,意味着多次探索和巨大的资金投入。现有的探测器不仅造价昂贵,收集数据也十分低效。

科幻总是这么丰满,可现实却总是如此骨感。如何将开头的科幻变为现实呢?试想有一天,数据收集终端可以变得如同沙粒般微小,并且能散布于地球的各个角落,那么以上所有这些问题,也都将不再成为问题。我们的整个地球,就如同一个巨大的显示屏,每个沙粒般微小的终端,就是这巨型荧屏的一个像素点。在中央计算器的监控下,每一个坐标的物理量(GPS坐标、温度、湿度、速度、光强、磁场强度等)信息都尽收眼底。这些惊人的大数据,都可以被全球的中心云服务器实时监控、追踪和分析。

而实现这样终极信息化的单元,正是被我们称之为"智能微尘"(Smart Dust)的极度微小、高度集成的传感器系统。这听起来像是个让人毛骨悚然的没有任何隐私的世界——智能微尘简直是CIA(美国中央情报局)监控全人类的最好工具,无论你走到哪里,你的一举一动都可能被某个掌控了这个中心云服务器的人记录和观察。

但是,任何事情都是一把双刃剑,最坏的世界某种程度上也是最好的世界。因为地球上每一点的物理量都是被动态监控的。天气预报、地质勘探、地震预报、洋流检测等,都将极度准确,手到擒来。宇宙勘探,只需要发射一个装满智能微尘的炸弹,令其爆破后的尘埃遍布星球表面,整个地表的形貌测绘、地质勘探将轻而易举;灾难搜救,利用智能微尘使整个海洋包括海洋中的所有物体能被清晰3D成像,搜救工作就变成一个形状模式识别的过程,只需要计算机即可定位;地质勘测也是同样,如果这样的智能微尘能被注入深层的地下,那么资源检测和地震预报也都变得极度信息化。人类个体,也会因智能微尘而获得诸多福利——疾病检测将变得非常容易,因为当服用了智能微尘胶囊之后,整个消化系统都能被清晰成像,治病问诊从此变成一件轻松之事。在理想的世界中,智能微尘能给人们的生活带来极大的便利,因为这是终极的信息化。

智能微尘是什么

智能微尘这个概念最先在1992年被提出,20世纪90年代开始由美国国防部高级研究计划局(DARPA)出资研究。这一概念的愿景,是由一系列具备通信模块的微型传感器来组成一个分布于环境中的监测网络。每一个监测单元就是所谓的"微尘",成千上万的微尘被散播在环境之中,相互之间用自组织方式构成无线网络,来收集环境数据(温度、气压等)。收集到的数据则通过"微尘"的通信模块传向终端的计算机(或者云服务器)来分析和处理。所以如果要实现这个总体的微尘网络,单个的微尘必须具备这几个功能模块:传感器模块,用于采集环境数据;通信模块,用于无线传输数据;电源模块,用于供电和自充电;微处理器模块,用于控制和调度所有的模块。

有这样一个有趣的类比,如果将要监测环境变量的区域想象成人类的皮肤,那么可以将每个智能微尘想象成人类的单个触觉神经元。神经元能够收集触觉信号,同时也能够相互连接成网络来传递这些信号。外界的刺激在神经元之间相互传递,最终被传向大脑,由大脑来分析处理触觉信号,得到触觉的意识。一个智能微尘网络也有非常相似的架构:被进行环境变量采集的区域,就好比是人的皮肤;而每个智能微尘,就好比是每个神经元;中央云服务器,就好比是人的大脑。数据由微尘采集并传输,直至最终到达云服务器。

近几年很火的一个名词叫物联网(Internet of things),目标是把常见的设备都连入互联网。拿家用设备来说,电灯、冰箱、炉灶、门窗、空调等全部智能化,连入互联网,让家居信息化和可自动控制化。而智能微尘的网络(Internet of Smart dust),可以被视为一种终极的物联网。要实现这个微尘传感器网络的宏图,核心技术还是制作出单个的微尘。这实际上是一个"麻雀虽小,五脏俱全"的微型计算机单元。前面提到了必须具备传感器、通信、电源、微处理器四大模块。而这些所有的模块加在一起,要达到灰尘的尺寸,也就是几十微米。这其中的难度可想而知。

加州大学伯克利分校的 Pister 和 Kahn 教授提出智能微尘这一概念时给出的结构示意图。基于当时的技术水平,这个所谓的"微尘"体积定位有 5 立方毫米之大。可以看得出,这是一个巨型灰尘,距离"微尘"还有很大的差距。从图中可以看出,它的确具有传感器、通信、电源、微处理器四大模块。底部占用体积最大的是电池模块。大块头的蓄电池上面,是一个太阳能电极板,用来将太阳能转化为电能进行充电。在这个基座上,放置有微传感器芯片和微处理器芯片。通信芯片有两个,分别用于信号的发射和接收。

即便整个系统的大部分体积被电池所占用,但电池的总体尺寸还是只有毫米级别。由于电池本身很小,因此要求整个系统有极低的功耗。不间断工作一天的功耗指标要求不超过 10 微瓦。这是什么概念呢? 在 10 微瓦的功耗下,普通的芯片会因为电量不足而停止运算。正是因为这样苛刻的要求,传统的通信模块不能被采用,传统的控制软件也不能直接使用。整个系统都需要重新优化设计来满足低功耗要求。

低功耗只是一个挑战,要做出我们目标中的微尘,5 立方毫米是远远不够的。还需要进一步激进地推进微小化,让整个系统的尺寸达到微米的级别,也就是再缩小至将近千分之一。这真是一个知易行难的事情——按照如今每两年面积缩小一半来计算,实现这个尺寸需要 20 年的时间。当然,低功耗和微尺寸只是指做出单个原型微尘所要达到的目标。要实现工业化大批量生产智能微尘,让它们能够真正分布于我们的环境之中,还需要满足更为苛刻的工业化生产条件,这包括了低生产成本、稳定性良好、抗环境干扰、工作寿命长等条件。

正是因为这些苛刻的工程条件,导致了在其概念提出 20 年后的今天,智能微尘仍滞留在理论阶段。随着近年半导体技术、传感器技术、大数据和云计算的飞速发展,这项理论逐渐有了脱离科幻小说层面的可能,未来几十年,它或许能够成为可以实现的革命性新产品。

智能微尘的微型化传感器中心

　　智能微尘的目标是采集环境的温度、压力等物理量信息。要实现这个功能,当然离不开传感器。那么,什么是传感器呢? 传感器(transducer / sensor)不是一个新概念,传感器渗透于我们生活中的方方面面。传统的水银温度计就是一个简单的传感器,它将温度信号直接转化为可视可读的数字。比如生活中最常见的感应水龙头,手接近时会自动出水,这是因为控制电路里有红外线传感器,能够检测人体的红外线强度来判断手是否接近了水龙头。再拿我们最熟悉的声光控灯来说,为什么这种灯能够白天保持熄灭,夜晚只有人走近发出了声响才点亮呢? 这是由于控制电路里有检测光强的光敏电阻和检测声音的蜂鸣片,控制电路执行了简单的判断逻辑:如果光变弱,而且有声音,就点亮电灯。光敏电阻、蜂鸣片也是传感器的例子。

　　传感器的作用,通常是把不是电学的物理信号化成电学信号。为什么要这么做呢? 这是科学技术中一个经典的化未知问题为已知的技巧——因为电路技术是最为成熟的工程技术之一,一旦将一个非电路信号变为电路信号,所有的电路信号处理、功率放大、控制、自动化、计算机技术等都能较为轻松地与它对接。实现声控的蜂鸣片,是使用了能够将声音振动转化为电荷信号的压电元件;实现红外控制的红外探测器,是使用了吸收红外线后聚集电荷的铁电材料;实现光控的光敏电阻,则是利用了光照后电流变化的特殊半导体材料。

　　常见的传感器,都是将力、热、光、速度、加速度、磁场、声波等其他的物理信号转化为电信号的工具。它们是电路系统和我们所生活的充斥各种各样物理信号的五彩缤纷世界的接口。传统的传感器,例如前面提到的蜂鸣片、红外传感器等,体积都相对比较大,通常在毫米甚至厘米量级。在系统的集成和微型化日新月异的今天,这些大体积的传感器,会带来诸多问题。体积过大,意味着更重,成本更高,功耗更大。如今的系统对于体积和功耗的要求越来越苛刻,这也导致了对于微小传感器的更高要求。如今的大多数电子系统中都集成了形形色色的传感器,这些传感器当中,绝大多数都是采用了微型传感器。

　　智能手机就是一个最好的例子。你可能只关心智能手机是不是用起来很酷、很方便,但不去在乎它里面是怎样的结构和原理。其实如果你拆开一个智能手机,会发现里头布满了各式各样的传感器,其中的大多数还是微传感器。拿一部 iPhone 6 来说,其中有加速计、陀螺仪、电子罗盘、指纹传感器、距离与环境光传感器、MEMS 麦克风、气压传感器等。正是这些传感器的存在,使得 iPhone 能够具备导航、横屏图像转置、计量步数、根据环境光调节屏幕亮度、检测指纹等一系列传统计算机所不具备的功能。最新的 iPhone 6s 还推出了 3D touch 功能,这个功能也是基于显示屏后面的压力传感器阵列，使得手机能够识别手指的按压力度。这些新型的传感器，具有体积小、重量轻、功耗低、可靠性高、工作速度快等优势。这些微传感器，特征尺寸能达到微米甚至纳米级别。微米是多大呢？头发丝的直径是 20 ~ 400 微米。这么小的传感器，怎么制作呢？如今已经工业化的方法是通过被称作微机电系统（MEMS）的技术。

拿在智能手机中非常常见的加速度计来说，它就是一个微机电系统，它的功能是探测外部的加速度输入。这就是为什么智能手机能知道我们是否晃动了手机，晃动有多剧烈。比方说如果你使用微信的"摇一摇"功能，那么这个加速度计就能告诉手机芯片"用户晃动了手机"，然后让芯片根据这个信号输入执行相应地功能—匹配其他用户进行聊天。如果我们把这个加速度计拆开，放到扫描电子显微镜下，会看到像梳子一样的结构，这是它检测加速度的核心部件。这实际上是一系列并联的电容，用来检测外部的机械运动。晃动手机会导致中心质量块的晃动，从而改变电容两个极板间空隙的大小，继而改变电容的数值，这个数值又能被后处理电路所读出。有了这样一一对应的关系，我们测量加速度时实际上是在逆向而来顺藤摸瓜，通过读入电流变化来计算输入的加速度是多少。所谓检测手机的加速度，实际上是检测了加速度计中心悬挂的质量块移动导致的电容变化。

通过这个例子我们可以看到，这里所谓的"机械"，完全没有我们概念中的机械的样子。我们习惯见到的机械，都是由齿轮、连杆等构成的精巧的联动机构，而这里的"微机械"仅仅是一个外形独特的能够运动的质量块而已。是不是大跌眼镜？其实这是所有工程技术的特点——最好的方案应该是能够解决问题的最简单方案。我们这里的目标是检测加速度，根据牛顿第二定律，大的质量块能够放大加速度的效应，所以，这里只需要一个简单的质量块就够了。

制作这样的一个微机电系统，通常需要用到复杂的半导体加工工艺，从个空白的硅晶片开始，一步一步利用可控的化学反应添加薄层，或者将薄层刻蚀成想要的形状。经过几十甚至上百道工序之后，这样的系统就制作好了。这的确听起来很复杂，并且复杂的制作流程意味着造价不菲。那么，在成熟的工业技术中是如何降低造价的呢？别忘了，这个传感器总共只有一滴水的大小，而一个硅晶片却有盘子那么大。这意味着完成一套工艺，我们能同时做出成千上万个传感器——又是工业技术中的另一个技巧，通过量产来降低成本。

仅仅是一个传感器就需要这么复杂的工序。当我们讨论理想中的智能微尘的时候，目标是在相同的尺寸上集成进去好几个传感器。如果要实现监测环境变量的功能，我们需要将温度传感器、加速度计、磁强计、气压传感器等集成在一个十几微米大小的尺寸上；从这里你就能看出挑战是多么的巨大。这意味着要设计更复杂的机械结构，用更复杂的材料和更复杂的加工技术。至少从现在的技术来讲，这个目标是一个几乎不可能实现的任务。

智能微尘的微处理器

微处理器是智能微尘的另外一个重要模块,它是简化版的 CPU(中央处理器),需要运行低功耗的操作指令,来指挥数据的采集和传输。这部分的微小化技术相对成熟,主要依赖于传统的 CMOS 电路加工技术。这部分的微小化离不开整个半导体工业微小化的技术支撑。

在半导体工业界,有一个著名的摩尔定律(Moore's law),这个定律是由英特尔(Intel)创始人之一戈登·摩尔(Gordon Moore)提出来的。这是一个经验规律:每 18 个月芯片的集成度会提高一倍,即实现相同的性能,只需要一半面积的芯片。曾几何时,计算机还基于笨重的电子管,这直接导致了一台计算机要占据一间房子的大小,而且功耗巨大。而如今的计算芯片,已经只需要依赖高度集成的晶体管:在指甲盖大的芯片上,就可以集成进去上 10 亿个晶体管。这也就是为什么如今的计算机会更小、更轻,反而计算能力更为强大。如今一个能装在口袋里的智能手机的运算能力,已经远远超过当年将人类送上月球的占地 200 平方米的大块头计算机。"同样的计算能力,当年人们用它将人类送上了月球,而如今你却用它来拿小鸟砸猪(手游'愤怒的小鸟')"说的就是这样的强烈对比。这都是微型化带来的天翻地覆的变化。

如果仅仅只是将半导体工业的微型化看作面积减半的数字游戏,那可就是过于低估了其中的难度。每一个新的技术的研发,都是巨大的资金投入和技术挑战——尤其是在特征尺寸缩减到纳米级别的今天。每一次的面积缩减的过程,都需要通过许许多多的科学研究和技术创新才能实现。

智能微尘的通信模块

就通信模块来讲,低功耗还是首要需求。并且智能微尘要能实现双向无线电数据传输,同时具备发射和接收模块——因为每一个微尘都要向相邻的微尘发射或者接收数据,所以仅有一种模块是不够的。

由于微尘本身体积很小,如果基于传统的射频通信,天线的尺寸也没法做大。这意味着微尘要在很短的波长工作,由于波长和工作频段成反比,这意味着需要使用高频段,从而消耗更多的功耗——这对于智能微尘来讲是致命的。另一个方法是采用光波通信。这种通信方法不单外围电路简单,而且能够低功耗地工作在短波长,非常适合智能微尘这样的极小结构。为了保证光波能被有效地接收和反射,"天线"被设计成了独特的立方角形的结构。这种结构能够被微机电系统技术加工出来。

在网络互连方面,采用分布式网络。就是说数据被传向临近的微尘节点,逐级到达云服务器,而不是直接一次性地传达到云服务器,这一方面会降低通信模块的设计要求,只需要满足短距离传输即可,另一方面也能降低工作时的功耗要求。文章前面做过微尘网络和人的神经系统的类比,实际上人的神经系统也是类似的工作机制,信号从来不是直接转向大脑的,而是先由感知到输入的神经元传向临近的神经元,逐级扩散,直至信号到达大脑。

智能微尘的电池微型化

最后一个,也是最棘手的一个模块,就是电池模块。要实现智能微尘这个极微小的系统,需要各项技术的共同进步,电池的微小化就是当前最大难题和技术瓶颈。如果不能实现电池的微型化,很难将整个系统做得更小。如今的电子系统,例如智能手机或者平板电脑,如果你打开后盖,会发现里面绝大部分的容积被电池所占据。这是因为如今的锂电池,储能密度远远达不到微小化的要求。要追求更高密度的储能电池,就需要新的电池技术和材料,例如氢氧燃料电池、石墨烯电池,甚至核电池等。

当然,另外一个路径是可以减少微系统本身的功耗,这样极少的电量也能支持很长的时间。如今的微处理器功耗在 100 瓦量级,要实现执行一条指令平均只需 12 皮焦能耗的智能微尘目标,需要在性能上进行折中,使用计算更慢的处理器,另外一方面,系统集成,甚至更底层的优化改进设计,可能达成这一目标。例如可以让微尘在大多数时间处于休眠状态,只在执行任务收集数据的时候才被唤醒。这意味着一切软件系统,都需要针对智能微尘,进行自下而上的重新设计。

另外,理想的智能微尘,能够散布于环境中,长时间检测物理参数。由于散布的微尘数量巨大,使得人为充电几乎不可能。所以理想的智能微尘,应当具有一个自充电模块,例如吸收太阳能、昼夜温差的热能,或者环境中电磁波的能量,甚至地表振动的能量。就目前的技术来讲,做出这样的模块尚没有成型的技术,更别说将这样的模块做到微米大小。

智能微尘系统级别的挑战

假设 20 年后的今天,我们的各项技术突飞猛进,前文提到的四个模块都能够做到很小,智能微尘可能还是不能实现的,因为各个系统分别很小不意味着组装之后的系统还能够同样的小——系统互联和封装同样不是一个简单的工作。

要实现这两个目标,有两个解决问题的途径。一是研究出微米尺度下的互联和封装技术,能够把这些系统简单快速地组装在一起,并封装在起保护作用的外壳里。之所以要求简单快速,是因为我们的目标绝不仅仅是制作一两个这样的微尘,而是千千万万个,所以制作工艺一定要能够规模化。想要规模化的组装微尺度下的模块绝不是一件易事。需要封装成型的原因是要监测环境变量,微尘要能够抵御环境的侵蚀,例如防水、防晒、防寒,等等。一个电路系统如果不封装就投放环境的话很可能会立刻失效。第二个方法当然是从根本上解决问题,就是将这四个模块同时生产加工在一个单晶硅上,这样不但免去了互连的需求,而且封装技术也变得更为简单。这种技术的最大难点在于不同模块的生产工艺往往差别很大,想要同时加工在一个半导体芯片上意味着要么改动某些模块的设计,要么改良加工工艺使之能够同时生产电路、传感器、通信模块和电池——这从目前的技术水平来讲,是很难想象短期内就能够实现的。当智能微尘的硬件实施成为可能,软件方面的数据处理也将面临极大的挑战。智能微尘不仅每个单元体积极小,网络的节点数目也十分庞大。这对控制软件、网络技术和数据处理都提出了更高的要求。

如果解决了互联网的难题,我们就能通过云服务器来收集智能微尘网络所采集的数据了。这样新的难题又来了。如今的移动终端仅仅限于个人电脑、平板电脑和手机,即便这样,我们已经需要面对海量的数据整理、存储和分析,也就是现在的热门话题——大数据。当我们制作了智能微尘网络之后,网络节点数目一下增加了好几个量级,每个节点又会全天候不间断地采集环境物理量,这样我们面对的,不仅仅是大数据,而是超级大数据。因为数据量将随着数据终端的数目,呈现超大规模的增长。这么大的数据量我们现有的云技术能够应对吗?如何处理这么大规模的数据?这些都将是棘手的难题。

智能微尘的应用展望

说了这么多智能微尘面临的挑战,并不意味着这项技术就被判了死刑。与此相反,如果我们热情地去追求这个终极目标,可能带动半导体技术、传感器技术、通信技术、电池技术、网络技术和大数据技术等一系列科技的进步,这个美好的蓝图可能会逐步变为现实——人类科技的进步史,就是一部迎难而上的历史。

在现阶段的技术水平,物联网和可穿戴产品初露端倪。而智能微尘,其实可以看作是终极的物联网。智能微尘是下一代的超级物联网的数据采集终端。它所带来的无限应用可能和美好的信息化未来世界,使得这个概念成为下一代技术中一颗耀眼的明珠。这种微尘传感器网络,能够被用于气象预报、地质检测、灾难救援、无人监控、医疗应用、外星探测以及军事情报收集等领域。但是实现这项愿景却面临诸多就目前的技术来讲极度困难的课题:它需要微处理器、通信模块、传感器模块、电源模块的高度单片集成,需要先进的高密度电池技术和自充电技术,还需要新的面向庞大节点数目的网络传输技术和大数据技术等。

　　智能微尘这个概念的实现,需要整个工业界的推动和进步,不是一两个创新能够实现的,也不是一朝一夕的事。回顾人类科技走过的一百年,又有谁能在计算机刚刚诞生的时候预见到如今集成度如此之高的智能手机系统、可穿戴设备和物联网系统呢? 又有谁能预见到如今的互联网技术和大数据技术呢? 历史的车轮滚滚向前,科技的进步日新月异,没有做不到,只有想不到。相信在几十年后,这项技术能够被推向工业化,让"给全球每一个沙粒一个 IP 地址,构建感知和连接一切的世界"的美好梦想成为可能。

参考译文:

Smart Dust—the Tool of Information

　　In 2060, when you are sitting in the Tesla Model Z, and the tram is suspending in the air, moving automatically to destination A at subsonic speed. Robot in A riots and destroys infrastructure there, causing many casualties. You're one of the rescuers and you're in a state of anxiety. You turn the apple ring 5 in your hand and think, "I want to see the distribution map of local wounded people." The ring senses your thoughts, and then a 3-D virtual reality interface emerges in the air in front of you—it's a bird's eye view of the street of place A, with red dots marking the location of the critically injured. As you slide your finger in the air, the interface enlarges the location of the most seriously injured person immediately, showing a series of physiological indicators, the injured area and the estimated remaining time of rescue in detail. Then you think, "I need a plan." "The Voice" has not yet finished, the map in front of you automatically updated to show a road map, pointing out the best rescue route. But you don't know, in the era of information shortage decades ago, the work of search and rescue has never been so efficient.

　　The searching and the rescuing job of MH370 is seriously implementing, for the power of the black box of the lost aircraft can only maintain its operation for only one day. If electricity runs out, the black box will become a perpetual mystery sunk in the deep Indian Ocean. Finding an airplane in the vast Indian Ocean is as difficult as finding an embroidery needle in a football field. The information and data collected by rescue teams' detectors at a time are very limited, which makes it time-consuming and difficult for tracing the location of the aircraft in the deep Indian Ocean through the way of blanket scanning. Even if the topographic map of the seabed can be constructed over an enough period of time, the data can not be real-time. Moreover, ocean currents can also lead a movement of the target

black box, thus it is also possible to miss the target even if the whole ocean landscape is formed by line scanning.

The 8.0 earthquake, which affected more than half of China, made the city ruined, traffic disrupted, roads damaged and the work of disaster relief extremely difficult. The priority of rescue is to locate the surviving casualties in the ruins as soon as possible so as to search and rescue wounded people in time. During this process, timely and effective information is also important. If there is no clear goal, rescue work will get half a result with twice effort. Traditional non-informationized manpower search and rescue are difficult to realize accurate positioning. It is common that many wounded can not be found in time for the first time, thus missing the golden opportunity to rescue.

Nearly 10 years of waiting, and 5 billion kilometers of flight, the New Horizon probe arrived over the Pluto on schedule, taking the first photograph of the planet that lies in the deepest of the solar system, unveiling its mystery for the world. It takes 4.5 hours to send back an HD picture from the deep of the solar system to the surface of the earth. The data transmission rate is only 1Kbps. To get more information about the surface geomorphology and geological characteristics of this distant planet, it means more explorations and huge investment. Existing detectors are not only expensive but also inefficient to collect data.

Science fiction is always so full, but the reality is always so skinny. How to turn the science fiction mentioned before into reality? Imagine one day when data collection terminals can become as small as sand and can spread around each corner of the earth, then all of these problems will no longer be a problem. Our entire planet is like a huge display screen, and each terminal, which is as tiny as sand grain, is a pixel of this screen. Under the control of the central calculator, the physical information of each coordinate (GPS coordinates, temperature, humidity, speed, light intensity, magnetic field intensity, etc.) can be seen thoroughly. These amazing big data can be monitored, tracked and analyzed in real-time by global central cloud servers.

The ultimate informationization unit is what we called Smart Dust, a very small and highly integrated sensor system.

It sounds like a creepy world without any privacy—Smart Dust is the best tool for the CIA to monitor the entire human beings, and wherever you go, your actions may be recorded and observed by someone who controls the central cloud server.

However, everything is a double-edged sword, and the worst world is, to some extent, the best. Because every physical quantity on the earth is monitored dynamically, weather forecasting, geological exploration, earthquake prediction, ocean current detection, and other data will be extremely accurate and available. When doing space exploration, it only needs to launch a bomb that filled with Smart Dust to make the dust spread all over the planet's surface after the explosion, which will make it easy to survey and map the surface topography and to do the geological exploration. When conducting disaster search and rescue, Smart Dust can create a clear 3D image of the whole ocean, including all objects in the ocean, therefore the work of search and rescue becomes a process of recognizing shape pattern that only needs a computer to locate. Geological exploration is the same as disaster search and rescue. Resource exploration and earthquake predication will also be extremely informative if such Smart Dust can be filled into the deep underground. Individuals will also benefit from the Smart Dust—disease detection will become very easy, because when people take Smart Dust capsules, the whole digestive system will be clearly imaged, thus making it easy to treat diseases.

In the ideal world, Smart Dust can make great convenience to people's lives, because it is the ultimate informatization.

What is Smart Dust?

The concept of Smart Dust was first proposed in 1992 and has been studied by the Defense Department's Advanced Research Projects Agency (DARPA) since the 1990s. The purpose of this concept is to form a monitor network distributed in our life by a series of micro-sensors with communication modules. Each monitoring unit is the so-called "micro-dust". Thousands of micro-dusts are scattered in the environment, and form self-organized wireless networks among them to collect environmental data (temperature, pressure, etc.). Then the collected data are sent to the terminal computer (or cloud server) through the "dust" communication module to be analyzed and processed. So if we want to make this overall dust network a reality, single dust must possess these functional modules: sensor module for environmental data collection, a communication module for wireless data transmission, power module for power supply and self-charging, and microprocessor module for controlling and scheduling all modules.

An interesting analogy is that if the area where environmental variables are to be monitored is imagined as human skin, then each Smart Dust can be imagined as a single tactile neuron in humans. Neurons can collect tactile signals, and they can also connect to form a network to transmit these signals. External stimulation is transmitted among neurons, and is eventually transmitted to the brain, in which tactile signals will be processed, thus tactile awareness will be got. A Smart Dust network also has a very similar system: the area

where environmental variables are collected is like human skin, each Smart Dust is like a neuron, and the central cloud server is like the human. Data is collected and transmitted by dust, and finally, it will reach the cloud server.

One of the most popular terms in recent years is the Internet of things that aims to connect common devices with the Internet. Take household equipment, for example, all the intelligent lights, refrigerators, stoves, doors and windows, air conditioning, etc. are connected to the Internet to make home informationalized and automatic controlled. To realize the macro-vision of the dust sensor network, the core technology is to produce single dust. This is actually a "sparrow, although small, five internal organs" computer unit. The four necessary modules of the sensor, communication, power supply, and microprocessor are mentioned above. Together, all these modules have to reach the size of dust, which is only tens of microns. The difficulty can be imagined.

It is a structure drawing given by Professor Pister and Professor Kahn of the University of California, Berkeley when they proposed the concept of Smart Dust. Based on the technology at that time, the expected volume of this so-called "dust" is 5 cubic millimeters. It can be seen from the drawing that this is huge dust that is far from "micro-dust". As can be seen from the drawing, it does have four modules: sensor, communication, power supply, and microprocessor. The battery module occupies the largest volume at the bottom. Above the large batteries is a solar energy electrode plate, which is used to convert solar energy into electricity for charging. Above this base is micro sensor chip and a microprocessor chip. And two communication chips are used to transmit and receive signals.

Even though most of the volume of the whole system is occupied by batteries, the overall size of batteries is only millimeters. Because the battery itself is very small, the whole system is required to have very low power consumption. The power consumption for operating a whole day is required to not exceed 10 microwatts. What is this mean? At 10 microwatt power consumption, ordinary chips will stop computing because of power insufficiency. It is for such requirements that the traditional communication module can not be used directly as well as the traditional control software. The whole system needs to be re-optimized to meet the requirements of low power consumption.

Low power consumption is just a little part of the challenge. Five cubic millimeters is not enough to create the dust in our target. There is also a need for further efforts in miniaturization to make the size of the entire system reaches the micron level, that is, to shrink to nearly one-thousandth of now. It's a thing that is easy to understand but difficult to do—it will take 20 years to achieve based on the fact that the area is halved every two years.

Of course, low power consumption and micro dimension only refer to the goal of making single prototype dust. To realize large-scale production of Smart Dust and to make them truly distribute within our environment, more demanding industrial production conditions need to be met, including low production costs, good stability, anti-environmental interference ability, long-serving life, and other conditions.

It is because these demanding engineering conditions that Smart Dust still staggers in the theoretical stage after its concept was put forward 20 years before. With the rapid development of semiconductor technology, sensor technology, big data, and cloud computing in recent years, this theory has gradually shown its possible to break away from the science fiction and may become a revolutionary new product that can be realized in the coming decades.

The aim of Smart Dust is to collect physical information such as temperature and pressure of the environment. The sensor, of course, is indispensable to achieve this function. So, what is the sensor? The transducer / sensor is not a new concept. Sensors exist in all aspects of our lives. The traditional mercury thermometer is a simple sensor, which directly converts the temperature signal into a visible and readable number. Another example is the sensor faucet, which is common in our life. When hands approach, water will automatically come out from the sensor faucet. The reason of this phenomenon is that there are infrared sensors in the control circuit, which can detect the infrared intensity of the human body to determine whether the hand is close to the faucet. The sound and light-controlled lamp is an example, too. Why can this lamp be extinguished during the day, only when people approach at night and make a sound can it be lighted? This is because its control circuit has a photoresistor to detect the intensity of light and a buzzer to detect sound. The control circuit performs a simple judgment: if the light becomes weak and there is sound, lighten the lamp. Photoresistors and buzzers are also examples of sensors.

Normally, the role of the sensor is to convert physical signals into electrical signals. Why do we do this? This is a classic technique in science and technology to turn unknown problems into known ones. Circuit technology is one of the most mature engineering techniques. Once a non-circuit signal is converted into a circuit signal, all circuit signal processing, power amplification, control, automation, and computer technology can be easily connected to it. Sound controlled buzzer uses piezoelectric components that can convert sound vibration into charge signals. An infrared detector that can conduct infrared

control uses ferroelectric materials that collect charges after absorbing infrared rays. The photosensitive resistor that achieves light control uses special semiconductor materials that happens current changes after illumination.

Common sensors are tools that can convert other physical signals such as force, heat, light, speed, acceleration, magnetic field, and sound wave into electrical signals. They are the connection point between the circuit system and the colorful world with all kinds of physical signals that we live in.

Traditional sensors, such as the buzzer and infrared sensors mentioned above, are relatively large, usually in the millimeter or even centimeter scale. Today, with the integration and microminiaturization of the system, these large volume sensors will cause many problems. Oversize means heavier weight, higher cost, and greater power consumption. Nowadays, the requirement of volume and power consumption is becoming more and more strict, which leads to higher requirements for microsensors. Most of today's electronic systems contain various sensors, among which are microsensors.

The smart phone is the best example. You may only care about whether smart phones are cool and convenient to use, but don't care the structure and principle of smart phones. In fact, if you disassemble a smart phone, you will find that there are various sensors in it, most of which are micro sensors. Take iPhone 6 as an example, it cnotains an accelerometer, gyroscope, electronic compass, fingerprint sensor, distance and ambient light sensor, MEMS microphone, barometric sensor and so on. It is because the existence of these sensors that enables the iPhone to have a series of functions that traditional computers do not have, such as navigation, horizontal screen image transposition, steps measured, screen brightness adjustment according to surrounding light, and fingerprints detection. The latest iPhone 6S also launches the 3D touch function, enabling the mobile phone to recognize the pressing strength of fingers, which is also based on the pressure sensor array behind the display screen. These new sensors have the advantages of small size, light weight, low power consumption, high reliability, and fast working speed.

These microsensors can reach the micron or even nanometer scale. How big is the micron? The diameter of the hair is 20 to 400 microns. How to make such a small sensor? The method that has been industrialized today is through a technology called microelectromechanical systems (MEMS). The accelerometer, which is very common in smart phones, is a kind of micro electromechanical system, whose function is to detect input acceleration from outside. That's why smart phones can know whether we shake the phone and how violent the shake is. For example, if you use the function of the "shake" of

Wechat, the accelerometer can tell the chip that the user wobble the phone and let the chip perform the corresponding function according to this input signal to match other users to chat.

If we disassemble this accelerometer and watch it under the scanning electron microscope, we will see a comb-like structure, which is the core component for the function of acceleration detection. This is actually a series of parallel capacitances that are used to detect external mechanical movements. Shaking the phone will cause the center mass to shake, thus changing the gap between the two plates of the capacitance, and then changing the figure of the capacitance, which can also be read by the post-processing circuit. With this one-to-one correspondence, we actually measure the acceleration in reverse, calculating the input acceleration by reading the current changes. The so-called acceleration detection of mobile phone is actually measuring the capacitance change caused by the movement of the mass hung at the center of the accelerometer.

From this example, we can see that the so-called "machinery" here does not look like the machinery in our concept at all. The machines we are used to see are all elaborate linkage mechanisms consist of gears, connecting rods, and other components, and the "micro-machine" here is only a mass block with a unique shape that can move. Is it a big surprise? In fact, this is a characteristic of all engineering technology — the best solution should be the simplest one that can solve the problem. Our goal here is to detect the acceleration. According to Newton's second law, large mass blocks can amplify the effect of acceleration, so here just a simple mass block is needed.

To make such a micro-electromechanical system, a complex semiconductor processing is usually needed. Starting with a blank silicon wafer, adding thin layers step by step through chemical reactions, or etching thin layers into desired shapes, such a system will be made after dozens or even hundreds of such processes. It does sound complicated, and the complex production process means high costs. Then how to reduce the cost in mature industrial technology? Don't forget that the size of this sensor is the same as a drop of water, but a silicon wafer has the size of a plate. This means that we can make thousands of sensors at the same time when completing a set of technology. It's another technique in industrial technology, which reduces costs through mass production.

Just a sensor alone requires such a complex process. When we discuss the ideal Smart Dust, the goal is to integrate several sensors in the same size. If we want to achieve the function of monitoring environmental variables, we need to integrate temperature sensors, accelerometers, magnetometers, barometric sensors, and other parts into a size of more than

a dozen microns, from which you can see how great the challenge is. This means that there is a need to design more complex mechanical structures and use more sophisticated materials and more sophisticated processing techniques. This goal is almost an impossible task at least in terms of current technology.

Microprocessor of Smart Dust

Microprocessor, a simplified CPU (central processing unit), is another important part of Smart Dust. It needs low-power operating instructions to command data acquisition and transmission. The miniaturization technology of this part is relatively mature, which mainly depends on the traditional CMOS circuit processing technology. And it is inseparable from the technical support of the whole semiconductor industry.

In semiconductor industry, there is a famous Moore's Law, which was proposed by Gordon Moore, one of the founders of Intel. This is a rule of experience: the degree of chip integration will double every 18 months, that is, only half the area of the chip is needed in order to achieve the same performance. Once upon a time, computers were also based on bulky electron tubes, which directly led to the house-size and the huge power consumption of a computer. Today's computing chips, however, rely only on highly integrated transistors: on chips with nail size can integrate more than a billion transistors. That's why today's computers are smaller and lighter, but is more powerful. Today, the computing power of a phone that can be put in pocket is far exceed the 200 square meters large computer that sent humans to the moon before. "With the same computing power, we used it to send humans to the moon years ago, and now we use it to take small birds to hit pigs (the game of "angry birds"), which is such a strong contrast. These are the incredible changes caused by miniaturization.

If the miniaturization of the semiconductor industry is merely regarded as a digital game of halving the area, then you underestimate the difficult. The development of every new technology needs huge financial investment and has great challenge on technology, especially today when the characteristic size is reduced to the nanometer level. Every area reduction process needs a lot of scientific researches and technological innovations to achieve.

Communication Module of Smart Dust

As for the communication module, low power consumption is just the primary requirement. And the Smart Dust should be able to achieve bidirectional radio data transmission, and at the same time have modules of transmission and reception- because each dust has to send to or receive data from the adjacent dust, so only one module is not enough.

Because of the small size of the dust itself, the antenna can not be enlarged if it is based on traditional radio frequency communication. This means that the dust has to work at very short wavelengths. Because the wavelength is inversely proportional to the frequency of the work, this means that the high-frequency section is needed to consume more power, which is fatal to Smart Dust. Another method is to use light wave communication. This kind of communicating method is not only simple in peripheral circuit, but also can work at short wavelength with low power consumption, which is very suitable for Smart Dust structure. In order to ensure that light waves can be effectively received and reflected, "antenna" is designed as a unique cubic structure, which can be processed by micro-electromechanical system technology.

In the aspect of network interconnection, a distributed network is adopted. That is to say, the data is transmitted to the nearby dust nodes and reaches the cloud server step by step, but not directly conveying to the cloud server at one time. On the one hand, this will reduce the design requirements of the communication module, only need to meet the requirement of short-distance transmission. On the other hand, it can also reduce power consumption. In fact, the human nervous system is similar to the dust network. Signals never transmit directly to the brain, but first from the sensory that senses the input signal to the adjacent neurons, and then spread step by step until the signal reaches the brain.

Battery Miniaturization of Smart Dust

The last and most difficult module is the battery module. To realize the Smart Dust system, a collective progress of various technologies is needed. The miniaturization of batteries is the biggest problem and technical bottleneck at present.

If the miniaturization of batteries can not be realized, it is difficult to make the whole system smaller. If you open the back cover of today's electronic systems, such as smart phones or tablet PCs, you will find that most of the volume is occupied by batteries. This is because the energy storage capacity of today's lithium batteries is far less than the requirements of miniaturization. To pursue a higher density of energy storage batteries, new battery technologies and materials are needed, such as hydrogen-oxygen fuel cells, graphene cells, and even nuclear cells.

Of course, another way is to reduce the power consumption of the micro-system itself, so that very little power can also support a long time. Today's microprocessors' consumption is 100 watts. Only an average of 12 pico smart dust is needed to achieve the goal for executing an instruction, which is necessary to compromise the performance and use slower processors. On the other hand, system integration and even lower-level optimization and improvement may achieve this goal. For example, dust can be dormant for most of the time, only waking up when collecting data. This means that all software systems need redesign the Smart Dust.

In addition, the ideal Smart Dust can be dispersed in the environment, detecting physical parameters for a long time. Because of the huge amount of scattered dust, it is almost impossible to recharge it artificially. Therefore, the ideal Smart Dust should have a self-charging module, such as absorbing the heat energy of solar energy and the temperature difference between day and night, the energy of electromagnetic wave in the environment, and even the energy of surface vibration. As for the current technology, there is no technology to make such a module, let alone to make such a module in micron size.

Challenges of the System of Smart Dust

Assuming that 20 years later, our technology is advancing quickly, the four modules mentioned above can be made very small, but the Smart Dust can still not be realized, because the small size of each system does not mean that the assembled system can still be small - system interconnection and packaging is not a simple task, either.

In order to achieve these two goals, there are two ways. One is to develop interconnection and packaging technology in the micron scale, which can assemble these systems simply and quickly and encapsulate them in protective enclosures. The reason why we want to be simple and quick is that our goal is not only to produce one or two dust, but tens of thousands of them, so the manufacturing process must be on a large-scale. It's never easy to assemble micro-scale modules on a large scale. The reason for encapsulating is to monitor environmental variables, and dust should be able to resist environmental erosion, such as waterproof, sun protection, cold protection and so on. If a circuit system is put into the environment without encapsulation, it is likely to fail immediately. The second way, of course, is to solve the problem fundamentally by manufacturing the four modules on a monocrystalline silicon at one time, which not only eliminates the need of interconnection, but also simplifies the packaging technology. The biggest difficulty of this technology is that the production process of different modules often varies greatly. To manufacture on a semiconductor chip at one time means that either changing the design of some modules or improving the fabrication process to enable simultaneous production of circuits, sensors, communication modules and batteries, which is difficult to achieve in the short term on the basis of the current technical level.

When the hardware implementation of Smart Dust becomes possible, the data processing of software will also face great challenges. Smart Dust is not only very small in size per unit, but also has a large number of network nodes, which puts forward higher requirements the software controlling, internet technology and data processing.

If we solve the problem of the Internet, we can get the data collected by the Smart Dust network through the cloud server. Then a new problem is coming again. Today's mobile terminals are limited to personal computers, tablet PCs and mobile phones. Even so, we have to face massive data sorting, storage, and analysis, which is the current hot topic—big data. When we made the Smart Dust network, the number of network nodes increased by several orders of magnitude, among which each node will collect the environmental physical quantities all the time, therefore, what we faced are not only big data, but super big data. Because the amount of data will increase at a very large scale with the number of data terminals. Can our existing cloud technologies deal with such a large amount of data? How to deal with such large-scale data? This will be a difficult problem.

Prospect of Smart Dust

Saying so much about the challenges of Smart Dust does not means that the technology is "condemned to death". On the contrary, if we pursue this ultimate goal enthusiastically, which may lead to a series of technological advances on semiconductor technology, sensor technology, communication technology, battery technology, network technology, and big data technology, then this wonderful blueprint may gradually become a reality - the history of human science and technology progress is a history of facing difficulties and dealing with difficulties.

At today's stage of technology, the Internet of Things and wearable products are emerging. Smart Dust, in fact, can be regarded as the ultimate Internet of Things. Smart Dust is the next generation of data acquisition terminal of the super Internet of Things. The infinite application possibilities and a beautiful informative future world brought by it make this concept a dazzling pearl in the next generation of technology. This dust sensor network can be used in weather forecasting, geological detection, disaster rescue, unmanned surveillance, medical applications, alien detection, and military intelligence collection.

However, to realize this vision, there are many problems that are extremely difficult in terms of current technology: it requires highly monolithic integration of microprocessors, communication modules, sensor modules and power modules, advanced high-density battery technology and self-charging technology, as well as new network transmission technology, and large data technology with a large number of nodes.

The realization of the concept of Smart Dust requires the promotion and progress of the whole industry, which can not be realized by one or two innovations, nor overnight. Looking back on the past hundred years of human technology, who can foresee today's highly integrated smart phone system, wearable devices and Internet of things system when the computer first emerged? Who can foresee today's Internet technology and big data technology? The wheel of history is rolling forward, and the progress of science and technology is changing quickly. It is believed that after a few decades, this technology can be industrialized, making it possible to "giving every grain of sand an IP address to build a world of perception and connectivity".

单词释义：

1. 亚音速	subsonic speed
2. 分布图	distribution map
3. 鸟瞰街景	bird's eye view of the street
4. 生理指标	physiological indicators
5. 绣花针	embroidery needle
6. 冥王星	Pluto
7. 高清照片	HD picture/ high-definition picture
8. 地貌	geomorphology
9. 美国国防部高级研究计划局	the Defense Department's Advanced Research Projects Agency（DARPA）
10. 微型传感器	micro-sensors
11. 微处理器	microprocessor
12. 太阳能电极板	solar energy electrode plate
13. 微传感器芯片	micro sensor chip
14. 微处理器芯片	microprocessor chip
15. 通信芯片	communication chip
16. 传感器技术	sensor technology
17. 水银温度计	mercury thermometer

18. 感应水龙头	sensor faucet
19. 红外线传感器	infrared sensors
20. 声光控灯	sound and light controlled lamp
21. 光敏电阻	photoresistor
22. 蜂鸣片	buzzer

译文解析：

1. 汉译英中词性转换。

原文：是 A 地的鸟瞰街景，其中的红点标注着危急伤员的位置。

译文：Tt's a bird's eye view of the street of place A, with red dots marking the location of the critically injured.

解析：原文中形容词"其中的"在译文中变为介词"with"。

原文：眼前的地图自动更新显示出一个线路图

译文：the map in front of you automatically updated to show a road map

解析：原文中形容词"眼前的"在译文中译为介词词组"in front of"。

原文：因为数据量将随着数据终端的数目，呈现超大规模的增长。

译文：Because the amount of data will increase in a very large scale with the number of data terminals.

解析：形容词"大规模的"转换成名词"scale"。

2. 长句的翻译。

原文：你在空气中滑动手指，界面立刻放大到了一个最严重伤员的位置，详细地显示出他的一系列生理指标、受伤的部位和预估的救援剩余时间。

译文：As you slide your finger in the air, the interface enlarges the location of the most seriously injured person immediately, showing a series of physiological indicators, the injured area and the estimated remaining time of rescue in detail.

解析：译文中用"as"引导，体现原文中句子间隐含的时间关系，即"当滑动手指，界面放大"。英语属形合语言，多连接词，呈树状结构，使用谓语动词"showing"，连接两个分句，符合英语形合的表达习惯。

原文：救援队所拥有的探测器，每次只能收集非常有限的信息和数据。这决定了如果想以地毯式扫描印度洋海底来追寻飞机的去向，必定会是一个漫长而艰苦的过程。

译文：The information and data collected by rescue teams 追寻飞机的去向词 the location of the most seriously injured person immediately, showing a series of physiollocation of the aircraft in the deep Indian Ocean through the way of blanket scanning.

解析：原文中的"收集"是主动形式，在译文中译成被动含义的"collected by"，符合英语多被动的表达习惯。译文中的引导词"which"的使用将两个分句连接起来，符合英语树状结构的特点。形式宾语"it"的使用起到平衡句子结构，避免句子结构混乱的作用。

原文：如果解决了互联网的难题，我们就能通过云服务器来收集智能微尘网络所采集的数据了。

译文：If we solve the problem of the Internet, we can get the data collected by the Smart Dust network through the cloud server.

解析：原文中第一个分句中"解决了"前并没有表明主语，但根据句意我们可以知道，主语是"我们"，在汉语中经常出现无主语句的情况，通常根据上下文来暗示主语。英语中添加"we"做主语，将主语显性的表明。原文中"采集"是主动语态，译文中翻译成了被动语态"collected"，符合汉语多主动，英语多被动的特点。

原文：它需要微处理器、通信模块、传感器模块、电源模块的高度单片集成，需要先进的高密度电池技术和自充电技术，还需要新的面向庞大节点数目的网络传输技术和大数据技术等。

译文：Tt requires highly monolithic integration of microprocessors, communication modules, sensor modules and power modules, advanced high-density battery technology and self-charging technology, as well as new network transmission technology and large data technology with a large number of nodes.

解析：原文三次重复使用"需要"，在译文中并没有译成三个"need"，而是使用连接词"and"和"as well as"，这符合汉语多重复，英语少重复的特点。英语一般注重表达的多样性，力戒无意义的重复，一般尽量避免重复使用相同的词，常用的手法有替换同义词或省略等。这里采取了转换同义词的方式。

原文：每一个监测单元就是所谓的"微尘"，成千上万的微尘被散播在环境之中，相互之间用自组织方式构成无线网络，来收集环境数据（温度、气压等）。

译文：Each monitoring unit is the so-called "micro-dust". Thousands of micro-dusts are scattered in the environment, and form self-organized wireless networks among them to collect environmental data (temperature, pressure, etc.).

解析：原文中可以看到每个短句间都是用的逗号，但在译文中，我们把第一个逗号处理成了句号，分为两个句子翻译，如果按照原文用逗号，可能译文过长，不利于读者理解，因此将其分为两个句子来翻译。

原文：有这样一个有趣的类比，如果将要监测环境变量的区域想象成人类的皮肤，那么可以将每个智能微尘想象成人类的单个触觉神经元。

译文：An interesting analogy is that if the area where environmental variables are to be monitored is imagined as human skin, then each Smart Dust can be imagined as a single tactile neuron in humans.

解析：译文中使用连接词"that"引导表语从句，形成树形结构，符合英语形合的特点。

原文：近 10 年的等待，50 亿公里的飞行，新视野号探测器如期飞抵冥王星上空，首次拍下了这颗位于太阳系最深处行星，为世人揭开了它的神秘面纱。

译文：Nearly 10 years of waiting, and 5 billion kilometers of flight, the New Horizon probe arrived over the Pluto on schedule, taking the first photograph of the planet that lies in the deepest of the solar system, unveiling its mystery for the world.

解析：原文中的小短句在译文中通过增加连词"and"和使用非谓语动词的形式，将原文小短句之间的关系显性地表现出来，体现了英文的形合特点。

原文：试想有一天，数据收集终端可以变得如同沙粒般微小，并且能散布于地球的各个角落，那么以上所有这些问题，也都将不再成为问题。

译文：Imagine one day when data collected terminals can become as small as sand and can spread around each corners of the earth, then all of these problems will no longer be a problem.

解析：英语是形合语言，语言呈树形结构，而汉语是意合语言，语言呈竹形结构，根据英汉这两种差异，译文通过使用连接词"when"、"and"将汉语的小句连接起来，形成一个长句。符合英文表达习惯。

原文：宇宙勘探，只需要发射一个装满智能微尘的炸弹，令其爆破后的尘埃遍布星球表面，整个地表的形貌测绘、地质勘探将轻而易举

译文：When doing space exploration, it only needs to launch a bomb that filled with

Smart Dust to make the dust spread all over the planet's surface after the explosion, which will make it easy to survey and map the surface topography and to do the geological exploration.

解析: 第一个 it,在原文中,第二个小分句的主语省略了,在译文中我们添加了形式主语 it,代指"宇宙勘探",这里体现了汉语多无主语句多特点。第二个 it,在译文中作为形式宾语,代指后面的"survey and map the surface topography and to do the geological exploration"。

原文: 正是因为这样苛刻的要求,传统的通信模块不能被采用,传统的控制软件也不能直接使用。

译文: It is for such requirements that the traditional communication module can not be used directly as well as the traditional control software.

解析: 中文原文中有三个小分句,但在译文中,变成了一个长句,中间添加连接词"that"、"as well as"连接,这符合英语形合的特点,英语中多用连接词使句子连贯,形成树形结构,而汉语属意合语言,较少使用连接词,句子呈竹形结构。第二,原文中"不能被采用"和"不能被直接使用"这两个表达意思相同,在译文中用"as well as"来表达这种重复关系,省略了后面相同的表达,符合英语强调简洁,尽量避免重复的特点。

第八章 延寿抗衰

我的博士生导师 GaryRuvkun 教授有不少写人生物教科书的重量级研究：除了发现可以调控基因表达和个体生长发育的非编码小 RNA（核糖核酸）——这恐怕是他的最得意之作——Ruvkun 实验室还首先报道了胰岛素类生长因子信号通路在衰老过程中的调控作用和机制。当时刚加人实验室的我徘徊在衰老研究领域的大门前，考虑是否要在博士期间致力于对衰老的研究。不过最终，还是对非编码小 RNA 的兴趣占了上风，于是我对这些基因组的"暗物质"开始了研究。不过，在实验室耳濡目染衰老研究组的成果让我一直对这个领域充满兴趣，至今都还是我在自己研究课题之外的一个关注点。"衰老—长寿"领域无论是八卦流言还是正经的科学理论和发现都多到侃上几天几夜恐怕也聊不完，在这篇文章中我只能挑几段最有启发性的故事，希望引发读者更多的思考和阅读。

徐福的故事、辟谷的历史记载，还有荒诞的炼丹皇帝等，无论是传说还是事实，都反映了几千年来人们对"长生不老"的无限向往，同时也说明了人类对衰老的无知。在现代科学高度发展的今天，衰老是很多科学家研究的课题。不过，红葡萄酒抗氧化保健品、羊胎素等现代流行的抗衰老概念和产品就真的是长生不老的"灵丹妙药"吗？怀疑和科学的态度是最好的评判标准，还是先让我们看看科学家都在研究什么怎么研究的吧。

吃葡萄不吐葡萄皮可以让人长寿？

在众多对衰老和长寿的研究中，饮食和长寿有着最悠久深远的联系。近百年来，科学家在模式生物从酵母到秀丽隐杆线虫到果蝇到小鼠中的研究中都发现：少吃有助于长寿。一个有代表性的例子是 1986 年加州大学洛杉矶分校的科学家做的实验。他们给饲养的老鼠定量饮食，比较对照组（敞开吃，每周饮食摄入大约 115 千卡热量）和每周定量只摄入 85、50、40 千卡热量的三个实验组（四组老鼠都同样保证摄入足够的维生素和微量元素）。结果发现，摄入热最少的"节食组"的老鼠不仅身材苗条而且吃得越少越长寿。一般老鼠的寿命只有两年，但是节食最猛的老鼠创造了它们的长寿纪录——活了四年半。同时，与对照组相比，"节食组"老鼠的免疫系统功能也明显增强，患癌症的比例下降（Weindruch et al. 1986）。

观察到了这个现象还不够，爱刨根问底的科学家最关心的问题是：为什么降低卡路里摄入能够延长寿命？随着生活水平的提高，人们如今能享受到更多美食，可是很多研究结果却又让人担心美食中的热量（单位：卡路里）会影响寿命，难道为了长寿就必须放弃飘香诱人的美味吗？如果知道了低卡路里摄入能延长寿命的生物学机制，并对自身摄入的热量进行控制，我们说不定就可以随心所欲地大快朵颐，还能"长生不老"，岂不两全其美。

科学家们对酵母的研究揭示了一些关键性的答案。酵母是单细胞生物，因此衡量它"长寿"的标准与绝大多数动物不同。科学家通常用细胞分裂的次数而不是它们存活的时间来反映酵母的"寿命"，这也被称为"生殖寿命"。麻省理工学院（MIT）的 Leonard Guarente 研究组再次证实，生长在低糖（糖是酵母主要的卡路里来源）培养基上的酵母比正常培养基上的酵母更"长寿"。在此基础上，他们通过大规模筛选实验来研究到底是什么因素让这些低卡路里摄入的酵母"长寿"。结果发现，一个叫作 Sirtuin（Sir2）的蛋白发挥了关键性的作用：酵母缺失 Sirn2 蛋白后，即便生长在低糖培养基上也不能"长寿"（Lin et al. 2000）。这听起来不妙，似乎与"在大快朵颐中长生不老"目标背道而驰？别急，这个结果启发了 Leonard Guarente 研究组做了个反方向的实验。这次，他们利用基因编辑的方法提高了酵母中 Sir2 蛋白的数量。你猜怎么着？生长在正常培养基上的过量表达 Sin2 蛋白的"超级"酵母，几乎与低卡路里摄入的正常酵母同样"长寿"（Lin et al. 2000）这太棒了！也就是说如果能增加 Sir2 蛋白的量，不节食也可以长寿？！

Si2 究竟是何方神器？原来，Sir2 蛋白在一种化学小分子烟酰胺腺嘌呤二核苷酸（Nicotinamide adenine dinucleotide，简称 NAD）的辅助下，可以改变一些蛋白特定的化学修饰来调控它们的生物活性。几个被 Sir2 调控的关键性靶蛋白在保护基因组的稳定性、细胞内环境平衡等许多生物学过程中都可能发挥功能。虽然 Sir2 与抗衰老之间具体的机制还远不够清楚，但它潜在的抗衰老功能足以让大家兴奋不已。

在酵母的实验中，科学家利用基因编辑来增加 Sir2 蛋白的数量，但在人身上实现基因编辑还存在不少技术上的不成熟和伦理上的争论（参见相关文章：《基因编辑：一场由酸奶引发的新革命》）。既然增加数量有困难，科学家转而寻找方法来增强 Sir2 蛋白活性。2003 年，哈佛大学的 David Sinclair 研究组在《自然》杂志上发表了一篇文章，报道能够增强 Sir2 生物活性的多酚类小分子物质，比如白藜芦醇（resveratrol）。"喂食"酵母白藜芦醇就可以像减少卡路里摄入一样延长它们的生殖寿命。最重要的是，白藜芦醇在体外培养的人类细胞系中也同样能发挥效应（Howitz et al. 2003）。听起来太神奇了！

葡萄特别是葡萄皮中含有丰富的白藜芦醇。也许不是所有人乐意为了长寿而"吃葡萄不吐葡萄皮",不过红葡萄酒在长时间酿造过程中获得了葡萄皮中的白藜芦醇成分。这似乎解释了为什么酷爱红葡萄酒的法国人虽然喜好高脂肪食物而依然健康长寿,甚至总体上患心脏病的比例还比较低。在这篇白藜芦醇学术文章(Howitz et al. 2003)之前,红葡萄酒已经被认为有诸多保健功效,比如抗氧化、降低心血管疾病和癌症的发病率等。这些依据不仅给葡萄酒业带来效益,还创造了无数新商机,比如提取的白藜芦醇作为保健品登上了货架。一些研究抗衰老药物的科学家也成立了自己的公司,名利双收。

这么说我们在和衰老的千年之战中又扳回一局?可惜,事实并非如人所愿。在更多的科学家进行大量研究实验后发现,不仅 Sir2 在多种生物中对延长寿命的功效和机制都非常有争议,多数对白藜芦醇功能的神话般宣传更是言过其实。最糟糕的一个例子是,康州大学健康中心(University of Connecticut Health Center)的 Dipak Das 曾被认为是白藜芦醇和衰老研究领域的领军人物之一。但是,在他发表的一百余篇报道白藜芦醇功效的文章中,后来被发现有大量被捏造和篡改的数据。到目前为止,已经有数十篇论文被正式撤回,而更多的论文还在调查之中。这个科学界的丑闻更是增加了人们对 Sir2 和白藜芦醇的质疑。2011 年,21 个作者联合署名文章,澄清缺乏足够的实验证据支持白藜芦醇作为营养保健品(Vang et al. 2011)。白藜芦醇还被认为具有抗氧化功能,但是科学家通过严格的临床实验,发现没有任何证据说明这些流行的抗氧化产品对人体具有宣传的保健功能,摄入过量甚至还可能有害。所以,即便细胞氧化损伤是衰老的原因之一,延缓衰老也并非吃一粒抗氧化胶那么轻而易举。

不过,东方不亮西方亮。Sir2 的辅助因子——NAD——是个在新陈代谢、基因表达调控等许多生物学过程中都非常重要的多功能小分子。对 Sir2 的研究和很多从其他切入点开始的新研究都表明,提高细胞中 NAD 的含量对神经退行性疾病有明显改善,另外,有些实验还说明提高 NAD 的量可以减缓衰老。NAD 给大家带来的抗衰老新希望到底是否会成为下一种生物黑科技,未来会发现什么,还让我们拭目以待。这个故事至此已经充分体现了各种生命现象以及人类疾病的复杂性,衰老作为其中一例,远非一个饮食触发开关就能解释或是控制的那般简单。"研究"一词的英文"research"拆开来就是"re""search",意思是不断反复地寻找。在这不断研究的过程中,科学家们只有本着严谨无偏见的精神来设计和分析实验,才能一步步接近事实真相,开辟人类新知识的源泉。与人类生活和健康息息相关,并存在高度功名利益诱惑的衰老研究领域,更应如此。

对"吃"的继续深度研究

白藜芦醇神话破灭让许多喜好吃的食客大失所望,难道"长生不老"和"大饱口福"真的是鱼和熊掌不可兼得？先别急下定论。虽然"节食有助于长寿"法则在多种模式生物中一再被验证,拥有严谨和怀疑精神的科学家们并没有完全信服这条规律在我们人类身上同样适用。在研究衰老的模式生物里,和人类关系最相近的是小鼠,虽然同属哺乳类动物,而鼠类毕竟和灵长类动物(比如人类猩猩和猴子)有非常多的不同之处。那为什么不研究灵长类动物呢？因为且不说来自动物保护组织的反对,灵长类的自然寿命短则五六年,长则百年,是个耗资巨大、极其考验耐心和毅力的实验。

尽管困难重重,科学家还是迈出了研究灵长类动物衰老的第一步。从 20 世纪 90 年代末起,威斯康星州国家灵长类动物研究中心(Wisconsin National Primate Research Center,简称 WNPRC)和美国国家衰老研究所(National Institute on Aging,简称 NIA),都选择了恒河猴作为研究对象,各自开始了长达二十余年的跟踪研究。

2009 年,威斯康星州国家灵长类动物研究中心(WNPRC)率先在《科学》杂志上发表了研究成果(Colman et al. 2009)。还没来得及仔细阅读他们的统计数据和结论,我的目光就先被他们的第一张图吸引住了。对照组猴子无精打采、身上毛发所剩无几,而低卡路里摄入的节食组猴子精神焕发皮毛锃亮。看上去差别真不小！

阅读完了全文才发现统计数据和结论与图片给我的第一印象相差径庭:对照组和节食组的猴子其实有同样长的寿命！尽管节食组的猴子体重明显低,患几种典型的老年性疾病(糖尿病、癌症和心血管疾病)的比例也有所降低,但它们却死于其他原因,并没有更加长寿。美国国家衰老研究所(NIA)的结果随后在《自然》(Nature)杂志上发表(Mattison et al. 2012)。他们的研究说明,不仅节食组与对照组的猴子有类似的寿命,节食组在典型老年性疾病导致的死亡率上也与对照组没有显著区别。而后一结论与之前威斯康星州国家灵长类动物研究中心(WNPRC)的研究报道不同,美国国家衰老研究所(NIA)在文中解释道:不同研究所猴子的饲养条件(尤其是食物)的差别可能是导致不同结论的原因。总结起来,这个长达几个世纪、耗资巨大的实验虽然仍有值得讨论的地方,但都指出了"节食有助于长寿"法则在灵长类动物里并不完全适用。另外,人类节食是否能延长寿命还完全没有定论。这对于那些喜爱美食的人们而言,是个大大的好消息。但是,衰老和长寿机制的复杂性,以及在每个生物种中的特异性,对科学家而言无疑是个巨大的挑战。

逆境中绝处逢生

"故天将降大任于斯人也,必先苦其心志,劳其筋骨,饿其体肤,空乏其身,行拂乱其所为,所以动心忍性,曾益其所不能。"孟子用这一段话告知天下人:艰苦环境的磨砺让人奋发成长,成就大业。但他可能想不到的是,这一段哲理在一定程度上,居然放之"生物四海"而皆准。

前面我们提到了"饿其体肤"（降低卡路里摄入）在一些生物中延长寿命的例子。很有趣的是，这些长寿的个体通常具有更强的耐受力，能够抵御不利环境以及疾病。于是，科学家提出了一种假说：减少卡路里可能是一种轻微的压力，刺激机体产生防御性反应来增强抵抗不利环境和疾病的能力，从而延年益寿。注意了，这不是精神上的压力（酵母、线虫们还没有进化出能让它们为食物发愁的高级神经系统），而是指不利的环境压力，比如食物的缺乏、过高或低的温度、有害化学物紫外线等。

孟子的《告天下》在线虫身上的体现着实让人瞠目结舌。不过一毫米长的小小线虫有一套特殊本领：发育中的线虫幼虫在遇到非常不利的环境时，比如食物缺乏或虫族过度拥挤，会暂停正常的发育程序而进入"dauer"幼虫模式。逆境中的"dauer"幼虫不仅拥有耐受各种不利环境的极强能力，似乎也停止了衰老的过程。于是，这些极端坚忍、长寿的 dauer 可以熬过艰苦的岁月，等待环境好转时重新恢复正常发育模式。

除了减少卡路里摄入，科学家发现很多种不利的环境刺激都反而增强个体的耐受力，并且延长寿命。但在这里，特别需要强调的是个"度"的问题。轻微的压力能够刺激和提高耐受力乃至延长寿命，但超过了最适当的"度"，不利环境就会降低个体耐受力和生存力。这说明了生物个体自我调整和修复的强大能力，但每种环境刺激最适宜的"度"，在不同物种甚至是不同个体之间都存在差异，这也解释了为什么衰老研究在生物科学领域中都是争议最多也极富挑战性的。

"返老还童"术

更加奇葩的"长生不老"是所谓"返老还童"。小学时我爱看的一部书《自然未解之谜》中一篇文章讲述了极罕见的"返老还童"现象，这至今是人类未解之谜：几位老人重新长出乌发或者新牙齿的奇事，直到现在都让我印象极深。而人们想象世界中的"返老还童"就更加离奇：荣获三项奥斯卡大奖的 2008 年美国大片《本杰明·巴顿奇事》（The Curious Case of Benjamin Button）就讲述了这样一个返老还童的怪人本杰明·巴顿。他以一个耄耋老人的形象降临于世。童年在养老院长大的"小"本杰明副老态龙钟模样，与纯真可爱的小姑娘黛西初识便一见如故。而时光在他的身上倒流，本杰明越来越年轻、越来越强壮有力。几十年后，逆生长的本杰明帅气十足（布莱德·皮特饰演），与正是花容月貌的黛西在各自人生的中点重遇，共度了一段最甜蜜的时光。而当岁月慢慢爬上了黛西的额头，她曾经的恋人本杰明却继续逆生长成了小孩童模样，最终"倒长"成一个小婴儿在黛西怀中与世长辞。这一段穿梭在倒流时光岁月中的温暖和短暂的爱情让我唏嘘不已，由衷地觉得如此的返老还童真不是件快乐的事。

还是让我们回到现实世界。至今，真正生物学意义上的"青春永驻"还是人类未实现的愿望，就更不说如此"返老还童"了。遗憾的是，像"小"本杰明一样老态龙钟的儿童倒是真实存在的。这些不幸的儿童患有一种十分罕见的先天性遗传病,被称为"早衰症"(Hutchinson-Gilford Progeria syndrome),他们无论在外表还是身体功能上都像老人一样屡弱且易患例如心血管类的老年性疾病。2003 年,科学家终于通过不懈努力找到了疾病的"元凶":原来,一个负责支撑细胞核结构的重要蛋白(LMNA)发生了变异(Eriksson et al. 2003)。细胞核就像是细胞的控制中心,贮存着最重要的生命遗传信息。在 LMNA 变异的细胞中,细胞核的支撑骨架出现问题,变得形状不规则。这些病变的细胞无法对遗传物质进行正常的储存、保养和读取,整个机体于是大厦将倾。虽然早衰症极其罕见而且难治愈,但是科学家们并没有放弃。他们不仅在努力寻找治疗早衰症的药物,而且把研究早衰症特例作为一个切入点,来不断加深人类对自然衰老过程的认识。

启发和展望

尽管在 21 世纪的今天,完全字面意义上的"长生"和"不老"依然是不太现实的奢望:但是振奋人心的统计数据显示,在 2013 年世界上已经约有一半国家的人均寿命在 70 岁以上,其中近 20 个国家达到 80 岁以上。这相比 20 世纪初时的"人生七十古来稀",非常有力地显示了人类社会在 20 世纪取得的飞跃性进步。

"最美不过夕阳红,温馨又从容",记得我和姥姥一起看过的《夕阳红》节目让我觉得优雅的晚年生活是人生旅游中很幸福的一段。但是现实中衰老往往带给人们痛苦,因为衰老不仅指器官机能上的衰退,还和心血管疾病、糖尿病、癌症、神经退行性疾病等的发病成正相关。所以,如今大家希望的"长生不老"越来越注重的不仅是寿命,更重要的是健康、高质量的晚年生活,这也被称为"健康寿命"。

由于人均寿命的提高,人口老龄化成为当今社会面临的新考验。因此提高老年人的健康水平和生活质量,也就是延长"健康寿命",是当今科学研究的主要目标之一。值得一提的是,就在本文截稿前不久(2015 年年底),美国食品药品监督管理局(U. S. Food and Drug Administration, 简称 FDA)经过长时间的商讨后,终于通过了一项将 Metformin(二甲双胍)用于减缓衰老的人类实验。Metformin 是一个几十年前已经通过 FDA 批准的用于治疗二型糖尿病的药物。近年来的临床数据显示 Mefomin 不仅有效降低糖尿病人的血糖,还或许可以降低多种老年性疾病的发病率延长人们的寿命。FDA 近期批准的这个项目是从 2016 年开始,在 300 名非糖尿病的老年人身上试验 Metfomin 是否真的能够降低癌症心血管疾病和神经退行性疾病三个主要的老年性疾病的发病率。听起来这么神奇的药物,而且价格非常便宜,为什么我们不可以现在就都来尝试服用它来"长生不老"呢? 原来,虽 Mefrmin 在人类身上的安全性已经有足够的临床证据,它还从未在非糖尿病人身上使用过。同样,把衰老作为"病"来"治",这还是前所未有的全新思路。因此,FDA 能够批准这样一个"抗衰老"项目,是非常激动人心的一件大新闻。这说明证明延长"健康寿命"这个新"长生不老"理念已经不再是少数有钱人的奢望,而是正逐渐变成国家卫生健康部门的纲领方针、走进寻常百姓家实实在在的新理念。有趣的是,不仅仅各大生物公司和 FDA 这样的健康机构越来越注重"健康寿命",近年来以谷歌(于 2015 年划归于新成立的 Alphabet 旗下)为首的一些 IT(信息技术)公司也纷纷举起公共健康卫生的大旗。例如,2013 年谷歌投资成立的 Calico 公司(2015 年成为 Alphabet 公司旗下的子公司,和谷歌并行),专攻衰老的机制和找寻抗衰老的途径。在 21 世纪,人类是否能结合生物和信息等多学科的知识和技术,在"长生不老"上有新的突破,研发出新的生物黑科技? 让我们拭目以待。

参考译文:

Longevity and Anti-aging

There are many influential researches from my doctoral supervisor Gary Ruvkun edited into biology textbooks, his favorite one of which is the discovery of small non-coding RNA (ribonucleic acid) with the ability to regulate gene expression and individual growth and development. Except that, his favorite one is about the regulatory function and mechanism of insulin-like growth factor signal channel during the aging process, which was first reported by Ruvkun Lab. When I first became a member of the lab, standing in front of the gate of the aging research, I wondered whether I should devote myself into aging research during my Ph. D. study. However, I finally began to research the dark matter of the genom of small non-coding RNA out of my preference for it. While having been constantly influenced by the results of the aging research group, I have been fully interested in it and have payed close attention to it apart from other topics of my researches. The topics of "Aging and Longevity", whether gossip or formal scientific theories and discovery, are too much to finish in one day. In this passage, I have to excerpt several stories, which are the most thought-provoking, to better inspire readers' thought.

Whether the legends or the facts, such as the story of Xu Fu(an ancient legend about pursuing longevity), the historical records of Bigu(an ancient regimen for prolonging life) and the absurd story of emperors obsessed with Alchemy, all show people's yearn for longevity for thousands of years as well as their ignorance of aging. Nowadays, modern science has been highly developed, and aging is researched by many scientists. However, are some popular modern anti-aging concepts and products like brodeaux, antioxidant health

products and sheep placenta extract really the panacea of immortality? Since doubt and scientific attitude are the best evaluation criteria, we'd better firstly learn what and how the scientists are researching.

Longevity Can Be Realized by Spitting No Grape Skins While Eating Grapes

Of many researches of aging and longevity, the connection between diet and longevity is the most long-standing and profound. Over the last hundreds of years, the scientists have found that eating less, living longer through researches on model organism like yeast, caenorhabditis elegans, drosophilas and mice. There is a typical example of an experiment done by the scientists in University of California, Los Angeles in 1986. They feed the mice on a quantitative diet, which are divided into one control group (the mice can eat as much as they like with about 115 kilocalorie intake per week) for comparison and three experimental groups (with the intake of 85, 50, 40 kilocalorie per week). And it is assured that adequate vitamin and microelement will be provided for all these four groups of mice. The results show that the mice in "groups on diet" with less intake are not only fit but also long-lived. (In addition, the less they eat, the longer they lived.) Generally speaking, a mouse's lifespan is only 2 years, however, the mice who eat the least set a record of longevity that they have lived for 4.5 years. Meanwhile, compared with the control group, the immune system of the mice in "groups on diet" have obviously improved and the cancer rate has declined (Weindruch et al. 1986).

Such an observation is not enough. The biggest concern of the inquisitive scientists is: Why does lower calorie intake extend our lifespan? The improvement of living standards enables people to enjoy more delicious food. However, many studies have raised concerns that the calories in delicious food will shorten our lifespan. Knowing the biological mechanism in which low calorie intake can extend our lifespan, and controlling our calorie intake, we are not only might able to eat as much as we want, but also realize longevity and anti-aging.

Scientists' research on yeast has revealed some key answers. Yeast is single-celled, so the measurement of its lifespan is different from that of the majority of animals. Scientists usually measure its "lifespan", also called "productive life", by the number of times cells divide instead of how long they live. Leonard Guarente's team from Massachusetts Institute of Technology(MIT) confirmed once again that yeast grown in culture media with less sugar (sugar is the main source of calorie for yeast) lived longer than those grown in standard cultures. Based on this confirmation, they studied what factors on earth contribute to the longevity of the low-calorie yeast through large-scale screening experiments. It turns out that a protein named Sirtuin (Sir2) plays a key role: yeast without Sir2 does not live long even growing in low-sugar culture-media (Lin et al 2000). It seemingly runs in the opposite direction to the goal of "living longer while eating as much as we want", (which sounds not good). Don't worry. The results inspired Leonard Guarente's team to do an experiment oppositely. This time, they increased the amount of Sir2 in yeast by gene editing. Guess what? The "super" yeast, which overexpress Sir2 on normal culture medium, lived almost as long as the normal yeast with low calorie intake (Lin et al. 2000). This is great! That is to say, if one can increase the amount of Sir2, he can also have a long lifespan without dieting?!

What is Sir2 exactly? It turns out that Sir2 protein can change the specific chemical modification in some proteins to regulate their biological activity, with the auxiliary of a small chemical molecule called Nicotinamide adenine dinuckotide (NAD). Several key target proteins regulated by Sir2 may play a role in many biological processes, including protection of the genome stability and environmental balance in cells. Although the exact mechanism between Sir2 and anti-aging is far from clear, its potential anti-aging function is enough to excite us.

In the experiment of yeast, scientists increased the amount of Sir2 protein by gene editing, however, there is still a lot of technical immaturity and ethical debate (reference article: Gene Editing: A New Revolution of Yogurt). Since it is difficult to increase the amount of Sir2, scientists are looking for ways to increase the activity of Sir2 protein. In 2003, David Sinclair's team in Harvard University published a paper in Nature that small polyphenolic molecules can enhance Sir2's biological activity, such as resveratrol, which could be "fed" to yeast to extend their reproductive life which is similar to the effect of reducing calorie intake. Most importantly, resveratrol also works in human cells lines in vitro (Howitz et al. 2003), which sounds really amazing!

Grapes, especially grape skins, are rich in resveratrol. Perhaps, not everyone "spits no grape skins while eating grapes" for longevity. While there is resveratrol of grape skins in red wine during its long-last brewing, which seems to explain why the wine-loving French, despite of their preference for fatty foods, can still live longer and have low rates of heart disease in general. Before the appearance of this academic paper on resveratrol (Howitz et al. 2003), red wine had been linked to a number of health benefits, including oxidation resistance and reducing the risk of cardiovascular disease and cancer. It not only brings interest to the red wine industry, but also provide countless new commercial

opportunities, such as the extracted resveratrol is put into market as a kind of health product. Some scientists researching the anti-aging drugs have also found their own companies, gaining both fame and wealth.

That's to say, we pull back a game in the battle of millennium against aging? Unfortunately, the facts are unsatisfactory. After a lot of research experiments by more scientists, it became controversial of Sir2's efficacy and mechanism of prolonging life in many living organisms. Besides, most legend publicity of the efficacy of resveratrol are beyond the fact. Here is a worst example of Dipak Das in University of Connecticut Health Center, who was considered as one of the leading figures in the fields of resveratrol and aging. However, in his more than a hundred articles on the efficacy of resveratrol, it was later found that a large number of data was fabricated and falsified. So far, dozens of papers have been officially withdrawn, and there are still many papers under investigation. This scientific scandal has increased doubts about Sir2 and resveratrol. In 2011, 21 authors cosigned an article to clarify the lack of sufficient experimental evidence to support resveratrol as a nutraceutical (Vang et al. 2011). Resveratrol is also considered to have antioxidant function, but there is no evidence that this popular antioxidant product has the health function as it has been advertised, which may be harmful with excessive intake, after scientists' having conducted rigorous clinical trials. Therefore, delaying aging can't be easily realized by just eating an antioxidant capsule, even if cell oxidative damage is one of the causes of aging.

However, we can always find other ways. Sir2's cofactor, NAD, is a multifunctional small molecule that is important in many biological processes, such as metabolism and gene expression regulation. The study of Sir2 and many new studies starting from other entry points have shown that increasing the amount of NAD in cells has a significant improvement in neurodegenerative diseases. In addition, some experiments have also shown that increasing the amount of NAD can slow down aging.

Whether the new hopes of anti-aging brought by NAD will give birth to the next kind of bio-black technology, and what will be discovered in the future, we will see. This story has now fully reflected the various life phenomena and the complexity of human diseases. Aging, as an example, is not so simple that only a diet trigger switch can explain or control it. The English word "research" consists of "re" and "search", meaning to search repeatedly. In continuous researches, scientists can only design and analyze experiments in a strict and impartial spirit so that they can gradually approach the truth and open up new sources of human knowledge. This is especially true in the field of aging research, which is closely related to human life and health, and has a high degree of interest and temptation.

Continued In-depth Study of "Eating"

That the myth of resveratrol has shattered has disappointed many foodies who are fond of eating. Is it true that "longevity" and "eating as much as we want" can really not realized at the same time?

Don't jump to conclusions. Although the rule of "being on diet helps longevity" has been repeatedly proven in many model organisms, scientists with rigor and skepticism are not fully convinced that it can be applied equally to us humans. In the study of aging model organisms, mice have the closest relationship with human beings. Although they are also mammals, there are many differences between rodents and primates (such as human, orangutans and monkeys). Then why not study primates? Because not to mention opposition from animal protection organizations, the natural lifespan of primates is five or six years or even a hundred year. It is an expensive experiment that greatly tests our patience and perseverance.

Despite the large number of difficulties, scientists have taken the first step in studying the aging of primates. Since the late 1990s, the Wisconsin National Primate Research Center (WNPRC) and the National Institute of Aging (NIA) have chosen macaca mulatta as the research subjects and started their own tracking research for more than 20 years. In 2009, the Wisconsin National Primate Research Center (WNPRC) took the lead to publish their research results in the journal Science (Colman et al. 2009). Hardly when I read their statistics and conclusions carefully were my eyes attracted to their first picture. In the control group, the monkeys were listless and hairless, while the monkeys in low-calorie dieting group had a refreshing fur. It seems that there is a large difference between them!

After reading the whole text, I found that the statistics and conclusions were so different with what the picture firstly left me, because the monkeys in the control group and the diet group actually had the same lifespan! Although the monkeys in the diet group had significantly lower weight, and the possibility they may suffer from several typical senile diseases (such as diabetes, cancer and cardiovascular diseases) has also decreased, they have died from other causes and have not lived longer. The results of the National Institute of Aging (NIA) were later published on the journal Nature (Mattison et al. 2012). Their study showed that not only did the diet group have a similar lifespan to the control group, but the former one also had no significant difference in mortality from typical senile diseases compared with the latter one. The latter conclusion differs from the previous research reports from the National Center for Primate Research (WNPRC) in Wisconsin. The National Institute of Aging (NIA) explained in the report that the differences in feeding conditions (especially

food) among monkeys in different institutes may give rise to different conclusions.

To sum up, this costly experiment which has lasted for several centuries, although still worthy of discussion, points out that the rule of "being on diet helps longevity" is not fully applicable in primates. Furthermore, whether human being on diet can prolong life is completely inconclusive. This is great news for those who love eating. However, the complexity of the mechanisms of aging and longevity, as well as the uniqueness in each species, is undoubtedly a huge challenge for scientists.

To Live in Adversity

"When Heaven is about to place a great responsibility on a great man, it always first frustrates his spirit and will, exhausts his muscles and bones, exposes him to starvation and poverty, harasses him by troubles and setbacks so as to stimulate his spirit, toughen his nature and enhance his abilities. " By proposing such a paragragh, Mencius wanted to tell the people around the world that the harsh conditions stimulate people to strive to be strong and to make great achievements. But what he may not know is that this philosophy, to a certain extent, can be applied to all livings.

We have mentioned the example of "exhausts his muscles and bones" (reducing calorie intake) to prolong their lives in some organisms. Interestingly, these long-lived individuals are generally more tolerant and able to resist adverse environments and diseases. Thus, scientists have come up with the hypothesis that reducing calories may serve as a slight stress that stimulates the body to produce a defensive response to enhance its ability to resist adverse environments and diseases, thereby prolonging its life. Be aware, this is not a mental stress (the high-level nervous system has not evolved for yeast or nematodes to worry about food), but adverse environmental stresses such as lack of food, excessively high or low temperatures, harmful chemical, ultraviolet rays.

The embodiment of Mencius's "The World" on Nematodes is very astounding. However, a millimeter-long nematode has an unique ability that developing nematode larvae in a very unfavorable environment, such as lack of food or overcrowding of the zerg, will suspend normal development procedures and turn into the "dauer" larva mode. The "dauer" larvae in adversity not only have the strong ability to withstand various adverse environments, but also seem to stop aging. As a result, these extremely tough, long-lived dauers can survive the harsh conditions and return to normal development patterns after the conditions become better.

In addition to reducing calorie intake, scientists have found that a variety of adverse environmental stimulation enhance individual tolerance ability and extend its lifespan. However, what needs to be emphasized is "degree". Slight stress can stimulate and improve its tolerance and even extend its life, but exceeding the appropriate "degree" can reduce its tolerance and viability. This illustrates the strong ability of biological individuals to self-adjust and self-repair, but the optimal "degree" of each environmental stimulation varies between different species and even different individuals, which explains why aging research in the field of biological science is the most controversial and challenging.

The Magic Called "Rejuvenation"

The more exotic "Longevity" is the so-called "rejuvenation". An article in the book "The Unsolved Mystery of Nature" that I loved in elementary school tells the rare phenomenon of "rejuvenation", which is still the mystery beyond human beings' knowledge. It tells a weird story that several old men regrow black hair or new teeth, which have made me deeply impressed until now. And people's imagining of "rejuvenation" is even more bizarre: The American blockbuster "The Curious Case of Benjamin Button" in 2008 that won three Oscars tells about such a rejuvenating geek Benjamin Barton. He came to the world as an old man who seemed to be dying. The "young" Benjamin, who grew up in a nursing home and looked like in his old state, fell in love with the pure and lovely little girl Daisy at the first glance. While time was flowing back on him, Benjamin was getting younger and stronger. A few decades later, Benjamin (Brad Pitt played), who was in inverse growth and handsome, rencountered beautiful Daisy in the midst of their respective lives. And they spent a sweetest period together. When wrinkles slowly climbed up to Daisy's forehead, her ex-lover Benjamin continued to grow up and became a childlike child. Eventually, he reversely became a small baby passing away in the arms of Daisy. Their warm but ephemeral love story happened in the backflow time stunned me and gave me a sincere feeling that such rejuvenation is not a happy thing.

Let's return to the real world. So far, the true immortality of youth in the biological sense is still the unfulfilled desire of mankind, let alone such a "rejuvenation". Unfortunately, children like the "young" Benjamin truly exist. These unfortunate children suffer from a rare congenital genetic disease called "Hutchinson-Gilford Progeria syndrome". They are weak like the old not only in appearance but also in physical functions, and are prone to senile diseases, like cardiovascular diseases. In 2003, scientists finally found the "culprit" of the disease after unremitting efforts: It was a major protein (LMNA) responsible for supporting the nucleus structure that had mutated (Eriksson

et al. 2003). The nucleus is like the control center of the cell, storing the most important genetic information of life. In mutated cells, the supporting skeleton of the nucleus is problematic and becomes irregular in shape. The cells of these lesions are unable to properly store, maintain, and read the genetic material, and the entire body will be tilted.

Although Hutchinson-Gilford Progeria syndrome is extremely rare and difficult to cure, scientists have not given up. Not only are they trying to find drugs for treatment, but they also use the special case of Hutchinson-Gilford Progeria syndrome as a cutting point to continuously deepen human understanding of natural aging.

Inspiration and Outlook

Even in the 21st century, the "longevity" in the literal sense are still unrealistic expectations. However, the inspiring statistics show that in 2013, about half of the countries around the world have the average lifespan of 70 years. Nearly 20 countries have reached 80 years or more. This is a very powerful demonstration of the dramatic progress that human society has made in the 20th century, compared to the "A man seldom lives to be seventy years old" in the early 20th century.

There is a song singing:"The most beautiful scene is that the sunset is red, which is warm and calm. " I remember that the "Red Sunset" program that I have watched with my grandmother has made me feel that the elegance in one's old age is a very happy part of life's journey. But in reality, aging often brings people pain, because aging is not only a decline in organ function, but also positively related to cardiovascular disease, diabetes, cancer, and neurodegenerative diseases. Therefore, "longevity" that everyone hopes now means not only lifespan, but also health and high-quality life in the old age. This is also called "healthy lifespan".

As the increasing in average lifespan, the aging of the population has become a new test for today's society. Therefore, to improve health level and life standard of the elderly, namely to extend "healthy lifespan", is one of the main goals of scientific research today. It is worth mentioning that just before the deadline for this article (end of 2015), an experiment for human beings to slow down aging by using Metformin was finally passed by the U. S. Food and Drug Administration (FDA) after a long discussion. Metformin is a drug for the treatment of diabetes mellitus, type 2 approved by FDA decades ago. Recent clinical data show that Mefomin not only effectively reduces the blood sugar of diabetics, but also may reduce the incidence of various senile diseases so as to prolong people's lifespan. The project that FDA recently approved starts from 2016, to test whether Metfomin can indeed reduce the incidence of three major senile diseases, cancer, cardiovascular disease and neurodegenerative disease, in 300 non-diabetic elderly people. It sounds so magical, and the price is very cheap. Why can't we try to take it now to realize "longevity"? It turns out that although there is enough clinical evidence of Mefrmin for human safety, it has not been used on non-diabetics. And treating aging as a "sickness" is still a new idea that has never been seen before. Therefore, that FDA has approved such an "anti-aging" project is a very exciting piece of news. This shows that the extension of "healthy lifespan", a new concept of "longevity", is no longer the distant dream of a few wealthy people, but is gradually becoming the program guidelines of National Health and Wellness Department and a new concept entering the ordinary people's families. It is interesting that not only major biotech companies and health organizations (such as FDA), are paying more and more attention to "healthy lifespan", but some IT (information technology) firms led by Google (has belonged to the newly established Alphabet since 2015), in recent years, have also raised the banner of public health. For example, Calico, which was founded by Google in 2013 (in 2015, became a subsidiary of Alphabet, parallel with Google), specializes in aging mechanisms and ways to anti-aging. In the 21st century, can humans combine knowledge and technology including biology and information to make new breakthroughs in "longevity" and develop new biological black technology? We will see.

单词释义：

1. 基因表达   gene expression
2. 非编码小 RNA（核糖核酸）   non-coding RNA（ribonucleic acid）
3. 胰岛素类生长因子   insulin-like growth factor
4. 抗氧化保健品   antioxidant health products
5. 羊胎素   sheep placenta
6. 酵母   yeast
7. 秀丽隐杆线虫   caenorhabditis elegans
8. 果蝇   drosophilas
9. 加州大学洛杉矶分校   University of California, Los Angeles
10. 麻省理工学院   Massachusetts Institute of Technology（MIT）
11. 烟酰胺腺嘌呤二核苷酸   Nicotinamide adenine dinuckotide（NAD）
12. 靶蛋白   target proteins
13. 多酚类小分子物质   small polyphenolic molecules
14. 白藜芦醇   resveratrol

15. 康州大学健康中心	University of Connecticut Health Center
16. 神经退行性疾病	neurodegenerative diseases
17. 威斯康星州国家灵长类动物研究中心	Wisconsin National Primate Research Center（WNPRC）
18. 美国国家衰老研究所	National Institute of Aging（NIA）
19. 恒河猴	macaca mulatta
20. 早衰症	Hutchinson-Gilford Progeria syndrome
21. 美国食品药品监督管理局	U. S. Food and Drug Administation
22. 二甲双胍	Metformin
23.《本杰明.巴顿奇事》	The Curious Case of Benjamin Button

译文解析：

1. 科技英语中被动语态的广泛使用。

原文：值得一提的是. 就在本文截稿前不久（2015 年年底），美国食品药品监督管理局（U. S. Food and Drug Administation，简称 FDA）经过长时间的商讨后. 终于通过了一项将 Metformin（二甲双胍）用于减缓衰老的人类实验。

译文：It is worth mentioning that just before the deadline for this article（end of 2015），an experiment for human beings to slow down aging by using Metformin was finally passed by the U. S. Food and Drug Administration（FDA），after a long discussion.

分析：分析长句中的被动语态，首先提出句子主干，即"美国食品药品监督管理局通过了一项实验"，科技英语中常用被动语言，使译文更严谨，因此可译为"实验被美国食品药品监督管理局通过"。

2. 典故习语的翻译

原文：徐福的故事、辟谷的历史记载，还有荒诞的炼丹皇帝等，无论是传说还是事实，都反映了几千年来人们对"长生不老"的无限向往，同时也说明了人类对衰老的无知。

译文：Whether the legends or the truths，such as the story of Xu Fu（an ancient legend about pursuing longevity），the historical records of Bigu（an ancient regimen for prolonging life）and the absurd story of emperors obssessed with Alchemy，all show people's yearn for longevity for thousands of years as well as their ignorance of aging.

分析：汉语是高语境语言，英语则是低语境语言，这在典故的使用上格外明显。因此在英译时，应增译其隐含的历史背景，便于读者理解，也有利于中国文化的传播。

原文：东方不亮西方亮。

译文：We can always find other ways./All roads lead to Rome.

分析：对于习语的翻译，最好是能够对应上目标语中的相应习语。我找到了两个版本，一个是比较日常的说法，另一个是大家耳熟能详的习语。我认为两个都可以，毕竟这篇文章读起来很接地气，能够让读者看懂就可以。

3. 逻辑关系的判断

原文：2011 年，21 个作者联合署名文章，澄清缺乏足够的实验证据支持白藜芦醇作为营养保健品（Vang et al. 2011）。

译文：In 2011, 21 authors cosigned an article to clarify the lack of sufficient experimental evidence to support resveratrol as a nutraceutical (Vang et al. 2011).

分析：本句话隐含着一层逻辑关系，即作者联合署名是为了澄清一件事，这是目的关系，因此译为"to clarify"更合理。

4. 动名词性的转换

原文：这相比 20 世纪初时的"人生七十古来稀"，非常有力地显示了人类社会在 20 世纪取得的飞跃性进步。

译文：This is a very powerful demonstration of the dramatic progress that human society has made in the 20th century, compared to the "A man seldom lives to be seventy years old" in the early 20th century.

分析："有力地显示了"是动词词性，但科技英语中名词性结构更严谨，且使用"demonstration"名词性结构，使句子逻辑更清晰。

5. 长句的翻译

原文：我的博士生导师 GaryRuvkun 教授有不少写入生物教科书的重量级研究：除了发现可以调控基因表达和个体生长发育的非编码小 RNA（核糖核酸）——这恐怕是他的最得意之作——Ruvkun 实验室还首先报道了胰岛素类生长因子信号通路在衰老过程中的调控作用和机制。

译文：There are many influential researches from my doctoral supervisor Gary Ruvkun edited into biology textbooks, his favorite one of which is the discovery of small non-coding RNA (ribonucleic acid) with the ability to regulate gene expression and individual growth and development.

分析：对于长句的翻译，首先要理清句子逻辑关系，再按照英文的语言习惯重新排列组合成新句子，当然也可以使用分译法，但是我认为能翻译成一句话就尽量译成一句话，因为科技英语的特点也是长句居多。

第九章　纳米颗粒智能新药

药，一半天使一半魔鬼的双生子

生病是一件说起来就让人头疼的事情。生了病怎么办？看医生吃药吧。例如感冒发烧，吃过药后，烧退了，却又拉肚子了。这就是常见也不可避免的药物不良反应。细细说来是很吓人的。药，就像《圣斗士星矢》里的双子座黄金圣斗士撒加，兼有善与恶的双重人格，一半天使，一半魔鬼。下面花点篇幅说说药物的不良反应（Adverse Drug Reaction，简称 ADR）。对这个专业术语的解释是患者在服用药物时身体产生的与治疗无关、甚至是有害的作用。咱老祖宗早就总结了"是药三分毒"。完全没有不良反应的药物是不存在的——某些广告常常吹嘘的"无任何副作用"，不用怀疑，这全是美好的幻想。

可是，同一种药，不是所有人都会有同样的不良反应。更进一步说，即使在出现反应的人群中，个体和个体之间的反应差异也往往很大。医学上大致将不良反应归为六类：副作用、毒性反应、变态反应、继发反应、后遗效应和致畸作用。

副作用（side effect）指对治疗无关的不适应反应，一般比较轻微。比如说，吃了治病毒性感冒的抗生素，感冒症状减轻了，但是引起了耳鸣的不适反应。毒性反应（toxic action）是指大量或者长期用药对中枢神经、消化系统、循环系统及肝功能、肾功能产生的损害。有些病人长期需要服用药物来抑制癌细胞，这些药物往往对病人的肝脏造成不小的损伤。变态反应（allergic reaction）即过敏反应，是指人体受到药物刺激后，产生异常的免疫反应，从而导致生理功能障碍或组织损伤。经常听到医生会询问："对抗生素过敏吗？对某药物有过敏史吗？"常见的就是有些人服用了抗病毒的抗生素以后身上长红斑。这就是很典型的过敏反应。过敏反应与服药者的体质有关。多出现于少数过敏体质的人身上，具体的反应类型和严重程度也因人而异，很难预知。继发反应（secondary reaction）是由于药物治疗作用引起的不良后果。维基百科上举的一个例子就是"长期使用四环素类广谱抗生素会导致肠道内的菌群平衡遭到破坏，以至于一些耐药性的葡萄球菌大量繁殖而引起葡萄球菌假性肠炎。这样的继发性感染也称为二重感染"。后遗效应(residual effect)是指停药后血液中残存的药物成分产生的不良影响。有点像常说的"宿醉"(hangover),酒醉后,第二天醒来残留的酒精还是会让人头晕、犯困和没力气等不良反应。

致畸作用(teratogeneisis),顾名思义,指的是有些药物影响婴儿正常发育造成畸形。所以孕妇服药是要慎之又慎。举一个最近比较出名的药物不良反应事故,2014年马萨诸塞州剑桥(Cambridge,MA)的明星药厂 Biogen 的最畅销的多发性硬化症 Multiple sclerosis(MS)的治疗药物 Tecfidera,一名患者服用该药物后身上出现了罕见但是致命的进行性多灶性白质脑病(PML)。这名患者最后也死于 PML。此前,Biogen可是声名鹊起,股价节节攀升,这个事件导致 Biogen 的股票一夜狂跌 22%,从此一蹶不振。另外一起引起轰动的药物不良反应事件也发生在 2014 年。医药巨头 Novartis公开申明承认其日本分部隐瞒了白血病新药临床实验中患者出现不良反应的报告。新闻界发现足够的证据,证明 Novartis 日本分部删除了记录病人对新药产生不良反应的文档。即使面对外界舆论压力,Novartis 也没有公开出现反应的患者的具体人数。一些日本媒体猜测早期的临床实验中至少有 30 名患者出现了不同症状。Novartis 发言人同时也承认,在过去几年的对 10 种新药的开发中,至少有 1 万个对药物出现不良反应的案例被隐瞒或者从未公布。其中包括两种白血病药物 Gleevec 和 Tasigna 哮喘病药物 Xolair,帕金森病药物 Exelon 和用于防止器官移植排除的药物 Neoral。美国食品药品监督管理局(US Food and Drug Administration,FDA)明文规定药厂必须在发现药物不良反应的 15 天内上报案例。但是根据对 2004 年到 2014 年上报到 FDA 的 160万个案例的分析显示,如果患者对药物的反应不是致命的,制药商往往选择隐瞒数据。很多曝光于众的数据和新闻事件只是冰山一角而已。

我需要你:纳米颗粒智能新药

传统的制药行业在过去的一百多年内快速发展,无数药物问世,极大地提高了疾病的治愈率,功绩显赫。但是如前文所说的,与药效相伴相生的不良反应引起科学家和普通民众的担忧。有幸的是,人类拥有追求至臻完美的天性。科学界智慧的大脑早已着手开发智能新药来解决药物的不良反应。学术界达成共识的方法有:

1. 提高药物的溶解性:延长药物在血液中的循环时间从而提高药效和减低用药量。

2. 选择性释药:通俗地说就是让药物长有一双慧眼,在茫茫细胞的海洋中准确识别出病变细胞。病变细胞就像靶心一样,药物犹如飞镖,准确地射中靶心,人病变细胞内进部,一举清除病变细胞。

3. 热疗法:这是一个新的概念,和传统分子药物不同,热疗法局部提高病变组织和

155

细胞的温度杀死细胞。正常细胞和组织的微环境温度保持不变,不会受到影响。

那么,怎样才能把智能新药从理论变为现实呢?什么是打开月光宝盒的金钥匙?答案很可能就是纳米颗粒(Nanoparticles)!什么是纳米颗粒呢?下面的内容介绍的便是关于纳米颗粒的"之乎者也"。纳米颗粒,一种新型的微观材料,通常在纳米尺度。一个纳米是多少?举例来说,一个水分子是十分之一纳米,乳糖分子是一纳米,癌细胞是一万到十万个纳米,一个网球的大小是一亿纳米。大多数纳米颗粒的粒径在一到一百个纳米之间。一些常见的纳米颗粒,例如脂质体纳米颗粒(Liposome)、树枝状聚合物纳米颗粒(Dendrimer)、金纳米颗粒(Gold nanoparticle)、量子点(Quantum Dots)等。

纳米材料有什么与众不同的神奇魔力?简单一点解释就是,纳米材料是连接宏观材料和原子结构之间的桥梁。在宏观世界,材料的性质是均一的,不会随着材料尺寸的改变而改变。有意思的是,当材料的尺寸降到纳米级别后,材料表面的原子数量相比宏观材料增长了几个数量级,因此材料的表面积也增长几个数量级,随之带来许多有趣和意想不到的性质的改变。比如,黄金以宏观材料的形式存在时是金灿灿的色泽。但是当金以纳米材料形式存在时,随着粒径大小的改变,金纳米颗粒溶液就呈现出红,深蓝和紫等不同颜色。科学家们因此深深为纳米材料的性质着迷,孜孜不倦地探索着。

纳米颗粒凭着它微小的体积和可调控修饰的表面性质,在智能新药开发领域大显神通。首先来谈谈纳米颗粒是怎样提高药效,延长药物在血液中的循环时间,达到减低用药量的。学术界有一个专有名词(Drug delivery),主要说的就是利用高聚物或者脂质体纳米颗粒作为药物的载体和运输工具,有效增加药物的溶解度,延长循环时间和提高最终进入细胞内的药物含量。

在过去的十几年里,纳米颗粒的释药体系获得了巨大的成功,已经有不少应用于临床,比如说位于旧金山半岛福特斯城的明星药厂 Gilead 开发的用于治疗毒菌感染的脂质体纳米颗粒药物 Ambisom。Gilead 在新药开发领域可是大名鼎鼎,最早的成名作就是对乙肝、丙肝以及艾滋病的特效药开发。近几年这个创造力十足的药厂在纳米新药领域也在大展拳脚。

高聚物纳米颗粒主要是指具有抗菌或者抗癌功效的高聚物胶束(Micelle)。药物或者被包裹在胶束内部,或者配合在胶束表面。最广泛应用的高聚物之一是聚乙二醇(PEG)。PEG 有非常好的水溶性和生物兼容性。PEG 胶束纳米颗粒可以提高药物在人体内的溶解性,降低肾脏对药物的清除率,增强由细胞表面受体引导的药物进入细胞内部的过程。PEG 胶束纳米颗粒可以整体上提高药物在体内的循环周期,并降低用药量。脂质体胶束是由包裹了液体的磷脂双分子层形成的。磷脂双分子层结构和体内细胞膜很相似,因此具有很好的生物兼容性和降解性。另一类磷脂纳米颗粒药物是实心固体的脂质体。固态磷脂被融化后和药物分子均匀混合,通过乳化反应,冷却后形成实心的载有药物成分的纳米颗粒。

　　高聚物或者磷脂纳米颗粒已经在多种疾病的治疗中大展身手。举一个癌症治疗的例子。紫杉醇(Paclitaxel)是获得 FDA 批准的第一个来自天然植物的化学抗癌药,用于子宫癌、皮肤癌、肺癌的治疗。1960 年美国国家癌症研究所(NCI) 和农业部合作成立了一个项目:采集和筛选植物样品,从中找到可能有医用价值的天然化合物。植物学家亚瑟·巴克雷(Arthur Barclay)跑到华盛顿州的一个森林里,采集了 7 千克的紫杉枝叶和果实带回了研究所。紫杉树貌不惊人,通常生长在溪流岸边,深谷或者潮湿的山沟中。紫杉的材质很重,没有什么太大的用途,所以伐木工人都把它叫作"垃圾树",一般都是当柴火,或者砍了当篱笆桩子用。但是研究三角学院(另种译法为"研究三角园区",缩写 RTI)的化学家 Monroe Wall 和同事们偶然发现从紫杉原料中可以提取出一种活性物质。这种活性物质在肿瘤细胞测试中显示出了活性!"垃圾树"中竟然可以提取出抗癌药物。

　　在此之后,掀起了一场关于紫杉树中这种活性抗癌物质的学术研究热潮。科学家们开始对这种物质进行提纯、结构研究以及实现实验室化学合成的一系列实验。最终在 1993 年,医用紫杉醇由大名鼎鼎的美国百时美施贵宝（Bristol-Myers Squibb）公司开发上市。当年就获得了超过 9 亿美金的销售利润。紫杉醇的抗癌机理是阻碍癌细胞的微管稳定,从而导致癌细胞死亡。可惜的是紫杉醇的水溶性很差,通常需要先溶解在医用乙醇中(Taxol®)中,然后和聚氧乙烯蓖麻油(Cremophor® EL)混合静脉注射进人体来提高药物的溶解性。使用聚氧乙烯蓖麻油的最大弊端就是会诱发过敏反应。患者通常要提前服用抗过敏药物来预防因为注射紫杉醇带来的不良反应。2005 年,科学家们研发出包裹紫杉醇的天然高聚物纳米颗粒药物 Abraxane®。这种纳米新药提高了紫杉醇的溶解性,患者不再需要注射聚氧乙烯蓖麻油。同时,纳米颗粒载体也提高了药物从血液到癌症组织处的传输效率,大大提高了紫杉醇的药效。

　　纳米颗粒在对神经性疾病的治疗中也是功绩显赫。长久以来怎么使药物有效地到达中枢神经系统一直是个难题。血脑屏障(blood brain barrier),俗称 BBB 效应,是造成这个问题的重要原因。人体自发精心地设立了这个机能防止大脑受到体外异物和血液中携带的传染物的入侵。可是,这个机能无法将药物从异物和传染物中辨别出来,从而使得绝大部分药物也被挡,从而无法直达患处。为了达到治疗效果,必须加大用药剂量才能使一部分的药物输送到大脑,这就增大了患者产生不良反应的风险。高聚物和脂质体纳米颗粒能够有效地帮助药物穿过血脑屏障,使其能够进入大脑和中枢神经组织,从而发挥作用。美国俄亥俄州医学院的研究员,曾经发表的论文中,展示了他们用外表包裹了氨基酸的纳米颗粒,成功将抗癌药物阿霉素输送到脑部中枢神经系统的研究成果。

　　在医用的其他领域,比如说在对艾滋病的治疗中,高聚物纳米颗粒也被证实可以提高 HIV-1 病毒蛋白酶抑制剂药物的溶解性。在眼科疾病的治疗中,多数药物是通过眼药水滴液的方式作用于眼球。但是人总是不停地眨眼,在眨眼的过程中,眼内黏液也在不停移动,药物在进入角膜前就大量损失了,所以病人要隔几个小时就需要滴点新的药水。纳米颗粒能够将药物陷入眼黏液膜层中,延长药物在眼球中的停留时间而减少滴眼药水的次数和提高药效。

科学家们已经研发出几种方法来赋予药物一双识别病变细胞的慧眼,实现纳米颗粒药物的选择性释药。最常见的和最早使用的方法是配体和受体识别法。一把钥匙开一把锁,病变细胞上有"锁孔",纳米颗粒药物就是那把配合的"钥匙"。正常的细胞上没有那个"锁孔",纳米颗粒药物就不碰触正常细胞。具体说来就是病变细胞表面会大量表达和正常细胞不一样的物质。纳米颗粒的表面可以通过化学反应和病变细胞表面的物质发生作用,这样纳米颗粒从正常细胞和病变细胞的茫茫细胞海洋中就能一眼识别出病变细胞。纳米颗粒从此就有了火眼金睛,一眼就能识别出病变细胞,并结合富集在病变细胞表面。科学家们在设计纳米颗粒时,已经在其内部设定了"程序",纳米颗粒在接收到信号以后,就启动自我分解的模式,把药物准确地靶向释放到病变细胞内部。这个"程序"就是设计和调控高聚物内部的化学键。这些化学键被"编程"后或者能够被病变细胞分泌的酶降解,或者能够进入细胞内部,随着细胞内部溶液酸性的增强而降解,从而释放出来发挥药效。

另外一种选择性释药是利用癌细胞血管的渗透现象(blood vessel leakiness in cancer)。癌细胞附近的血管和正常细胞相比,空隙更大,组织结构无序。科学家们可以通过控制纳米颗粒大小,让纳米颗粒的尺寸刚好能自由地穿过癌细胞附近的血管,而无法进入和穿过正常细胞附近的血管。根据这种对血管壁孔径的选择,纳米颗粒绕过了正常细胞,只在病变细胞和病变组织附近集合。下面的章节也会说到一个运用这个原理实现靶向释药的真实例子。

最后提到的纳米颗粒"热疗法"也是非常有意思的。大多数的研究热点集中在对金纳米颗粒和磁性纳米颗粒上。原理其实不复杂,高温会引起酶的灭活、类脂质破坏、核分裂的破坏、产生凝固酶使细胞发生凝固,另外使细胞蛋白质变性。人体正常细胞能存活的温度在 37℃ 到 38℃ 之间。细胞在 39℃-40℃ 培养 1 小时会受到一定损伤,但仍有可能恢复。如果在 41℃-42℃ 中培养 1 小时,细胞损伤严重,温度升至 43℃ 以上时细胞就无法存活了。

那么纳米颗粒怎么实现"热疗法"呢?贵金属,比如说金、银和铂,有一种特殊的表面等离子共振现象(Surface Plasmon Resonance,SPR),SPR 现象使光子禁闭在纳米颗粒的小尺寸上,纳米颗粒表面因此产生很强的电磁场,从而增强了纳米颗粒的包括吸收和散射在内的辐射性能。金纳米颗粒得益于 SPR,可以大量吸收光能,并且把光能通过非辐射过程转化成热能。如果把金纳米颗粒召集到病变细胞和组织处,照射可吸收的光源,金纳米颗粒把光能量有效地转换成热能,提升病变细胞附近的温度。通过调整照射的光源能量,可以把病变细胞微环境的温度升高到43℃以上,让他们无法存活。相似地,磁性纳米颗粒,比如说铁,当处于交流磁场(alternating magnetic fields)下,将磁场能转换成热能。类似于金纳米颗粒的疗法,把磁性纳米颗粒聚集在病变组织处,在患者体外加上交换磁场,也会达到对病变细胞的热疗效果。

前路虽有荆棘,但未来不是梦

纳米颗粒新药是现代科技创造的新生事物。他带着自信、智慧和一身绝技来到世间,对病变细胞大声宣战:"疾病,我来了,我知道你在哪!"用独门绝技,完成了提高药效、降低药物剂量、实现靶向进攻和降低不良反应的使命。不过纳米颗粒智能新药作为新生儿,也非尽善尽美。学术界一直在不懈地进行研究和实验,以期更好地了解和完善纳米颗粒药物,解决关于纳米颗粒药物的疑虑。最后这节来谈谈纳米颗粒智能新药需要克服的一些问题。

第一个问题是纳米颗粒药物在进入人体后的分布问题。科学家开发和设计纳米颗粒新药的初衷是让所有的纳米颗粒药物都能准确地到达病变处。但是目前存在的问题是,纳米颗粒药物在进入人体这个庞大复杂的系统以后,有些迷路。纳米颗粒可以成功穿越很多人体设置的生物障碍,比如说本章开头提到的血脑屏障、还有很多外表皮层的结缔组织。正因为纳米颗粒这个特殊的穿越本领,除了如科学家期望的达到病变组织外,还分布到了人体的其他器官。由此带来的问题,一是兵力的分散,本来应该是全部的纳米颗粒军团与病变细胞作战,现在有一部分战士跑去了没有敌人的阵营。二是纳米颗粒药物很有可能对正常器官产生影响。已经有许多文献表明,纳米颗粒除了很大一部分集中在癌症或者病变组织以外,在人体的过滤器肺脏,人体解毒器肾脏,还有人体最大的免疫器官脾脏里都有大量分布。这不难理解,这些器官都是人体对抗外界侵犯的防御体系。当纳米颗粒药物进入人体后,这些器官会很警觉地把纳米颗粒药物假设为敌人,然后采取诱导围攻策略,然后试图防御。另外一些报道还发现纳米颗粒在脑部和心脏处也有少量分布。怎么帮助纳米颗粒药物成功到达战场,而不被人体的其他防御器官迷惑和围困,不少大学和科研机构的实验室都在进行此方面的科研攻关。

第二个问题是纳米颗粒极具活性的表面性质可能带来一些副作用。本章一开头就介绍过,纳米颗粒和普通的宏观材料在性能上有很大的不同。因此,这些纳米材料和组织器官之间的相互作用也和普通的宏观材料很不一样。目前关于纳米颗粒表面作用可能对人体带来的不良副反应,现在还没有特别深入而系统的研究。原因之一是许多正在进行临床实验和被 FDA 批准进行实验的纳米药物是可降解和生物兼容性好的材料,如前几章提到的可降解的高聚物或者脂质体纳米颗粒。但是有些研究人员还是担心纳米颗粒独特的表面性质可能会在释药的过程中产生一些难以预料的不良反应。这种反应是造成纳米颗粒副作用,比如说过敏反应的一个重要的机理 。

另一个问题是不能降解的纳米颗粒如何在释药后排出体外?FDA 有明文规定,任何注射人人体的诊断试剂都必须在合理的时间内离开人体。这项规定是为了确保诊断试剂不会过久停留在体内,造成不良反应,或者妨碍其他诊断测试。举个例子说,金纳米颗粒的线衰减系数(linear attenuation coefficient)比骨骼要高 150 倍。如果金纳米颗粒长期逗留在人体内,会使电脑断层扫描(CT)结果失真,造成错误诊断。目前的一个可能的解决方案是控制纳米颗粒的粒径,使其能够通过人体的肾过滤(renal filtration)或者尿排泄排出体外。这样就可能解决或者至少极大程度上降低纳米颗粒药物在体内停留时间过长而对人体造成的不良反应。所以在设计合适的纳米颗粒尺寸能够使其在完成任务后,顺利穿越内皮被排出也至关重要。有意思的是,纳米颗粒的形状和外表的化学修饰层也会对其能否顺利被排出体外产生影响。比如说球状的金纳米颗粒和棒状的金纳米颗粒被人体清除的效率就有所不同。对此,科学家们无疑需要进行更多的研究和改善。

纳米颗粒药物从实验室真正走上药品货架的道路虽不是长路漫漫无期,但确是布满荆棘。纳米颗粒智能新药是一个科技梦,为了实现这个梦,已经有无数科学家们在为之奋斗。实验室里无数个不眠的昼夜和书桌前挑灯夜战的汗水,正是这些不眠夜和汗水让纳米颗粒智能新药一步一步走近人们的生活。纳米颗粒智能新药的未来不是梦,我们拭目以待!

参考译文:

Intelligent New Drug—Nanoparticles

Drug—A Mixture of Angel and Devil

Illness is a head-aching thing. What should we do if we are ill? The first choice is to see a doctor and to have some pills. For example, if having a cold and a fever, you take medicine. Then the fever is gone, but you suffer from diarrhea. This is the common and inevitable adverse drug reaction. It's very frightening to say in detail. Drug, like the Gemini Gold Saint in the "Saint Seiya", has a dual personality of good and evil, which means that it is half an angel and half a devil. Let's take a moment to talk about Adverse Drug Reaction (ADR). The meaning of this terminology is the non-therapeutic and even harmful effects on body when the patient takes drug. The ancestors have already summarized this as "Every medicine has its side effect." Medicine bringing no adverse drug reaction do not exist. So, undoubtedly, it is a wonderful fantasy that some advertisements often boast their drugs without "no side effects".

However, not all people taking the same medicine will have the same adverse drug reaction. Furthermore, the reactions differ greatly among individuals. Medically, ADR is roughly classified into six categories: side effect, toxic reaction, allergic reaction, secondary reaction, residual effect and teratogenesis.

Side effect refers to the inadaptable response unrelated to treatment, which is often slight. For example, taking antibiotics for the treatment of viral cold can reduce the symptoms of cold, but it will cause the tinnitus. Toxic reaction refers to the damage of central nervous system, digestive system and circulatory system as well as liver function and kidney function caused by one taking a large number of medicine or taking long-term medicine. Some patients need to take long-term medicine to inhibit cancer cells, however, these drugs often cause a lot of damage to their livers. Allergic reaction, also called anaphylaxis, refers to the abnormal immune reaction after the human body is stimulated by medicine followed by physiological dysfunction or tissue damage. Doctors often ask: "Are you allergic to antibiotics? Do you have a history of allergy to certain drugs?" It is common for some people to have erythema after taking antiviral antibiotics. This is a very typical allergic reaction. It is relevant to the constitution of the taker. This reaction often happens to a few people with allergic constitution, and the specific type and severity are also various and unpredictable. Secondary reaction is the adverse consequences caused by medicine treatment. One example cited on Wikipedia is that "Long-term use of tetracycline broad-spectrum antibiotics will disrupt the balance of intestinal flora, leading to the proliferation of drug-resistant staphylococci which will cause staphylococcal pseudoenteritis. Such secondary infection is also known as double infection." Residual effect refers to the adverse effect of residual drug components in the blood after drug withdrawal. It's a bit like hangover. After you were drunk, the residual alcohol will still make you dizzy, sleepy and weak in the next day. Teratogenesis, as the name suggests, refers to some drugs that affect the normal development of infants and cause deformity. So pregnant women must be extremely careful to take drug.

Take an example of the recent famous ADR accident, Tecfidera is the best-selling drug for Multiple sclerosis (MS) of Biogen, a star pharmaceutical factory in Cambridge, MA. In 2014, a patient had a rare but fatal progressive multifocal encephalopathy (PML) and finally died of it. Compared with increasing popularity and a rising share price before,

Biogen had a sharp 22% drop in the stock overnight and collapsed.

Another sensational accident occurred in 2014. Novartis, a pharmaceutical giant, has publicly admitted that its Japanese division concealed ADR reports of a new leukaemia drug in clinical trials. The press found enough evidence that Novartis Japan had deleted documents recording patients' adverse reactions to the new drug. Even faced with public pressure, Novartis did not give the specific number of patients suffering ADR to the public. Some social Japan media speculated that at least 30 patients in early clinical trials had various symptoms. A spokesman for Novartis also acknowledged that over the past few years, at least 10, 000 cases of ADR in 10 new drug developments had been hidden or never made public. Among them, there are Gleevec and tasigna both used for leukemia, Xolair, the asthma drug, Parkinson's disease drugs Exelon and Neorala used to prevent organ transplant elimination. US Food and Drug Administration (FDA) had stipulated that drug manufacturers must report cases within 15 days after discovering ADR. However, according to the analysis of 1. 6 million cases reported to FDA from 2004 to 2014, manufacturers often choose to hide the data if a patient's adverse reaction is not fatal. Many of the data and news exposed to the public are just the tip of the iceberg.

I Need You: Nanoparticles Intelligent New Drug

In the past 100 years, traditional pharmaceutical industry has developed rapidly. Numerous drugs have been produced, which has greatly improved the cure rate of diseases. What an outstanding achievement! However, as mentioned above, the adverse reaction going with medicine efficacy should arise deep concern of scientists and the general public. Fortunately, human beings have the nature to pursue extreme perfection. Smart scientists have already been developing intelligent new drugs to deal with ADR. The methods that the academic community unanimously approved are:

1. Improve the solubility of the drug: prolong the circulation time of drugs in the blood to improve the efficacy and reduce the dosage.

2. Selective drug release: Generally speaking, let the drug have a pair of eyes to accurately identify the diseased cells in the vast ocean of cells. Drugs will enter into the diseased cells and remove them at one stroke like the darts hit the bull's-eye.

3. Thermal therapy: It is a new concept. Unlike traditional molecular medicine, thermal therapy can kill the diseased cells by locally raising the temperature of pathological tissues and cells. The micro-environment temperature of normal cells and tissues remains unchanged and will not be affected.

So, how can we turn intelligent new drugs from therapy to reality? What is the golden key to this? The possible answer is nanoparticles! Then what are nanoparticles? The following content is about them. Nanoparticles, a new type of micro material, are usually at nanometer scale. How much is a nanometer? For example, a water molecule is one tenth of a nanometer; a lactose molecule is one nanometer; a cancer cell is from 10,000 to 100,000 nanometers; and a tennis ball is one hundred million nanometers. Most nanoparticles are between one and one hundred nanometers in size. Picture 2 illustrates some common nanoparticles: Liposome, Dendrimer, Gold nanoparticle and Quantum Dots and others.

What is the magic power of nanomaterials? To be simple, they are bridges linking macroscopic materials and atomic structures. In the macrocosm, the properties of materials are uniform and will not change with the swift of material size. Interestingly, when the size of the material is reduced to the nanometer level, the number of atoms on its surface will increases several orders of magnitude compared with that of the macro material, so the surface area of the material also increases several orders of magnitude, which brings many interesting and unexpected changes in properties. Gold, for example, is golden when it's in the form of a macro material. However, when it exists in the form of nanomaterials, gold nanoparticle solutions will be red, dark blue, purple and other different colors with changing particle sizes. Scientists are fascinated by the nature of nanomaterials and are diligently exploring them.

Nanoparticles, with their tiny size and modified surface properties, have been widely used in the field of intelligent drug development. First, let us talk about how nanoparticles can improve drug efficacy, extend the circulation time of drugs in the blood, and reduce the dosage. In academia, there is a term called Drug deliver, which mainly refers to the use of high polymer or liposome nanoparticles as the carrier and transportation of drugs can improve the solubility of drugs, prolong the circulation time and increase the content of drugs that eventually enter the cell effectively.

Over the past decade, nanoparticles drug release system was a great success. There have many clinical applications, such as Ambisomo, a liposome nanoparticle drug used to treat viral infection. It is developed by Gilead located in Fortes City, San Francisco Peninsula. The company, a celebrity in the field of new drug development, made the earliest well-known contribution in the development of specific drugs—hepatitis B, hepatitis C, and AIDS. In recent years, this innovative pharmaceutical company has also made great efforts in the field of nanoparticles.

Polymer nanoparticles mainly refer to the polymer micelle with antibacterial or anticancer efficacy. The drug is either wrapped inside the micelle or attached to the surface

of it. One of the most widely used polymers is polyethylene glycol (PEG). It has excellent water solubility and biological compatibility. PEG micelle nanoparticles can improve the solubility of drugs in human body, reduce the clearance rate of drugs in the kidney, and accelerate the process of drugs entering into cells guided by cell surface receptors. As a whole, PEG micelle nanoparticles can improve the drug cycle in the body and reduce the dosage needed a time.

Liposome micelles are composed of phospholipid bilayer coated with liquid. The structure of phospholipid bilayer is similar to that of the cell membrane in the body, so it has good bio-compatibility and biodegradability. Another kind of phospholipid nanoparticle drug is solid liposome. The melted solid phospholipid is combined with the drug molecules uniformly. Through emulsification reaction, it forms solid nanoparticles containing drug components.

High polymers or phospholipid nanoparticles have gained extensive uses in multiple diseases. Take an example of cancer treatment. Paclitaxel is the first FDA-approved chemical anticancer drug from natural plants, which is applied in the treatment of uterine cancer, skin cancer and lung cancer. In 1960, the National Cancer Institute (NCI) and the Department of Agriculture joint hands to set up a project: collecting and selecting plant samples to find natural compounds featuring potential medical value. Arthur Barclay, a botanist, ran to a forest in Washington state, collected yew leaves and fruits weighed 7kg and brought them back to the Institute. Yew trees do not have attracting appearance and usually grow on the banks of streams, deep valleys or in damp gullies. They are so heavy and have no great use, so loggers call them "garbage trees", which means that the trees are usually used as firewood or fence post after being chopped. However, Chemist Monroe Wall and his colleagues from the Research Triangle Park (RTI), by accident, discovered that an active substance could be extracted from yews. This active substance shows activity in tumor cell tests! Anti-cancer drugs can be extracted from "garbage trees"!

Since then, there was an upsurge of academic research on this active anticancer substance from yew trees. Scientists make a series of experiments from refining the substance, studying its structure to realizing chemical synthesis in the laboratory. Finally, in 1993, medical paclitaxel was developed and listed by big-name Bristol-Myers Squibb which earned over 900 million dollars in profit on sales. The anticancer mechanism of paclitaxel is to prevent microtubules stability of cancer cells, which leads to the death of these cells. It's a pity, however, that the water solubility of paclitaxel is very poor. It usually needs to be dissolved in medical ethanol (Taxol ®), and then injected intravenously into human body with Cremophor® EL to improve the solubility of drugs. The biggest drawback of Cremophorl® EL is that it can induce allergic reaction. That's why patients usually take anti-allergic drugs in advance to prevent this caused by injection of paclitaxel. In 2005, scientists researched and developed Abraxane, a natural polymer nanoparticle drug encapsulating paclitaxel. This new drug increased the solubility of paclitaxel, so patients no longer need to be injected with Cremophorl® EL. Besides, nanoparticles also make paclitaxel more effective by improving the efficiency of delivering drugs from blood to cancer issues.

Nanoparticle is necessary in the treatment of neurological diseases. For a long time, it is a difficulty that how to make drugs target the central nervous system. Blood brain barrier, usually known as BBB effect, is a great threat. The human body has set up this function to protect the brain from the invasion of foreign bodies in vitro and infectious substances carried in the blood. However, this function can not distinguish drugs from foreign bodies and infectious substances, so most drugs are also blocked from reaching the affected areas directly. To achieve the therapeutic effect, we must increase the dosage only to make part of the drug deliver to the brain. But, meanwhile, it will increase the risk of adverse reactions in patients. High polymer and liposomes nanoparticles can effectively help drug entering the brain and central nervous tissue by passing through the blood brain barrier. Then the drug can work. A published paper written by researchers at Ohio Medical College had shown their research results of transporting doxorubicin, an anticancer drug, to the central nervous system of the brain with nanoparticles coated with amino acids.

In other medical fields, such as in the treatment of AIDS, polymer nanoparticles have also been proved to improve the solubility of HIV-1 protease inhibitors. And in the treatment of ophthalmic diseases, most drugs act on the eyeball through eye drops. However, people always blink. During this process, the mucus in the eye keeps moving, which leads to most drug lost before entering the cornea. That's why patients are given a few drops of drug every few hours. Nanoparticles can embed the drug in the mucus layer of eyes and prolong the residence time of the drug in the eyeball, then reduce the number of drops and improve the efficacy.

Scientists have developed several methods to give drugs sharp eyes that can identify pathological cells to realize the selective release of nanoparticle drugs. The most common and the earliest method is ligand and receptor recognition. One key can only open one lock. Pathological cell has a "keyhole" and nanoparticle drug is the "key". Without that "keyhole" in cells, nanoparticle drugs would not touch these normal cells. Specifically, the surface of pathological cells express a lot of substances different from normal cells. The surface of nanoparticles can react with the substances on the surface of diseased cells

through chemical reactions, so that nanoparticles can recognize the diseased cells at a glance from the vast ocean of cells. From then on, nanoparticles have piercing eyes finding the diseased cells instantly, and combine and enrich on these cell surfaces. When designing nanoparticles, scientists have set up a "program" inside them. After receiving the signal, these nanoparticles initiative the mode of self-decomposition and target drugs into pathological cells. This "program" is chemical bonds that design and control the inside of polymer. These bonds can be "programmed" or be degraded by enzymes secreted by diseased cells, or can enter into the cell interior, decompose with the increase of the acidity of the solution inside the cell, and be released to exert its efficiency.

Another selective drug release depends on blood vessel leakiness in cancer. Compared with the normal cells, the blood vessels near the cancer cells have larger gaps and disordered organizational structures. Scientists can control the size of nanoparticles making them freely pass through the blood vessels near the cancer cells rather than entering vessels near the normal cells. According to this choice of vascular wall pore size, nanoparticles bypassed normal cells and only gathered near diseased cells and tissues. The following chapter will also talk about a real-life example of achieving targeted drug release by this principle.

The last "thermal therapy" of nanoparticles is also very interesting. But, most researches focus on gold nanoparticles and magnetic nanoparticles. Actually, the principle is not complicated. High temperature will lead to enzyme inactivation, lipids destruction, nuclear division destruction, coagulase production to make cells coagulate, and denaturation of cell proteins. The temperature at which normal human cells can survive is between 37℃ and 38℃. Cells cultured from 39℃ to 40℃ for 1 hour will be damaged, but they are still possible to recover. If cultured between 41℃ and 42℃ for 1 hour, the cells will be seriously damaged, and will not survive when the temperature rises above 43 ℃.

So how do nanoparticles achieve "thermal therapy"? Precious metals including gold, silver and pins, have a special Surface Plasmon Resonance (SPR) phenomenon. The phenomenon confines photons to the small-sized nanoparticle, which generates a strong electromagnetic field on the surface of nanoparticles. Thereby enhance its radiation performance including absorption and scattering. Benefiting from SPR, gold nanoparticles can absorb a lot of light energy and convert it into heat energy through non radiation process. If being gathered to the pathological cells and tissues as well as being irradiated the absorbable light source, the gold nanoparticles can effectively convert the light energy into heat energy and improve the temperature around the pathological cells. By adjusting the light source energy, the micro-environment temperature of the diseased cells can be raised above 43℃, so they can not survive. Similarly, magnetic nanoparticles, such as iron, can convert magnetic energy into heat energy when they are in alternating magnetic fields. Similar to gold nanoparticle therapy, magnetic nanoparticles can be gathered in the pathological tissues and add exchange magnetic field in vitro, which also have the effect of thermal therapy on diseased cells.

Faced with Difficulties, Having a Promising Future

Nanoparticle new drug is a new thing created by modern science and technology. He came to the world with self-confidence, wisdom and unique skills. It fights against the diseased cells and speaks loudly: " I am here, I know where you are!" With these unique skills, the missions of improving efficacy, reducing drug dosages, achieving targeted attack and reducing ADRs have been accomplished. However, as a newborn, he is not perfect. So, the academic world has been researching and experimenting in order to better understand, improve and remove the doubts about nanoparticle new drug. Finally, this section will talk about some problems that need to be overcome.

The first obstacle is the distribution of nanoparticles into the body. The original intention behind the development and design of new nanoparticle drugs is that all of the drugs can reach the lesions accurately. The current problem, however, is that the drug get lost after entering the vast and complex system of the human body. Nanoparticles can successfully cross many biological barriers set by the human body, such as the blood brain barrier mentioned at the beginning of this chapter and many connective tissues in the outer cortex. This unique ability makes themselves reach diseased tissues as scientists expect and reach other organs of the body. Two problems are created from this. One is the dispersion of forces which were supposed to be the entire army of nanoparticles fighting the diseased cells. But now some soldiers have gone to the camp without enemies. Another is that nanoparticle drugs possibly have an impact on normal organs. It has been shown in a lot of literature that nanoparticles are not only concentrated in cancer or diseased tissues, but in lung-the filter, kidney-the detoxifier, and spleen-the body's largest immune organ. It is understandable that these organs are the body's defense system against external invasion. When nanoparticle drugs enter the body, these organs are alert to suppose these as enemies, then take an induced siege strategy, and try to defend against them. Moreover, some reports also found small amounts of nanoparticles in the brain and heart. Laboratories at universities and scientific research institutes are working on to help nanoparticle drugs make their ways to the battlefield without being confused or trapped by the body's other defense organs. The second one is that the highly active surface properties of nanoparticles may have some side effects. As mentioned at the beginning of this chapter, significant differences in performance are between nanoparticles and ordinary macroscopic materials. Therefore, the interactions between these nanomaterials as well as tissues and organs are quite different from those between ordinary macroscopic materials and tissues and organs. At present, there is no in-depth and systematic research on the adverse and side reactions that may be caused by the surface effects of nanoparticles. One of the reasons for this is that many nanodrugs in clinical trials and those FDA-approved for testing are biodegradable and bio-compatible materials, such as degradable polymers or liposome nanoparticles mentioned in the previous chapters. But, some researchers concern that the unique properties of nanoparticles may lead to some unexpected adverse reactions during the process of drug release. This kind of reaction is an important mechanism leading to the side effects of nanoparticles like allergic reactions.

Another question is how do non-degradable nanoparticles get out of the body after drug release? The FDA stipulates that any diagnostic reagents injected into the human body must leave within a reasonable time. The rule aims to ensure that these reagents do not remain in the body for too long. Otherwise, it will cause adverse reactions and hinder other diagnostic tests. For example, the linear attenuation coefficient of the gold nanoparticles was 150 times higher than that of bones. If the gold nanoparticles stay in the body for a long period, they can lead to a misdiagnosis by distorting the results of computed tomography (CT). One possible solution is to control the particle size of nanoparticles so that they can be discharged through renal filtration or urinary excretion. This may address, or at least significantly reduce, the adverse reaction caused by long residence of nanoparticle drugs in the body. So it is imperative that design the appropriate size of nanoparticles to pass through the endothelium and be discharged smoothly after completing their task. Interestingly, the shape of the nanoparticles and the chemical coating on their appearance can also have an impact on their smooth discharge. For example, the efficiency of removing spherical gold nanoparticles and rod-shaped ones by human body is different. In this regard, it is no doubt that scientists need more researches and improvements.

The way that nanoparticle drugs from laboratories to the market may not be a long one, but it is full of thorns. Nanoparticle intelligent new drug is a technological dream that countless scientists have been working towards. It is these sleepless nights in the labs and sweat that bring nanoparticle intelligent new drugs into people's life, step by step. The future of this is not a dream. Let us wait and see!

单词释义：

1. 毒性反应	toxic action
2. 变态反应	allergic reaction
3. 继发反应	secondary reaction
4. 致畸作用	teratogeneisis
5. 后遗效应	residual effect
6. 药物不良反应	Adverse Drug Reaction（ADR）
7. 热疗法	thermal therapy
8. 结缔组织	connective tissues
9. 癌细胞血管渗透现象	blood vessel leakiness in cancer
10. 血脑屏障	blood brain barrier
11. 高聚物胶束	polymer micelle
12. 乳糖	lactose
13. 丙肝	hepatitis C
14. 磷脂双分子层	phospholipid bilayer
15. 乳化反应	emulsification reaction
16. 阿霉素	doxorubicin
17. 表面等离子共振现象	Surface Plasmon Resonance（SPR）

译文解析：

1. 汉语是重意合的语言。

许多连接词都隐藏在上下语句之间。翻译成英语时需要找出前后的关系,比如转折关系,因果关系等。

原文：使用聚氧乙烯菌麻油的最大弊端就是会诱发过敏反应。患者通常要提前服用抗过敏药物来预防因为注射紫杉醇带来的不良反应。

译文：The biggest drawback of Cremophorl is that it can induce allergic reaction. That's why patients usually take anti-allergic drugs in advance to prevent adverse reactions caused by injection of Paclitaxel.

分析：前一句提到会诱发过敏反应,后一句又表明要服用抗过敏药物。因此我们得出这两个句子之间的联系:因为会诱发过敏反应,所以才要服用抗过敏药物。翻译时就要体现这种因果关系。又因为句子本身过长,所以独立成句,用 that 到会诱发过敏连接上下句,也可以体现科技英语文本严谨的特点。

2. 翻译时要注意情感的表达。

虽然科技文本追求的是客观,但某些地方也需要情感表达才能更准确表达源语言的含义。

原文: 所以孕妇服药要慎之又慎。

译文: So pregnant women must be extreme careful to take drug.

分析: "慎之又慎"表达了谨慎的程度比谨慎还要高。这就表明了孕妇服药是非常需要注意的。所以,个人觉得这里要体现谨慎的程度,故翻译为了"慎之又慎"表达了谨慎的程度比谨慎。

3. 汉语中经常出现长句。

一个句子中有很多逗号,翻译时不可能逐字逐句翻译,这样句子太过冗长。翻译时要注意句子整合,我在翻译这个句子时,采用了分句的形式。既保证了整段文字的连贯性,又避免英文句子过长造成理解困难。

原文: 科学家们在设计纳米颗粒时,已经在其内部设定了"程序",纳米颗粒在接收到信号以后,就启动自我分解的模式,把药物准确地靶向释放到病变细胞内部。

译文: When designing nanoparticles, scientists have set up a "program" inside them. After receiving the signal, these nanoparticles initiative the mode of self-decomposition and target drugs into pathological cells.

分析: 通过阅读句子,我们可以进行整合。前两小句放到一起,整个句子成分也是完整的。后三小句合为一句,内部隐含了一种时间上的递进关系。先接收信号,然后分解,最后靶向给药。逻辑清楚之后,就可以直接翻译了。第一句主语都为"科学家",所以共用一个主语。第二句有时间上的先后顺序,翻译时就可将其中一个动词变成非谓语的形式。

4. 翻译时有拆分句子就会有合并句子。

某些句子可以合到一起,让文本更精炼。在翻译这一句时,我发现了其中前后两小句具有相似性,所以选择了合并句子。

原文: 病变细胞就像靶心一样,药物犹如飞镖,准确地射中靶心,进入病变细胞内部,一举清除病变细胞。

译文: Drugs will enter into the diseased cells and remove them at one stroke like the darts hit the bull's-eye.

分析: 该句主要内容是:药物准确进入病变细胞内部,清除病变细胞。"病变细胞就像靶心一样,药物犹如飞镖"其意思就可理解为飞镖命中靶心就像药物命中病变细胞。因此可以进行整合,用 like 连接两个句子,精简句子,避免过于冗长繁琐。

5. 英汉两种语言存在很多差异。

其中一方面是汉语多用动词,一个句子中可以有很多动词,而英语是多名词,一个句子中只有一个谓语动词。翻译时要合乎目标语言的特点,善用词类转换法。

原文：长期使用四环素类广谱抗生素会导致肠道内的菌群平衡遭到破坏。

译文： Long-term use of tetracycline broad-spectrum antibiotics will disrupt the balance of intestinal flora.

分析：原文中共有三个动词,分别为:使用,导致,破坏。简化句子,"使用抗生素会破坏菌群平衡。"我们发现谓语动词是破坏,前面说过,英语一个句子中只有一个谓语动词,所以"使用"就要改变词性,最常见的就是变为名词。"长期"本身是副词,也随之变为形容词。这样句子就变得简单易懂。

6. 文本中多次出现成语俗语的表达。

这也是汉语语言的一大特色。针对成语俗语的翻译,我们可以合理选择直译也可以选择意译,无论哪种方法,都是要能准确真实地表达成语在文本中的含义。考虑到本文可能会有国外读者阅读,因此翻译时多选择了意译,且都采用常见词汇,便于理解。

原文：纳米颗粒凭着它微小的体积和可调控修饰的表面性质,在智能新药开发领域大显神通。

译文： Nanoparticles, with their tiny size and controllable surface properties, have been widely used in the field of intelligent drug development.

分析："大显神通"原指无所不能的力量,如今用来形容充分显示出高超的本领。放在语境中理解,就是纳米颗粒在智能新药领域被广泛应用,所以才能大显神通。因此选择翻译为"大显神通"原指无所不能的力量,用词上简单明了,含义上也完全可以理解。

7. 小标题的翻译技巧。

小标题的翻译应该通俗易懂,成分简单。几乎所有标题都是省略冠词等成分,只表达最核心含义。因为即使省略,大家也还是能够理解。翻译时也要遵循此特点。

原文：前路虽有荆棘,但未来不是梦

译文： Faced with Difficulties, Having a Promising Future

分析：这里其实有比喻的成分。"荆棘"指的是困难、难题,若是直译的话难免会造成理解上的错误,所以将引申义"困难"翻译出来。"梦"不是指做梦,而是指梦想。个人对这句话的理解是"前方虽有困难,但美好未来一样可以到来。"翻译时选择这里其实有比和"having"代替了两个动词"有"和"到来",且省略了主语"我们"。前后两句也比较并列,朗朗上口。

171

第十章　基因编程新发现

　　是什么样的生物新黑科技开启了基因组编程时代的大门？是怎样的发现让科学家与好莱坞明星同台，获得硅谷亿万富翁赞助的豪华版诺贝尔奖？又是怎样的事件触发全球科学家以及社会舆论对生物技术安全和伦理进行大讨论？以上这些问题的答案都是一个长到连生物学家都记不住的名字：成簇规律间隔短回文重复序列（clustered regularly interspaced short palindromic re peats）大家都干脆亲切地称呼由它英文首字母组合成的新词 CRISPR CRISPR 作为一种最新也最受瞩目的基因编辑技术，给人类遗传病的治疗带来了新的希望，而对 CRISPR 安全性的考虑以及对伦理上的挑战也同样是它成为万人关注焦点的重要原因。

　　CRISPR 富有传奇色彩的故事要从它诞生那一刻讲起：在 20 世纪 90 年代，科学家就发现的一小段很奇特的细菌 DNA 序列。这一段看上去无厘头的序列成为长达两个世纪的不解之谜，最终居然与乳品业的科学家为了提高酸奶发酵菌的抗性所做的研究不期而遇。科学家不懈探索终于揭晓谜底。科学家通过研究 CRISPR 意外地发现了它是细菌的"独门神功"，小小的细菌竟然也有一套免疫系统，令人刮目相看；而就在科学家不断深入了解它的机理的同时，生物技术科学家也获取了编辑基因组的新"杀手锏"……

　　那么，到底什么是 CRISPR？且让我从头说起它的前世今生。

CRISPR：细菌的独门神功

　　虽然有些讨厌的细菌会让人生病，但实际上绝大多数细菌都与人"和平共处"。另外，很多种细菌还是人类的朋友，它们被用于发酵工程，帮助人类生产食品、药物以及清洁能源。

　　细菌也会生病，它的敌人主要是被称为"噬菌体"的病毒。噬菌体侵染细菌后，会启动一套"黑客"程序，盗用细菌体的材料和能量来生产和包装数以万计的病毒。随着这些新病毒的释放，一个细菌体便化为乌有。这对主要依赖于细菌进行发酵的乳品业（比如酸奶和乳酪噬菌体工业）是个头疼的问题。

　　为了帮助细菌有效地抵抗噬菌体,科学家们进行了很多研究工作。他们发现了一个有趣的现象:在细菌与噬菌体的战役中,虽然噬菌体通常大获全胜,但是细菌并没有全军覆没。更重要的是,这些顽强存活下来的极少数细菌再次遇到同种的噬菌体时,能够非常有效地抵抗噬菌体的攻击。这些受到菌体病活下来的细菌究竟有什么"过人之处",能够造成酸奶和乳酪减产,战胜曾经的"常胜将军"噬菌体? 正是这个问题促动了生物学家们进行不断研究。在 2007 年,这个问题终于有了答案。

　　答案的线索可以追溯到1987 年,日本科学家们在几种细菌的基因组中发现了很奇特的一段重复序列。每个重复的单元是一小段序列接着一段它的反向序列在生物学上被称为 CA 间隔序列再接着一小段看上去很随机的序列(科学家称之为"间隔序列")这样的单元可以重复许多次,连成一串排在细菌的基因组里。

　　因为实在是令人费解,这篇论文当时并没有在科学界引起重视。在接下来的差不多 20 年中,随着基因组测序的普及,科学家们不断在多种细菌和古生菌的基因组中都发现了这个奇特的重复序列。这让科学家们更好奇了,这个奇特的重复序列也获得了自己的学术大名—成簇规律间隔短回文重复序列(Jansen et al. 2002)因为又拗口又难记,科学界都只称呼它的英文首字母的组合 CRISPR。几位西班牙科学家搜集了来源于几十种不同细菌基因组中的上千段 CRIPSR 的"间隔序列",在一个包含当时所有已知基因组的信息库里进行序列比对,找寻还有哪些生物可能具有类似的序列。你猜怎么着? 许多间隔序列居然和一些噬菌体基因组序列高度一致(Bolotin et al. 2005)太奇怪了,细菌用着一种奇特的方式在小心翼翼地收藏着敌方的信息。莫非,这些信息是用来对付噬菌体的?

　　为了证明这个猜想,来自丹麦 Danisco 生物制品公司(现被杜邦公司收购)的科学家进行了一组严格控制的实验。Horvath 和他的同事们选择了用于乳品发酵的嗜热链球菌作为研究对象,筛选出在噬菌体侵染后产生抵抗性的菌株,与噬菌体侵染前的不具备抵抗性的菌株进行对比。结果发现在产生抵抗性的菌株的 CRISPR 位点上,插入了一个新的重复单元。而这个单元的间隔序列恰好与用来侵染的噬菌体完全匹配!而当他们把这一个序列单元从有抗性的细菌基因组移除后,发现这些细菌就不再对同种噬菌体有抗性了。反过来,最直接、最震撼的证据是,当他们把这个序列单元插入到一个本来不具有抗性的细菌 CRISPR 位点后,这个改造过的细菌居然就能顽强抵抗这种噬菌体了! 2007 年的那个春天,这篇论文发表在最顶尖的《科学》(Science)杂志上(Barrangouetal. 2007),就此打开了生物科学界一扇新的大门。

　　这个革命性的发现给科学家带来了更多的疑问:最重要的问题包括,细菌到底是怎么利用噬菌体的这一小段 DNA 序列扭转战局、转败为胜呢? 在接下来的短短几年内,众多科学家发表了数百篇论文阐述 CRISPR 的机理。虽然在不同种细菌里具体机制有所差别,但最基本的原理都一致而且简单:细菌通过这一小段序列就能够准确定位噬菌体相对应的序列,然后咔嚓一刀把噬菌体的 DN 链剪断。被剪断基因组 DNA 链的噬菌体就像被敌方拔掉了大旗,再也无心进攻了。非常有趣的是,细菌对抗特定种类的噬菌体的本领是获得性的:每次和一种新的噬菌体交手后,细菌就会把敌方的信息收藏到 CRISPR 位点中,用于以后抵御同一种噬菌体,这是典型的反攻型战略。这个极其令人惊讶的发现让大家对细菌刮目相看。一直以来,大家都认为只有高等生物,比如人类,才拥有这种可以用来对付细菌和病毒的特异性免疫系统(当然,人类的免疫系统主要是由特异性的免疫细胞组成,与细菌的 CRISPR 系统完全不同)。而事实上,细菌在与病毒上亿年的持久战中,就进化了这一套独门神功。

编码基因组的 CRISPR 神奇搭档:"搜索引擎"+"剪刀手"

计算机版的生命科学论

每个领域都有不少专业名词让人望而生畏,不过你并不用担心接下来将要登场的,比如基因组、基因编辑之类的生物专业词。只要你对计算机稍有了解,就一定很容易理解我接下来用计算机的一些基本概念来类比解释生物学。其实说起来,发展到了基因组学时代的生命科学同样是一门信息科学,而生命科学与信息科学交融的一个典型的例子就是信息科技大佬谷歌公司,如今谷歌瞄准了新兴的医疗大数据浪潮,大步进军生命科学领域。

每种生物都有自己的一套基因组。基因组就像一段很长很长的源代码。比计算机的"0""1"字符稍稍复杂,基因组主要有"A""G""C""T"四种字符(它的生物学名叫作碱基)。人类的基因组由约 30 亿对碱基序列构成,包含着一个个体生长发育的所有信息。

自 1990 年启动的人类基因组计划(Human Genome Project)被称为"生命科学的阿波罗计划",这项伟大工程的任务只有一个:获得人类基因组全部源代码的序列。测序工程从 1990 年启动,2001 年完成草图,同年 2 月,两大生命科学领域最顶尖的杂志《科学》(Science)和《自然》(Nae)同时在封面报道了人类基因组计划草图这一里程碑式的工作。2003 年,科学家最终完成了 99% 的人类基因组序列测定。而有了人类生命的序列代码只是了解生命奥秘的第一步,因为基因组使用着一门对人类来讲完全陌生的高深的代码语言,只有破解出这门语言的单词拼法、语法结构才能读懂基因组这本"天书"。也就是说,遗传学和基因组学的主要任务就是研究清楚基因组中一段段代码是如何控制我们的身高体重、相貌性格,等等。

遗传学家的研究方法很直接,就是删除或者修改一段代码后,观察个体性状发生的变化,以此来推测这段代码的功能。听起来简单,操作起来却是极具挑战性的。首先,要在上亿个碱基序列中精确定位到某段代码,如果没有高效的搜索方法,这项工作就如同大海捞针。虽然科学家们已经发明了很多技术来实现基因编辑,但是几乎没有一种方法是既准确高效,又简单低成本的。而 2011 年,科学家在逐渐了解到 CRISPR 的分子学机制后立刻意识到,细菌的这套本领真是大自然给基因编辑领域的一份馈赠。于是几个小组快马加鞭地研究如何利用 CRISPR 完成基因编辑。为什么 CRISPR 让科学家们如此兴奋? 他们最终如何妙用细菌的"神器"来编码基因组呢?

基因编辑决胜法宝之一:搜索引擎

首先, CRISPR 系统的非凡之处在于它的精确定位系统。要在一本 30 亿字的"天书"里准确定位到一个目标绝非易事。如果用于搜索的"关键词"太短,那么在基因组的多个位置都可能出现同样关键词,这在基因编辑上是非常可怕的错误:因为它可能导致对目标之外的代码进行改动。理想的搜索方式是对一段长度约为 20～30 个碱基对进行精确匹配,因为在 20 个碱基对的序列四种字符随意排列组合的种类就有 4 = 110 种。从概率上计算,同样的一个序列大约在一万亿长的碱基对序列里才会出现一次,因中,"A"T 在 30 亿长的人类的基因组中出现完全同样序列的概率极小,也就是说,出现命中目标之外的几率是极小的。可惜的是,以往基因编辑的"搜索工具"要么用于搜索的"关键词"太短,要么搜索时的匹配精准率还不够高,而 CRISPR 几乎满足最理想化的搜索方式。这也难怪,在细菌在与病毒上亿年的苦战中,细菌只有掌握极其快速准确的定位才能成功切割噬菌体的基因组,同时不出现错切自己的基因组这样的"乌龙事件"。在这样的生命对决中进化而成的 CRISPR 神功,它的精确定位功能对生物学家而言实在是天赐之喜。

基因编辑决胜法宝之二:神奇"剪刀手"

精确定位到目标还只是成功的第一部分,接下来如何对代码进行修改也大有讲究。要知道,基因组是所有生命的"蓝图",包含着一个生命体几乎所有的信息。因此每个细胞都拥有极其严格的保护机制,以确保这些最为珍贵的生命代码不被轻易更改,这就像对文档加锁保护。生物学家的解决方法就在找到需要修改的位点后切断 DNA(脱氧核糖核酸)链,这样就像解除了保护,同时激发细胞启动修复程序。有趣的是,细胞通常使用的有两套修复方案,分别被生物学家用来做不同的编辑功能。

一种是"模糊修复",这种相对简单的修复方案就是重新搭接上被切断的 DNA 链,但是在搭接的过程中经常会导致 DNA 链上个别代码的更改。而在基因组的功能区域,哪怕是单个字符的缺失、插入或者更改都可能导致整个一段代码失效。因此,生物学家恰好利用了"模糊修复"极易出错的这一特点来让这一段代码失效,这种技术也被称为"基因敲除"。另一方面,在一些基础研究和绝大多数临床应用中,都需要对代码进行精确编辑。于是,科学家就要利用细胞的精准修复功能。由于绝大多数物种的基因组都有双数对的、分别来自父亲和母亲的完整代码本,在其中一个代码本的DNA 链受到损伤后,细胞会完完全全地按照另外一套代码本的序列进行修复。虽然

来自父母亲的代码本不完全相同,但是在绝大多数位点都是一致的,所以细胞使用这套"精准修复"保证重要信息的完整性。有趣的是,细胞按照另外的代码本进行修复的这个特性也给基因编辑带来了机会。科学家想出办法给细胞提供一段人造的代码本,这样细胞忠实地按照科学家提供的样本修复后的 DNA 序列就正正好好是编辑后的序列的巧妙吧!

好了,讲了这么一大通,让我们重归 CRISPR 正题。神奇的是,细菌的 CRISPR 恰恰具有这两大重要系统的功能,即搜索引擎和一个"剪刀手"。这样一来,这对绝妙搭档不仅是细菌战胜噬菌体的法宝,也变成了生物学家编辑基因组的利器。能够把 CRISPR 用于基因编辑,大功属于加州大学伯克利分校 Jennifer Doudna 实验组和瑞典于默奥大学 Emmanuelle Charpentier 实验组。在 2010 年到 2012 年的几年中,她们合作发现了细菌里最简单的 CRISPR 系统,在此系统上又进一步合并了其中的搜索元件。于是,大大简化后的 CRISPR 系统只需要个被称为 CAS9 的蛋白和一段序列:这段序列就是搜索神器,引导 CAS9 的蛋白和这一段序列,这段序列就是搜索神奇,引导 CAS9 蛋白这个"剪刀手"来实现定点剪切功能。同时,科学家发现这套 CRISPR 系统在细菌以外的生物体也可以正常工作。这一系列研究都为 CRISPR 用于高等生物基因编码铺平了道路。最终,2013 年年初,CRISPR 转化为生物工具,来自麻省理工学院的张锋组和哈佛大学的 George Church 组同时报道了利用 CRISPR 实现基因组编辑的新技术。他们在高等生物(比如人类)的细胞中表达 CAS9 蛋白和一段靶向人类基因组的引导序列。这样 CAS9"剪刀手"就可以准确地在人类基因组的靶向位点剪切,来实现"基因敲除"。同时,科学家也可以提供一段额外的代码本来可以达到精确编辑。于是一个蛋白和一段 DNA 序列就这样开启了人类基因组的新编程时代。

一夜成名后的荣誉与挑战

CRISPR 基因编辑技术自 2013 年初被发表后立刻成为基础科学以及医疗领域的新宠。与此同时, CRISPR 的发现者和技术发明者也获得了科学界的最高认可,几乎包揽了除诺贝尔奖(目前还没有,让我们拭目以待)以外的大大小小各种奖项。

特别值得一提的是,同时在阐述 CRISPR 基本机制和促进技术转化过程中做出重要贡献的两位女科学家:加州大学伯克利分校的 Jennifer Doudna 和瑞典于默奥大学的 Emmanuelle Charpentier 得到了格外让人心动的 2015 年生命科学突破奖(Breakthrough Prizes)。这是由硅谷巨头 Sergey Brin(谷歌创始人之一)与前妻 Anne Wojcicki(23 BandMe 生物基因公司创始人之一), Mark Zucker. berg(脸书董事长及总裁)及夫人 Priscilla Chan, Yuri Milner(俄罗斯风险投资家)还有 Art Levinson(苹果公司董事长及基因泰克董事长)联合发起的鼓励基础科学研究的最新奖项。有如此强大的赞助商阵容,难怪生命科学突破奖的奖金高达 300 万美元,由此得到了"豪华版诺贝尔奖"的别名(要知道,诺贝尔奖奖金仅 10 万美元)。该奖项如同奥斯卡奖一样隆重,参加者身着盛装,电影明星卡梅隆·迪亚兹(Cameron Diaz)和本尼迪克特·康伯巴奇(Benedict Cumberbatch)也受邀出席。

基础科学领域的新星

　　前面提到过,遗传学家需要通过改变基因组的遗传代码来研究它的功能。而传统的基因编辑技术相对耗时长、成本高、成功率低。CRISPR 的出现,给基础生命科学研究带来了革命性的变化。举个例子,以前研究人员要想获得一个新的转基因小鼠模型,通常需要花一两年的时间,经过干细胞—嵌合体—转基因小鼠多道程序。而CRSPR 极大地简化了这个过程,把时间缩减到了三个月。更酷炫的是,科学家发挥无限的想象力在原有的 CRISPR 神奇剪刀上,不仅创造出了特异性更高、可调控的"超级剪刀手",还有一系列大大拓展"剪刀手"原先功能的新技术。比如,用来增强或者减弱特定基因功能的元件,还有用来显示特定序列在基因组上位置的标记元件等。这些巧用 CRIPSR 的新技术就像雨后春笋一样,成为生命科学家在基因编辑、基因表达调控和细胞生物学等多个方向研究中的新工具。

临床应用:希望与挑战

　　我们的基因组就像一段长长的代码,在生命的繁衍过程中不免出现极少数代码在拷贝时出错,而人类的很多疾病都是由某些特定代码的错误导致。相信大家对现在逐渐流行的"精准医疗"的概念已经有所耳闻。简单来讲,这个过程就是对个体进行基检测,先抓出在基因源代码层面上导致疾病的"罪魁祸首",然后再进行针对性的治疗。除了药物治疗等方法,在基因水平上的治疗相当于对源代码进行矫正。这种方法最为直接,但潜在的危险性也最大。

　　CRISPR 技术在基础研究中大显身手,而它最受关注的潜力还是在治疗疾病中的应用。如果能用这种既简单又高效的技术进行基因治疗,它可能会成为人类健康的大功臣。理想很丰满,现实很骨感。总结起来,CRISPR 技术距离实际临床应用的目标还有差距,原因主要有三:

　　第一,我们在上一节里曾经提到"从概率上计算,CRISPR 出现命中目标之外的可能是极小的"。到底有多小呢?科学家在不同系统中进行反复实验后发现,原来CRISPR 在进行目标搜索时允许引导序列上极个别位点的不完全匹配,导致有可能命中目标之外的序列,但是在临床上的基因定点治疗需要万无一失!因此,众多科学家不断努力提高 CRISPR 的精准度以期服务于临床。

第二,基因治疗需要精确修复机制,可是在大多数修复模板存在的情况下,细胞还是会启动模糊修复模式。这不仅无法治疗疾病,甚至可能导致基因功能完全丧失,造成更严重的后果。如何提高精准编辑的效率,仍然是科学家们需要攻克的难关。

第三,虽然 CRISPR 在基因编辑的大家族里算起来是效率最高的一个,但是要达到理想的治病效果,CRISPR 的效率还需要大大提高。

总之,CRISPR 技术已经是人类向前迈进的一大步,给基因治疗带来了新的希望。自 CRISPR 出现的短短两三年内,它在技术上成熟和完善的速度以及受关注的程度都让人瞠目。我们有理由相信,尽管目前还不完美,但是完善后的 CRISPR 或许是未来的完美基因编辑技术,将有希望真正成为临床上基因治疗的福音。

完美婴儿? 科学、伦理与法律

自然和生命,很久以来被认为具有某种神奇的魔力而令人敬畏。这种敬畏之心在一定程度上来自人们对很多生命现象未解之谜的无知。但是,科学技术的飞速发展,生命科学家一步步揭示生命原理,人类蓦然意识到自己不仅可以理解生命,甚至可以对一个生命体的生长发育进行一定的干预和控制。此时,就像一个在心理上完全没有准备的人突然获得了超能力,他一时间手足无措,正如同人类拥有超强大超乎想象的新科技的这一历史瞬间。此刻,基因编辑技术的迅猛发展和巨大应用潜力对人们的生命观和健康观,更重要的是对整个社会的伦理和法律体系都提出了前所未有的挑战。

2015 年 3 月,《MT 科技评论》(MIT Technology Review)上发表了一篇评论性文章《定制完美婴儿》(Engineering the Perfect Baby),副标题是:科学家正在发明可以编辑未来孩子们基因的技术,我们是不是应该立刻阻止它的发展,否则就太晚了?

一石激起千层浪:国际顶尖学术期刊《自然》《科学》紧接着刊登了评论文章,呼吁科学家暂停对人类胚胎基因组编辑的研究。恰恰此时,来自中国中山大学的黄军就研究组在 2015 年 4 月的 ProteinCe 杂志上发表了对利用 CRISPR 技术编辑人类胚胎的研究发现,结果显示 CRISPR 技术在精确编辑的应用上还不够完美。虽然他们使用的是在体外受精过程中出问题的胚胎,这些胚胎是不能发育成人的,但是批评者纷纷评论这个实验相当于人工修改和制造人类,一些新闻头条报道"基因编辑人类胚胎的传闻变成事实"。这无疑在社会舆论对人类胚胎基因编辑正在进行激烈辩论的当口,投下了一颗重磅炸弹。

虽然目前我们对人类基因组的了解也很有限,距离制造出真正的"完美婴儿"还非常遥远。但是,公众的担心不是没有道理。毕竟,一个受精卵包含着生长发育成一个人的全部遗传信息。也就是说,人类如今确确实实已经走到了改造自身基因的历史性的门槛前,只是在层层法律、伦理和社会舆论的压力限制下没有研究组敢"越雷池一步"。

2014年,一项仅在美国的社会调查显示:近一半的社会群众认同把基因编辑技术用于人类胚胎来降低重大遗传病的风险,而只有极少数(15%)的民众接受用这项技术让宝宝变得更聪明。由于这两种不同的应用其实用的是同种技术,而假设只有其中一种是合法的,这将对法律规则的实施和监督上提出了更高的要求和挑战(当然,前提是未来真有一天,我们了解了基因到底如何控制人们的智力水平来让宝宝变得更聪明)。

我们在技术、法律、道德体系和心理等各个方面上,到底有没有准备好跨进这个改造自身基因的新时代?截至2015年初秋本文截稿时,我们对人类胚胎的基因编辑还没有一个在社会层面的统一规则,甚至就连CRISPR的核心发现和技术发明人的群体内部,都有着相对保守和开放的不同声音。但是,人类科技和文明的车轮飞速向前行驶,它从未停歇给我们准备的时间。而回顾人类的发展历史,大多数飞越性的进步并非人们的精心筹划,却是由于新文明和新科技与传统的知识体系、伦理和法律相摩擦冲撞,引发人类新的思考,这才有了新的哲学、伦理观和法律体系。在这个生命科学大放异彩的世纪,从克隆技术、转基因,到器官工程、试管婴儿的一系列新生物技术在最初出现时,甚至至今都仍存在争议和谴责的声音,而它们给人类健康和生活带来的进步和希望是毋庸置疑的。

科学技术是把双刃剑,而如果因为畏惧它可能导致的问题就囿于当下,就不会有人类文明的发展进步。只有科学家联合所有社会力量,用理性、包容的态度评判,用严格、有效的法律规范管理,才能推动新科技造福人类。尽管目前人类胚胎的基因编辑还尚未形成一个统一规则,但是广泛的社会关注说明,人类已经迈出了解决问题的第一步。(注:2015年5月18日,美国国家科学院(NAS)和国家医学院NAM宣布,将为人类胚胎和生殖细胞基因组编辑制定指导准则。)

参考译文:

New Findings of Genetic Programming

What kind of biological new black technology has opened the door to the era of genetic programming? How did the discovery allow scientists to stand on the same stage as Hollywood stars and win the "Nobel Prize in Luxury Edition" sponsored by Silicon Valley billionaires? What kind of incident triggered global scientists and public opinion to have a big discussion on biotechnology safety and ethics? The answer is a long name that even biologists can't remember: clustered regularly interspaced short palindromic repeats. Everyone simply and kindly refers to the combination of its initials in English. New word

called CRISPR, as the latest gene editing technology, brings new hope to the treatment of human genetic diseases, and the consideration of CRISPR safety and ethical challenges also make it an important reason for million people to focus.

The legendary story of CRISPR starts from the moment it was born: in 1990s, scientists discovered a very strange piece of bacterial DNA sequence. This seemingly nonsensical sequence has became a mystery for two centuries, in the end, it unexpectedly met with research by dairy scientists to improve the resistance of yogurt fermentation bacteria. Unremitting exploration by scientists finally revealed the answer. Scientists have unexpectedly discovered that it is a bacterium's "exclusive magic power" through research on CRISPR. Even some small bacteria have an immune system, which is impressive; while scientists continue to understand its mechanism, biotechnology scientists also gain the new "trump card" for editing the genome …So what exactly is CRISPR? Let me talk about its past and present life from the beginning.

CRISPR: the Unique Power of Bacteria

Although some nasty bacteria can make people sick, most of them actually live in peace with people. In addition, many kinds of bacteria are friends for people. They are used in fermentation projects to help humans produce food, drugs and clean energy.

Bacteria can also get sick, and their main enemy is a virus called a "phage." After the bacteria phage infects the bacteria, it starts a "hacking" program that steals the material and energy of the bacterial body to produce and package tens of thousands of viruses. With the release of these new viruses, a bacterial body vanishes. This pair of dairy industries (such as yogurt and cheese The bacteria phage industry) is a headache. To help bacteria effectively resist phage, scientists have done a lot of research. They found an interesting phenomenon: in the battle of bacteria and bacteria phages, although the bacteria phages usually won, but the bacteria were not wiped out. What's more, these very few stubbornly survived bacteria can very effectively resist phage attack when they encounter the same type of phage again. What are the "excellent" of these bacteria in the dairy fermentation industry that survived by bacterial disease, can they be poisoned and cause yogurt and milk cooling to reduce the production of the phage that once defeated the "The Blow"? It is this problem that has prompted biologists to continue their research. In 2007, this question was finally answered.

The clue to the answer goes back to 1987, when Japanese scientists found a strange repeat in the genomes of several bacteria. Each repeating unit is a small sequence of sequences followed by its reverse sequence (biologically called the CA interval sequence followed by a small seemingly random sequence (scientists call it a "spacer sequence") It can be repeated many times, forming a string in the bacteria's genome.

Because it was puzzling, this paper did not attract much attention in the scientific community at the time. Over the next 20 years , with the popularity of genome sequencing, scientists continued to find this strange repeat in the genomes of a variety of bacteria and archaebacteria. This has made scientists even more curious, and this strange repetitive sequence has also gained its own academic name-clustered regular interval short palindrome repeated sequences (Jansen et al. 2002) because it is difficult to remember and difficult to remember, the scientific community only calls it The combination of the first letter of CRISPR. Several Spanish scientists collected thousands of "spacing sequences" of CRIPSR from the genomes of dozens of different bacteria, performed sequence comparisons in an information database containing all known genomes of the time, and looked for other organisms that might have similar sequences. Guess what? Many "spacer sequences" are actually highly consistent with some phage genome sequences (Bolotin et al. 2005) are too strange, the bacteria carefully collected the enemy's information in a strange way Could this information be used against bacteria phages?

To prove this conjecture, scientists from Danisco Biologicals (now acquired by DuPont) in Denmark conducted a set of tightly controlled experiments. Horvath and his colleagues chose to use Streptococcus thermophilus fermented in dairy products that was selected as the research object, and strains that developed resistance after phage infection were selected and compared with those that did not have resistance before phage infection. As a result, CRISPR was found in the resistant strain At the site, a new repeat unit was inserted. The spacer sequence of this unit exactly matched the phage used for infection! And when they removed this sequence unit from the resistant bacterial genome, they found these Bacteria are no longer resistant to the same phage. In turn, the most direct and shocking evidence is when they inserted this sequence unit into a non-resistant bacterial CRISPR site, the transformed bacteria could resist the phage tenaciously! In the spring of 2007, this paper was published at the top "Science" (Barrangouetal. 2007) opened a new door of the biological sciences.

This revolutionary discovery has brought more questions to scientists: the most important questions including, how exactly do bacteria use this small piece of DNA sequence of phage to reverse the war and turn defeat into victory? In the next few years, hundreds of papers were published by many scientists explaining the mechanism of

CRISPR. Although the specific mechanism is different in different types of bacteria, the basic principle is the same and simple: the bacteria can accurately locate the corresponding sequence of the phage through this small sequence, and then cut the DN chain of the phage with one click. The phage with the genetic DNA strands cut off seemed to be pulled off the banner by the enemy and no longer attacked intently.

It is very interesting that the bacteria's ability to fight specific types of phages is acquired: each time they fight a new type of phage, the bacteria will collect the enemy's information in the CRISPR site for future defenses against the same kind of phage. Bacteria phage, this is a typical counter-offensive strategy. This extremely surprising discovery made everyone look at bacteria. It has been thought that only higher organisms, such as humans, have this specific immune system that can be used against bacteria and viruses (Of course, the human immune system is mainly composed of specific immune cells, and bacterial CRISPR The system is completely different). In fact, bacteria have evolved this unique skill in the protracted battle with viruses for billions of years.

Genome-encoding CRISPR Magic Partner: "Search Engine" + "Scissors Hands" Computer Science

There are a lot of professional terms in every field that can be daunting, but you don't need to worry about what will come out next, such as biological terms such as genomes, gene editing, etc. As long as you have a little understanding of computers, it will be easy to understand that I will use some basic concepts of computers to explain biology by analogy. In fact, life science, which has developed into the era of genetics, is also an information science, and a typical example of the integration of life sciences and information science is the information technology leader Google Corporation. Now Google is targeting the emerging wave of medical big data. Great stride in the field of life sciences.

Each organism has its own set of genomes. The genome is like a long piece of source code. Slightly more complicated than the "0" and "1" characters of a computer. The genome mainly has four characters: "A", "G", "C", T (its biological name is called base). The human genome consists of about 3 billion pairs of bases. The base sequence is composed of all the information about the growth and development of an individual.

The Human Genome Project, which was launched in 1990, is called the "Apollo Project of Life Sciences". There is one task of this great project: to obtain the entire source code sequence of the human genome. The sequencing project started in 1990 and completed the sketch in 2001. In February of the same year, two of the top magazines in the life sciences, Science and Nae, simultaneously reported on the cover the milestone of the draft human genome project Work. In 2003, scientists finally completed the sequencing of 99% of the human genome. And the sequence code of human life is just the first to understand the mystery of life Step, because the genome uses a high-level code language that is completely unfamiliar to human beings, only by cracking the word spelling and grammatical structure of this language can the genome book be read. In other words, the main task of genetics and genetics is to study how the code in the genome controls our height, weight, physical appearance, and so on.

The research method of geneticists is very straightforward. After deleting or modifying a piece of code, observe the changes in individual traits to infer the function of the code. It sounds simple, but it is very challenging to operate. First of all, it is necessary to accurately locate a certain code in the hundreds of millions of base sequences. Without an efficient search method, this work is search for a needle in a haystack. Although scientists have invented many technologies to achieve gene editing, almost no method is accurate and efficient, and simple and low-cost. In 2011, after gradually understanding the molecular mechanism of CRISPR, scientists immediately realized that this ability of bacteria is really a gift from nature to the field of gene editing. So several groups are rushing to study how to use CRISPR to complete gene editing. Why is CRISPR so exciting for scientists? How can they ultimately use bacterial "artifacts" to encode the genome?

One of the best tips for gene editing: search engine

First, what makes the CRISPR system extraordinary is its precise positioning system. It is not easy to accurately locate a goal in a 3 billion-word "A sealed book". If the "keyword" used for searching is too short, the same keyword may appear in multiple locations in the genome, which is a terrible mistake in gene editing: because it may cause changes to code outside the target. The ideal search method is to accurately match a length of about 20 ~ 30 base pairs, because there are $4 = 110$ kinds of random permutations and combinations of four characters in a sequence of 20 base pairs. From a probabilistic calculation, the same sequence will only appear once in a trillion-trillion-long base-pair sequence, because "A" T has a very small probability of appearing exactly the same sequence in the 3 billion-long human genome. In other words, the chance of appearing outside the hit target is extremely small.

Unfortunately, in the past, the "search tool" for gene editing was either too short for "keywords" for search, or the accuracy of matching was not high enough, and CRISPR almost met the most ideal search method. It's no wonder that in the hundreds of millions of

years of hard battle between bacteria and viruses, bacteria can only successfully cut the phage genome without mastering extremely fast and accurate localization, and there is no such thing as an "bug" that wrongly cuts their own genome. The CRISPR magic that evolved in such a life showdown is a godsend to biologists for its precise positioning function.

The second magic weapon of Gene editing decisive : magic "scissors"

Pinpointing the goal is only the first part of success, and how to modify the code next is also very particular about it. You know, the genome is the "blueprint" for all life, containing almost all the information of a living body. Therefore, each cell has an extremely strict protection mechanism to ensure that these most precious life codes cannot be easily changed, which is like locking a document to protect it. The biologist's solution is to cut the DNA (DNA) strand after finding the site that needs to be modified, which is like removing protection and stimulating the cell to start the repair process. Interestingly, the cells usually use two sets of repair schemes, which are used by biologists for different editing functions.

One is "fuzzy repair". This relatively simple repair scheme is to rejoin the severed DNA strands, but often the individual codes on the DNA strands are changed during the overlapping process. In the functional area of the genome, even the deletion, insertion or modification of a single character may cause the entire code to fail. Therefore, biologists just use the "fuzzy repair" feature that is extremely error-prone to invalidate this piece of code. This technique is also called "gene knockout". On the other hand, in some basic research and most clinical applications, precise editing of the code is required. Therefore, scientists must use the precise repair function of cells. Since the genomes of most species have complete pairs of complete code books from fathers and mothers, after the DNA strand of one code book is damaged, the cells will completely follow the sequence of the other set of code books Make repairs. Although the codes from parents are not exactly the same, they are consistent at most sites, so cells use this set of "precise repairs" to ensure the integrity of important information. Interestingly, this feature of cells repairing according to another code book also opens up opportunities for gene editing. Scientists have come up with a way to provide cells with a piece of artificial code book, so that the cells faithfully repair the DNA sequence according to the sample provided by the scientist is exactly the cleverness of the edited sequence !

Well, after talking so much, let us return to the subject of CRISPR. Amazingly, bacterial CRISPR has exactly the functions of these two important systems, namely the search engine and a "scissor hand". In this way, this wonderful pair is not only a magic weapon for bacteria to defeat phages, but also a weapon for biologists to edit the genome. The ability to use CRISPR for gene editing belongs to the Jennifer Doudna experimental group at the University of California, Berkeley and the Emmanuelle Charpentier experimental group at Ume? University, Sweden. In the years from 2010 to 2012, they collaborated to discover the simplest CRISPR system in bacteria, and further incorporated search elements in this system. Therefore, the greatly simplified CRISPR system only needs a protein called CAS9 and a sequence: this sequence is the search artifact, the protein that guides CAS9 and this sequence, and this sequence is the magical search that guides CAS9 protein, the "scissors" "Hand" to achieve the fixed-point cutting function. At the same time, scientists have found that the CRISPR system works properly in organisms other than bacteria. These series of studies have paved the way for CRISPR to be used for genetic coding of higher organisms. Finally, in early 2013, CRISPR was transformed into a biological tool. The Zhang Feng group from the Massachusetts Institute of Technology and the George Church group from Harvard University also reported on a new technology for genome editing using CRISPR. They express CAS9 protein and a guide sequence that targets the human genome in cells of higher organisms, such as humans. In this way, CAS9 "scissors hands" can accurately cut at the targeted site of the human genome to achieve "gene knockout". At the same time, scientists can provide an extra piece of code that could have been edited precisely. So a protein and a DNA sequence started the new programming era of the human genome.

Honors and Challenges After One Night to Be Star

CRISPR gene editing technology has become a new favorite in basic science and medical science immediately after its publication in early 2013. At the same time, CRISPR's discoverers and technology inventors also received the highest recognition in the scientific community, and almost won all kinds of awards in addition to the Nobel Prize (not yet, let us wait and see).

It is particularly worth mentioning that two female scientists who have made important contributions in explaining the basic mechanism of CRISPR and promoting technological transformation: Jennifer Doudna of the University of California, Berkeley and Emmanuelle Charpentier of Ume? University Heartbreaking 2015 Breakthrough Prizes. This is by Silicon Valley giant Sergey Brin (one of the founders of Google) and his ex-wife Anne Wojcicki (one of the founders of 23 BandMe BioGene), Mark Zucker. berg (Facebook chairman and

president) and his wife Priscilla Chan, Yuri Milner (Russian venture capitalist) and Art Levinson (Apple's chairman and Genentech chairman) co-sponsored the latest awards to encourage basic scientific research. With such a strong lineup of sponsors, it is no wonder that the Life Science Breakthrough Award has a prize of up to $ 3 million, which has earned it the alias of the "Deluxe Nobel Prize" (you know, the Nobel Prize is only $ 100, 000). The award was as grand as the Oscars. The participants were dressed in costumes, and movie stars Cameron Diaz and Benedict Cumberbatch were invited to attend.

Rising Star in Basic Science

As mentioned earlier, geneticists need to study the function of the genome by changing the genetic code of the genome. The traditional gene editing technology is relatively time-consuming, high cost, and low success rate. The emergence of CRISPR has brought revolutionary changes to basic life science research. For example, before researchers wanted to obtain a new transgenic mouse model, it usually took one or two years to go through multiple procedures of stem cell-chimera-transgenic mouse. CRSPR greatly simplified this process, reducing the time to three months. The time was reduced to three months.

What's even cooler is that scientists have exerted unlimited imagination on the original CRISPR magic scissors, not only creating a more specific and adjustable "super scissors hand", but also a series of greatly expanded original functions of the "scissor hand" new technology. For example, elements used to enhance or weaken the function of specific genes, and marker elements used to display the position of specific sequences on the genome. These new technologies using CRIPSR, just like spring calculations, have become new tools for life scientists in the fields of gene editing, gene expression regulation, and cell biology.

Clinical Applications: Hopes and Challenges

Our genome is like a long piece of code. During the reproduction of life, a small number of codes inevitably make mistakes in copying. Many human diseases are caused by errors in certain codes. I believe that everyone has heard about the concept of "precision medicine" that is gradually becoming popular. To put it simply, this process is based on individual detection, and the "culprit" that causes the disease at the level of the source code of the gene is first captured before targeted treatment. Except for drugs and other methods, treatment at the genetic level is equivalent to correcting the source code. This method is the most direct, but also the most potentially dangerous.

CRISPR technology has played an important role in basic research, and its most interesting potential is its application in treating diseases. If this simple and efficient technique can be used for gene therapy, it may become a major contributor to human health. Ideal is full, the reality is very skinny. In summary, there are still three gaps between the CRISPR technology and the actual clinical application goals for three main reasons: First, we mentioned in the previous section that "probably, the probability that CRISPR appears to hit the target is extremely small." How small is it? Scientists have conducted repeated experiments in different systems and found that CRISPR allows incomplete matching of very few positions on the guide sequence when performing target search, resulting in the possibility of hitting sequences other than the target, but in clinical Gene-targeted therapy on this site needs nothing to lose! Therefore, many scientists continue to strive to improve the accuracy of CRISPR in order to serve the clinic. Second, gene therapy requires precise repair mechanisms, but in the presence of most repair templates, cells still initiate a fuzzy repair mode. This not only fails to treat the disease, it may even lead to the complete loss of gene function, with more serious consequences. How to improve the efficiency of precise editing is still a difficult problem for scientists to overcome. Third, although CRISPR is considered to be the most efficient gene in the large family of gene editing, in order to achieve the desired therapeutic effect, the efficiency of CRISPR needs to be greatly improved.

In short, CRISPR technology has been a big step forward for human beings, bringing new hope to gene therapy. In just two or three years since the emergence of CRISPR, its speed of technological maturity and perfection and the degree of attention it has attracted people's attention. We have reason to believe that although it is not perfect yet, the perfect CRISPR may be the perfect gene editing technology in the future, and it will hopefully become the gospel of clinical gene therapy.

The Perfect Baby? Science、Ethics and Law

Nature and life have long been considered awesome with some magical magic. This awe is partly due to people's ignorance of many unsolved mysteries of life phenomena. However, with the rapid development of science and technology, life scientists have revealed the principle of life step by step, and humans have realized that they can not only understand life, but can even intervene and control the growth and development of a living body. At this moment, like a person who was completely unprepared psychologically suddenly gained superpowers, he was at a loss for a moment, just like this historical

moment when human beings possessed super powerful and imaginative new technology. At this moment, the rapid development and huge application potential of gene editing technology pose unprecedented challenges to people's outlook on life and health, and more importantly, to the ethics and legal system of the entire society.

In March 2015, a review article, Engineering the Perfect Baby, was published on the MIT Technology Review, with the subtitle: Scientists are inventing technologies that can edit the genes of future children, Should we stop its development immediately, or it will be too late? This has sparked heated discussion: The top international academic journal Nature and Science immediately published a review article calling on scientists to suspend research on human embryo genome editing.

Just at this time, Huang Jun from Sun Yat-sen University in China published a study on editing human embryos using CRISPR technology in the April 2015 issue of Protein Cell. The results showed that CRISPR technology was not perfect for precise editing applications. Although they used embryos that had problems during in vitro fertilization, these embryos were not able to develop into adults, but critics have commented that this experiment is equivalent to artificially modifying and manufacturing humans. Become a fact. " This undoubtedly dropped a blockbuster when public opinion was engaged in a fierce debate on human embryo gene editing.

Although our knowledge of the human genome is currently limited, it is still far from creating a true "perfect baby". But public fears are not without reason. After all, a fertilized egg contains all the genetic information that grows into a person. In other words, human beings have indeed reached the historical threshold of transforming their own genes, but no research group dares to "step over the thunder pool" under the pressure of legal, ethical, and social public pressure.

In 2014, a U.S. social survey showed that nearly half of the public agrees that gene editing technology is used in human embryos to reduce the risk of major genetic diseases, and only a very small number of people (15%) accept it makes your baby smarter. Since these two different applications are used with the same technology, and it is assumed that only one of them is legal, this will place higher requirements and challenges on the implementation and supervision of legal rules (of course, provided that there is a real future One day, we learned how genes control people's intelligence to make babies smarter.)

In terms of technology, law, moral system and psychology, are we ready to step into this new era of transforming our own genes? As of the writing of this article in the early autumn of 2015, none of our gene editing of human embryos has been in society. The unified rules at the level, even the core discovery of CRISPR and the community of technology inventors, have different sounds that are relatively conservative and open. However, the wheels of human technology and civilization are moving fast, and it never stops to give us time to prepare. Looking back at the history of human development, most of the leapfrogging progress is not carefully planned by people, but because of the friction and collision between new civilizations and new technologies and traditional knowledge systems, ethics and laws, which has triggered new thinking of humankind. New philosophy, ethics and legal system. In this century of life sciences, in the new are from cloning and genetic modification to organ engineering and test-tube babies were still controversial and condemned when they first appeared, and there is no doubt that they bring progress and hope for human's health and life.

Science and technology are double-edged swords, and if we fear the problems that it may cause, we will not have the development of human civilization. Only when scientists unite all social forces, judge with a rational and inclusive attitude, and manage with strict and effective laws and regulations, can we promote new technologies to benefit humanity.

Although a unified rule has not yet been formed for the gene editing of human embryos, widespread social concern indicates that human beings have taken the first step to solve the problem. (Note：On May 18, 2015, the National Academy of Sciences (NAS) and the National Medical College [NAM announced that it will develop guidelines for human embryo and germ cell genome editing.)

单词释义：

1. 成簇规律间隔短回文重复序列　　clustered regularly interspaced short palindromic repeats

2. DNA 序列　　DNA sequence

3. 噬菌体　　phage

4. 反向序列　　backward sequence

5. 间隔序列　　spacer sequence

6. 嗜热链球菌　　streptococcus thermophilus

7. 菌株　　bacterial strain

8. 碱基　　basic group

9. 人类基因组计划　　Human Genome Project

10. 干细胞　　stem cells

11. 转基因小鼠　　transgenic mouse

12. 胚胎　　　　　　　　　embryo

13. 器官工程　　　　　　　organ engineering

14. 试管婴儿　　　　　　　test tube baby

15. 生殖细胞　　　　　　　germ cell

译文解析：

1. 词类转换。

由于英汉两种语言的表达方式不同,翻译时不能对号入座,应根据具体情况和上下文在译文中转换原文的某些词的词类,使译文更加通畅。

原文: 千百万山区的人民终于摆脱了病毒的入侵。

译文: Million of the people in the mountainous areas are finally off the invasion of the virus.

分析: 英语中的某些介词本身具有方向性和动态感,往往可以表示某种动作行为的进展状态,起到动词的作用,可以将其翻译为汉语动词。在这里 off 不译为介词含义而是译为动词"摆脱",是词类转换的翻译方法。

原文: 人们完全不懂在基因工程方面所承担的责任。

译文: we are ignorant of the duties we undertake in gene engineering.

分析: 在这里 ignorant 是形容词本意为:愚昧的、无知的。但是如果在本句中译为它的形容词意思,会使句子变得不通顺,因此我们能采用词类转换的翻译方法,将其译为它的动词意思,译为"不懂"。会使原文更加通顺,便于理解。

2. 视角转换的翻译方法。

视点转换指译者在充分理解原文的基础上,突破语言外壳,改变对原文的意思的思索方式,从另一个角度来表达原文的含义,使语言更加流畅。包括正说和反说转换、相对性转换、语态转换、句子成分转换。

原文: 基因和大脑功能面前,人人平等。

译文: genes and brain functions are no respecter of persons.

分析: 在这里采用了正说反译的翻译方法,no respector :尊重的人,驱离的人,但是原文并没有采用否定的词语,译文用否定词语来表示,更加地道。

3. 逻辑关系的调整。

在翻译时,仅仅掌握单个单词的词义是不够的,要彻底的理解原文,分析上下文的逻辑关系是翻译的关键。一句英文的意义大致可以分为三个相互制约的层次:词汇意义、语法意义、主题关系意义。错误的翻译是表层词义的堆砌,句子的意义必须深入到语法和主题。在挖掘句子的逻辑关系的基础上进行调整和变通。

原文：人们忽然关心起环境问题，因而我们必须从新的角度来看国际关系。对于基因问题的关心不仅使富国和穷国之间的关系更加紧张，而且还可能引起大范围的种种矛盾问题。

译文：Man's sudden concern for the genes has introduced a new dimension into international relations. It has heightened tensions between rich and poor nations ; it has introduced a widene range of issuses for potential conflict.

分析：在这一句中，我们没有完全按照句子的顺序，逐字逐词的翻译，而是按照句子内在的逻辑关系来翻译，使表达更加流畅通顺。

原文：科学技术是把双刃剑，而如果因为畏惧它可能导致的问题就囿于当下，就不会有人类文明的发展进步。只有科学家联合所有社会力量，用理性、包容的态度评判，用严格、有效的法律规范管理，才能推动新科技造福人类。

译文：Science and technology are double-edged swords, and if we fear the problems that it may cause, we will not have the development of human civilization. Only when scientists unite all social forces, judge with a rational and inclusive attitude, and manage with strict and effective laws and regulations, can we promote new technologies to benefit humanity.

分析：在此句中采用了介词短语后置，来更好的翻译长难句，不会使句子过长，造成头重脚轻，因此也需要根据上下文的意思来采取合适的翻译方法。

原文：我们在技术、法律、道德体系和心理等各个方面上，到底有没有准备好跨进这个改造自身基因的新时代？截至 2015 年初秋本文截稿时，我们对人类胚胎的基因编辑还没有一个在社会层面的统一规则，甚至就连 CRISPR 的核心发现和技术发明人的群体内部，都有着相对保守和开放的不同声音。但是，人类科技和文明的车轮飞速向前行驶，它从未停歇给我们准备的时间。

译文：In terms of technology, law, moral system and psychology, are we ready to step into this new era of transforming our own genes? As of the writing of this article in the early autumn of 2015, none of our gene editing of human embryos has been in society. The unified rules at the level, even the core discovery of CRISPR and the community of technology inventors, have different sounds that are relatively conservative and open. However, the wheels of human technology and civilization are moving fast, and it never stops to give us time to prepare.

分析：我们在这一句中，按照原文的逻辑顺序，理清逻辑关系，采取直译法即可。

第十一章　身体里的"大内密探"

　　香港有部喜剧电影《大内密探零零狗》，说的是在庄严的紫禁城里，有按照生肖排列的十二个大内密探，由先皇成立，只听候皇帝差遣。在宫内十二人惯常潜伏宫中各处收集情报，他们自小苦练武功，在宫内负责保护皇帝安全。灵灵狗则是十二密探中最特别的一个，他认为科技定会胜过功夫，发明了无数科技产品。故事就围绕这位高科技的大内密探在多次行动中使用他的发明创造而展开。

　　随着现代医疗的发展，对疾病信息收集成为准确诊断和治愈疾病至关重要的一个环节。人体就如电影说的那个庄严的紫禁城，疾病相关的信息隐藏身体各处。传统的诊断手段是在体外进行检测，比如说心电图扫描。或者是收集一些体内的样品血液和尿液，然后进行各种化验。要是有电影里配备了各种高科技产品的"大内密探"，进出人体各个组织器官，听候医生差遣，收集各种疾病信息，实时汇报体内疾病状况，那真是太妙啦！人体内的"大内密探"是不是天方夜谭？人体内的"大内密探"是谁，他们身在何处？就让我们来细说一下身体内的"大内密探"。

纳米颗粒："大内密探"身份揭秘

　　首先让我们来公布这位大内密探神秘的身份——纳米颗粒。纳米颗粒有怎样的十八般武艺能在身体里自由穿越和识别收集信号呢？前面说到了纳米颗粒在智能新药开发领域的应用，介绍了纳米颗粒靶向进攻肿瘤细胞，可控式释药的技能，从而纳米颗粒充当了身体里"大内密探"的角色，在人体内四处游走和收集关于健康的情报。本节再进一步分析一下有关原理。

　　纳米颗粒的表面可以通过各种化学修饰，加入不同的化学基团、高聚物层抗体或者其他配合体。纳米颗粒通过这种特定的表面功能团和分子标记物之间的作用，实现靶向选择识别病变细胞。最常见的和最早使用的方法是配体和受体识别法。一把钥匙开一把锁，病变细胞上存在特有的锁孔，纳米颗粒表面有那把配合的钥匙。因此纳米颗粒就能智慧地识别出病变细胞。如前章所描述，癌症细胞表面会大量表达和正常细胞不一样的分子标记物（biomarker），比如说表面抗原、小分子。常见的配体有和抗原特定结合的抗体（antibody）、契合寡聚物（aptamer），或者是能和小分子形成化学键的配合基（ligand）。纳米颗粒的表面可以通过化学修饰加入各种化学基团。这些化学基团可以和抗体、契合寡聚物、配合基的化学基团反应形成牢固的化学键。这样，这些能识别癌症细胞表面分子标记物的配体，就分布在了纳米颗粒表面。它们能帮助纳米颗粒从茫茫细胞海洋中一眼识别出病变细胞。当纳米颗粒识别出病变细胞后，便结合并且富集在病变细胞表面。

另外一种选择性识别是利用癌症细胞血管的渗透现象（blood vessel leakiness in cancer）。科学家们可以通过控制纳米颗粒大小，让纳米颗粒可以自由地穿过癌症细胞附近的血管，而无法穿过正常细胞附近的血管。纳米颗粒富集在病变细胞处，通过表面或者抓附的某些分子标记物和病变细胞表面结合，从而纳米颗粒改变了发射的信号。这些信号（通常是磁性信号和荧光信号）能被体外的仪器准确地探测到，并进行采集分析，可以随时监控身体内千变万化的信息。纳米颗粒表面的高聚物可以防止纳米颗粒在血液循环的过程中黏附在一起，实现纳米颗粒在人体内长时间的循环，完成收集情报信息的使命。同时这层高聚物"隐形外衣"，也能帮助纳米颗粒逃过人体消化系统或者是免疫系统的攻击，以免被过早地排出体外。

谷歌 X 实验室的野心计划：纳米颗粒"Fitbit"腕表

目前，Fitbit 是在美国很风靡的一款运动腕表品牌，2007 年由 James Park 在旧金山市（San Francisco）创立。Park 兄弟可是为数不多的叱咤风云的亚裔创业者。而且相比比尔·盖茨，乔布斯还有马克·扎克伯格等辍学先锋派，Park 兄也是少数坚持把哈佛读完，拿到电脑科学本科学位的"异类"。可能亚洲文化中"读好书才是王道"的思想是让 Park 兄弟在哈佛坚持待完 4 年的原因吧。Fitbit 产品是可穿戴的无线设备运动腕表，可以测量人每天的运动量，比如说行走的步数、登的楼梯步数和检测睡眠质量。将收集到的数据显示在腕表的电子显示屏上。Fitbit 把运动量通过测量仪器数字化和测量仪器可穿戴化两个概念完美地结合，从而获得了巨大的商业成功。2015 年 6 月 Fitbit 的 IPO 首日，股价一日之内狂增 52%，成为 2015 年美国前十大首日 IPO 涨幅最大的公司之一。

健康检测仪器和疾病诊断设备的可穿戴化，是当下硅谷创业者们最推崇的想法之一和投资者们看好的新金矿。如果你来到硅谷，听听各种大大小小的创业大赛，关于可穿戴式健康检测和医疗诊断仪器开发的项目比比皆是。创业者们在这个新领域内奇思妙想，不断寻求突破和实现产业化。

谷歌公司（Google）声名享誉世界，总部位于硅谷的山景城（MountainView），是无数科技粉丝向往的朝圣之地。2010 年谷歌旗下成立了个秘密的部门，名为 Goolge X，主要致力于开发各种前沿的尖端技术。Google X 位于 Google 总部半英里外（Google 结构重组后，Google X 中的生命科学部门被重组为 Verily 公司），由 Google 的创始人之一，现任 Alphabet 的主席谢尔盖·布林（Sergey Brin）掌舵。由身兼科学家和企业家双重身份的阿斯特罗·泰勒（Astro Teller）掌管实验室的日常业务。神秘的 GoogleX 的研究项目被内部人员称为"射月计划"（Moonshots）。这个神秘部门在 2014 年年底对外界公布的几个研究计划包括："无人车计划"（Project Self- Driving Car）；可以用来运输邮件物品的飞行仪器的"翅膀计划"（ProjectWing，概念有点类似亚马逊的 AmazonPrime Air）；通过多个热气球为指定地区的人提供快速稳定的 Wi-Fi 网络连接的"潜鸟计划"（Project Loon）；增强现实头戴式显示器（augmented reality head-mounted display）的谷歌"眼镜计划"（Project Glass）；可以用来检测血糖水平的隐形眼镜的谷歌"隐形眼镜计划"（Project Contact Lens）。

Google X 的雄心壮志在于让科技以 10 为指数发展，让科幻小说中的情节变成现实，不得不膜拜这个充满天才和奇思妙想的实验室。自 2013 年起，Google 的科技野心也扩张到生命科学领域，在生命科学领域的一系列大动作，让人们感受到了这个天才公司的魄力。2014 年 Google X 对外宣称的纳米颗粒健康监测腕表就是一个如本章开头所描述的大内密探电影里的高科技产品。虽然说 GoogleX 的这个机会融入了纳米颗粒诊断和大数据（big data）的新兴概念，却也不免落俗地采用了硅谷创业概念的新宠"健康监测＋可穿戴式"结合的模式。但是大牌毕竟是大牌。Google 项目一公布，还是引起了外界极大的关注和不少的争议。下面就来解说一下 Google X 的纳米颗粒腕表，或者称为"纳米颗粒 Fitbit"。Google X 生命科学实验室（2015 年，Google 重组之后 Google X 的生命科学部门单独成立为 Verily 公司。）的负责人安迪·康拉德（Andy Conrad）曾在一次新闻采访中用了一个非常形象生动的例子：现有的健康和疾病监测手段，就像是一个人想要了解巴黎的文化，但是却只坐直升机一年飞到巴黎一次。现行的监测手段都不是实时、持续的，而是间断缺乏时效性的、可想而知，很可能遗漏了对于早期疾病诊断至关重要的信息的收集。

电影人早在 20 世纪 60 年代就描绘出了超炫的 21 世纪的纳米颗粒腕表的锥形。康拉德把 Google X 要开发的健康腕表比作著名科幻电影《星际迷航》（Star Trek）里斯波克（Spock）的神器便携式手持科学分析仪 Tricorder。Tricorder 方便易携，时常被斯波克先生拿在手中，功能强大，可以同时具有信号探测扫描数据分析和记录存储信号的三重功能。康拉德的心中，Tricorder 就是利用功能性纳米颗粒作为传感器，来监测体内细胞发出的信号，然后由腕表对信号进行数据分析，最后将健康状况息息相关的结果显示并储存。

康拉德的一生都在从事和健康监测有关的研究，他曾就职于美国最大的实验检测公司 LabCorp。一次和 Google X 创始人布林及他带领的天才科学家团队的谈话，改变了他的职业轨迹。天才冒险家被 Google X 的天才云集和资金雄厚的科研环境深深吸引,也决定加入这个神秘的组织,用自己的智慧经验和超炫科技改变世界,让《星际迷航》中的神器变为现实。康拉德也进一步解释了纳米颗粒健康监测腕表的技术原理和部分细节。表面功能化的纳米颗粒可以识别病变细胞表面的分子标记物(biomarker),区分病变细胞和健康细胞。

纳米颗粒主要靶向标记循环肿瘤细胞(Circulating Tumor Cells)。循环肿瘤细胞是从原发肿瘤(Primary Tumor)分离下来并且进入血液循环的一类癌症细胞。它们是癌症传播的种子,可以随着血液的流动,到达人体其他部位和器官,诱发新的肿瘤细胞的生长,造成癌症在身体内的转移和扩散。循环肿瘤细胞早在 1869 年就被托马斯·阿什沃思(Thomas Ashworth)观察和发现。一个世纪以后的 1990 年,Liberti 和 Terstappen 证实了循环肿瘤细胞存在于中早期疾病的发展过程,人们才开始意识到这类细胞的重要性。但是因为循环肿瘤细胞的数量非常之少,而且仅有 0.01 % 的循环肿瘤细胞能引发癌症转移和扩散,使得这类细胞非常难被检测到。直到最近 5 年,随着检测技术的提高,循环肿瘤细胞逐渐成为研究的热点。学术界也由此产生了用循环肿瘤细胞作为疾病诊断的新术语"液体切片"(Liquid Biopsy),以区别于传统的组织切片技术。

康拉德计划使用的纳米颗粒是表面功能化的磁性氧化铁纳米颗粒。选用氧化铁纳米颗粒的原因之是氧化铁纳米颗粒是 FDA 批准用于人体内的显像对比剂。制成口服药片的纳米颗粒,进入血液循环系统后,随着血液流动达到各处,捕捉和识别病变细胞。手腕上佩戴的表环产生磁场,把纳米颗粒召集到一处。腕表对召集在一起的"密探们"发问:"你们都看到了什么?找到癌症的藏身之处了吗?"表环内的分析设备分析出纳米颗粒"大内密探们"采集到的情报,得出人体当下的健康状况报告。接下来,腕表上的磁场被移去,纳米颗粒大内密探们又重新回到血液中去执行下一次的任务。

神秘一直是 Google X 实验室的一贯传统。纳米颗粒"健康监测腕表计划"也不例外。Conrad 在新闻发布会上,对外界公开这个近似科幻电影情节的研发计划前,几乎没有人知道 Google X 这个结合了众多当下硅谷最热门黑科技的伟大实验计划。为了执行这个近似科幻的任务,Google X 召集了一批在疾病诊断领域聪明卓越的科学家。和其他超现实的前沿想法-样,纳米颗粒腕表也遭到怀疑。学术界的一些专家们对 Google X 能否实现这个科幻计划持保留意见。凯斯西储大学(Case Westerm Reserve)的专家们就认为 Google X 假想的纳米颗粒药片一进人体内就会遇到麻烦。纳米颗粒或者是颗粒表面的功能基团很可能在进入血液循环之前就被胃和胃肠道消灭分解殆尽。

Google X 的科学家们显然也意识到了这个问题并且在积极寻求解决方案。虽然他们没有公布具体的计划,但是 Google X 已经秘密地和马萨诸塞州的一家创业公司

Entrega 进行合作。Entrega 开发了一种微小的含有纳米颗粒的贴片,这些小贴片被封装在胶囊内,在被体内的消化酶分解后,小贴片就会附着在肠壁上,在几个小时内逐渐释放出纳米颗粒,使它们能够顺利进入血液循环。虽然 Entrega 没有公开配方,,但是学术界已经发表了用卡波普(carbopol)、果胶(pectin)和羧甲基纤维素钠(Carboxymethylcelllulose sodium)的混合物制成的可降解胶囊的文章。胶囊外壳有一层乙基纤维素(ethyl cellulosel)保护层,可以防止胶囊和载有纳米颗粒成分的贴片被消化系统的酶过早地分解。Entrega 很可能是使用了相似的技术和配方。

康拉德对外界所描述的可以召集纳米颗粒"密探"的可穿戴式仪器,目前也只是一个构想。但是 Google X 已经在寻求高人指路。加利福尼亚州伯克利分校的 Steven Conolly 教授应邀,在 Goolge X 做了一场关于磁性纳米颗粒在老鼠体内显像的报告。目前 Conolly 的实验室可以用用和老鼠身体大小的扫描仪探测纳米颗粒,但是目前 FDA 还没有批准在人体内进行的磁性纳米颗粒的显像实验。新闻报道,飞利浦公司已经在德国汉堡的研发中心,开始进行用于人体的磁性纳米颗粒显像的扫描仪的研究。不过 Conolly 教授也严谨地提出,即使以后 FDA 批准在人体内使用磁性纳米颗粒,前期还是需要进行大量的实验来证实其准确性和可靠性。不过 Google X 的一位专供纳米颗粒的化学家 Bajaj 宣称,他的团队已证明在体外实验中磁性纳米颗粒可以准确靶向识别出目标,目前正准备在模拟假肢中进行实验。

美国的西北大学(Northwestern Uniersity)的"大牛"教授 Chad Mirkin 也同样对 Google X 的雄伟计划持保留意见。Mirkin 是纳米生物医疗领域无人不知无人不晓的人物,也是奥巴马总统的科学顾问之一。这位教授本人已经成立了三个纳米科技公司。其中一个公司 Nanosphere 成功商业化了以金纳米颗粒为探针的疾病检测技术。金纳米颗粒可从血液、唾液和尿液中快速检测出感染物。关于这项技术,Mirkin 的实验室已经发表了 2000 多篇学术论文,并且金纳米颗粒探针也实现了产业化,可以随时为美国上千所的医院提供货源。然而,目前有购买意愿的医院少之又少。Mirkin 同时也表示,Google 在超出公司自己专长的以外的生命科学领域的大力投资,是一个伟大又勇敢的举动。Google X 的纳米颗粒 Fitbit 计划目前看来需要很多年的投入,而且也没有人能够肯定这项计划能最终实现。开发道路上的挑战不容小视。纳米颗粒健康腕表不确定的未来让人紧张却又无比憧憬和兴奋,说不定未来某天 Google 这块腕表就戴在了你我的手腕上。

美国国防部高级研究计划局(Defense Advanced Research Project Agency,DARPA)也有和 GoogleX 类似的纳米科幻计划。DARPA 除了大力支持无人车的开发以外,也在近年投入了大量资金,支持纳米颗粒探针的实时建康监测仪器的科研项目。DARPA 的构想是通过静脉注射或者药片服用的形式使水溶胶颗粒进入士兵体内。水溶胶内载有 5 到 6 种不同的荧光纳米颗粒(fluorescent nanoparticles)。这些纳米颗粒表面已经功能化,有不同的配体可以识别体内不同的生物分子标记物。纳米颗粒同时也和荧光湮灭物(fluorescent quencher)绑定在一起。湮灭物可以吸收荧光分子被激发后释放的能量,使得荧光无法正常释放,也就是说当纳米颗粒没有和疾病相关的标记物结合前,纳米颗粒的荧光被湮灭物湮灭,所以不发光。然而,当纳米颗粒和相关标记物结合后,荧光湮灭物从纳米颗粒的表面断开,使得电子激发后的能量能正常地释放,产生可以被检测到的荧光。DARPA 想要靶向监测人体新陈代谢的产物。很多代谢产物是疾病的早期信号标志。每种荧光信号都对应一种特定的代谢物。通过体外荧光探测器对士兵身体内的荧光信号进行收集和分析,最后得出士兵每天健康状况的全面分析报告。如果某种代谢物的含量超常,极有可能是此种疾病的早期信号,可以采取相应的预防或者治疗措施。DARPA 的保密工作可以说是全美第一,目前这个项目的具体进展情况,外界几乎没有知晓。

其他有意思的发明:纳米水溶胶颗粒
戒烟贴片和纳米针胰岛素贴片

另外一项关于纳米颗粒医疗设备有意思的技术是戒烟贴片。这个技术是伊利诺伊大学香槟分校的 Matt Phar 研发的,目前已完成了原型机的开发。戒烟对吸烟者来说是一个痛苦的过程,而且复发率很高。很多戒烟者信誓旦旦地开始,结果要么没有毅力中途而废,要么无法严格遵守疗程而不了了之。

纳米颗粒水溶胶贴片主要用于医生对戒烟者服药的控制。贴片中含有大量载有戒烟药物的水溶胶纳米颗粒。电路设备和小型机械装置。电路设备主要用于远程遥控和数据上传;小型机械设备主要用于促使药物从水溶胶纳米颗粒中释放出来。患者将戒烟贴片贴在手臂上,贴片内的电路让医生可以通过电脑对贴片进行操作和控制。到了定点的用药时间,医生可以远程操作贴片,使贴片内部的机械装置作用,挤压或者定量释放酶,使水溶胶破裂或者分解,内部的药物从纳米颗粒中释放出来,穿透贴片层,渗透进皮肤。

通过使用戒烟贴片,即使患者忘记用药,医生仍然可以通过有效的控制和监测,使疗程能够准确正常地进行。每次的用药量和用药时间都可以通过电路设备上传到控制设备,并且被记录下来。同时,开发者也考虑将数据和智能手机联网,让患者通过手机了解和记录自己每天的用药量。开发这项技术的 Pharr 教授,在 2015 年 4 月旧金山的一次学术会议上做了相关的学术报告,并且展示了原型机的图片。Pharr 教授透露,目前他的团队正在积极地和制造商洽谈实现原型机的量产,同时他们也在和投资商进行合作,目标是在未来一两年内完成戒烟贴片的商业化。

纳米针胰岛素贴片是由北卡罗来纳州大学教堂山分校的一名华裔的教授 Zhen

Gu 和他的实验室研发的。顾教授的祖母因糖尿病去世,所以他一直致力于开发更有效的胰岛素注射技术来对抗糖尿病。糖尿病病人需要每天监控血糖水平并且一天多次注射胰岛素。这是个非常麻烦的过程,即使有些病人体内有胰岛素注射泵,也很可能过量或者低量注射。

顾教授开发的纳米针贴片只有指甲盖大小,贴片内有超过 100 个纳米针。当患者把贴片贴到皮肤上时,纳米针扎进血管,会产生短暂的刺痛感。纳米针内充满了载有胰岛素和酶的“小口袋”。当血糖浓度高出正常值时,血液中的葡萄糖会通过“小口袋”的细孔渗透进入。“小口袋”内的酶把葡糖糖转换成酸性物质,使“小口袋”分解,把胰岛素通过纳米针注射到血管内。“小口袋”的设计,可以使整个胰岛素在一个小时内缓释完成。

顾教授已经在 5 只老鼠的身上测试过纳米针胰岛素贴片。数据表明,贴片能成功控制小鼠体内的血糖含量长达九个小时。顾教授目前已经开始在猪身上进行实验,因为猪的皮肤和人的皮肤更加接近。这位科学家的最终目标是开发出医用的贴片,让糖尿病患者每隔两三天更换一次贴片,并且无痛简单准确地控制和调节他们的血糖含量。因这项发明,2015 年顾教授也获得了负有盛名的麻省理工学院 35 位年龄不超过 35 岁的创新者称号(MIT 35 innovators under 35)。

从能描绘出人体内实时健康状况的腕表,到智能控制准确释药的贴片,纳米颗粒在医疗仪器设备的开发应用领域功绩显赫。纳米颗粒就像可以飞檐走壁的“大内密探”一样,神通广大,无所不能。随着纳米技术的不断创新和发展,医疗仪器走向了微型化、智能化和可控化的研发方向。未来纳米颗粒医疗仪器的发展目的是更小、更快、更准和更智能。这个世纪也许就是纳米医疗革命的时代,让我们用一颗悸动的心和创新的头脑来迎接这场小小纳米颗粒带来的大大的科技革新。

参考译文:

The Secret Services in Our Body

On His Majesty's Secret Service is a comedy movie in Hong Kong, which talks about twelve secret services arranged according to the zodiac in the solemn Forbidden City. They were appointed by a deceased emperor, and thus only follow emperors' dispatch. Those twelve secret services fly low in order to collect information from anywhere in the palace. Their work in the palace is to protect the emperor as they've been practicing KongFu since childhood. The Dog is one of these twelve secret services, who is the most special one, and he thinks that technology will definitely surpass KongFu with a lot of technology products invented. The whole movie is about the Dog and how he used his technological inventions in many operations.

With the development of modern medicine, the collection of disease information plays an essential role in accurate diagnosis and the cure of disease. The human body is like the solemn Forbidden City in the movie with information related to diseases hiding all over the body. Our traditional diagnosis methods include testing in vitro, such as an electrocardiogram scan, or collecting some samples of blood and urine from the body, and then getting them analyzed. It would be wonderful if there is a same "secret service" equipped with high-tech products as in the movie, assessing various tissues and organs of our body, waiting for the doctor's dispatch to collect information about diseases and timely reporting the status of diseases. Does it sound impossible to have secret services in our bodies? Who and where are they? We will elaborate on this.

Nanoparticles: The Real Identity of the Secret Services

Let's start by revealing the real identity of these mysterious secret services—the nanoparticles. What are the reasons that enable the nanoparticles to move freely inside our body and identify and collect signals? The previous chapter has introduced the application of nanoparticles in the field of intelligent drug development and how they target tumor cells and controlled-release medicine, for which they have played the role of secret services in our bodies to assess through human bodies and collect health information. And this chapter will further analyze the relevant principles.

The surface of nanoparticles can be modified by various chemical methods, adding different chemical groups, high polymer layer antibodies and other complexes. Nanoparticles achieve targeted selection to identify diseased cells through the interaction between this specific surface functional group and molecular makers. The most common and earliest methods are ligand and receptor recognition. As an old saying goes, a key to opens a lock. There are unique keyholes on the diseased cells, and the matching keys are on the surface of nanoparticles. Therefore, nanoparticles can itelligently identify diseased cells. As described in the previous chapter, a large number of biomaker, such as surface antigens and small molecules, will be expressed on the surface of cancer cells, which are different from the normal cells. Common ligands include antibodies that specifically bind to the antigen, aptamers, and ligands that can form chemical bonds with small molecules. The surface of nanoparticles can be chemically modified to add various chemical groups, which can react with the chemical groups of antibodies, aptamers and ligands to form strong

chemical bonds. In this way, these ligands that can recognize molecular markers on the surface of cancer cells are distributed on the surface of the nanoparticles. Nanoparticles can easily identify diseased cells from the vast ocean of cells with their help. And after identifying those diseased cells, nanoparticles are bound and enriched on the surface of the diseased cells.

Another selection recognition is the use of blood vessel leakiness in cancer. Scientists can control the size of nanoparticles so that they can pass freely through blood vessels near cancer cells, but cannot through blood vessels near normal cells. Nanoparticles are concentrated in the diseased cells, and combined with the surface of diseased cells through some molecular markers on the surface or attached to the surface, so that the nanoparticles change the emitted signals. Theses signals (usually magnetic signals and fluorescent signals) can be accurately detected by external instruments and collected for analysis, which can monitor the ever-changing information in the body at any time. The high polymer on the surface of the nanoparticles can prevent the nanoparticles from sticking together during the blood circulation, so as to the long-term circulation of nanoparticles in our bodies can be realized and information can be collected. At the same time, with the help of this "invisible coat", nanoparticles can avoid attacks from the human digestive system or the immune system, so that they can stay inside our body for longer time.

Goggle X Lab's Ambitious Project: Nanoparticle "Fitbit" Watch

At present, Fitbit is a popular sports watch brand in the United States. It was founded in 2007 in San Francisco by James Park. The Park brothers are successful Asian entrepreneurs which are rare in the United States. And compared with Bill Gates, Jobs, Mark Zuckerberg and other avant-garde dropouts, the Park Brothers is one of the few "weirdos" that finished school in Harvard and got a degree in computer science. Perhaps the idea that "Study is the absolute principle" in Asian culture supported them to stay at Harvard for four years. The fitbit products are wearable wireless device sports watches that can measure people's daily exercise, such as the number of steps walked, the quantity of steps on the stairs and the quality of sleep. The collected data will be presented on the electronic display of the watch. Fitbit perfectly combined the two concepts of digitization of measurement equipment and wearability of measurement equipment, which led to great commercial success. On its first IPO day, the stock price of Fitbit soared 52%, so the Goggle company became one of the top ten companies with the largest IPO increase in the United States in June, 2015.

The wearability of health detection equipment and disease diagnosis devices is one of the most highly regarded concepts of entrepreneurs in Silicon Valley as well as a gold mine that investors think highly of. If you come to Silicon Valley and listen to various entrepreneurial competitions, you will find that there are a lot of projects about wearable heath testing and medical diagnostic equipment development. Entrepreneurs have wonderful ideas about this new area, constantly seeking breakthroughs and industrialization.

Google is a world-renowned company whose headquaters located in the Mountain View, which countless tech fans look forward to. In 2010, Google established a secret department called Google X, which is dedicated to the development of cutting-edge technologies. Google X is located half a mile away from Google headquaters (after the reorganization of Google, the life science department of Google X was reorganized into Veliry). Sergey Brin, one of the Google founders and the current chairman of Alphabet, is in charge of Google X. And Astro Teller, a scientist as well as an entrepreneur takes charge of the lab's daily work. The mysterious Google X research project is called "Moonshots" by insiders. Several research plans announced by the mysterious department to the outside world at the end of 2014 include: Project Self-Driving Car; Project Wing, a flying instrument that can be used to transport mail items (conceptually similar to Amazon's Amazon Prime Air); Project Loon, which provides fast and stable Wi-Fi network connections to people in designated regions through mutiple hot air balloons; Project Class, which focuses on augmented reality head-mounted display; and Project Contact Lens, which can be used to measure blood sugar levels.

Google X's ambition is to make technology development at an exponential rate of 10 and make the plot in science fiction a reality. You have to worship this lab full of genius and imagination. Since 2013, Google's technological ambitions have also expanded into the field of life sciences. A series of major actions in the field of life sciences have made people feel the courage of this talented company. The nanoparticles health monitoring watch that Google X announced to the world in 2014 is a high-tech product as what has been described in the movie at the beginning of this chapter. Although this product of Google X incorporates the emerging concepts of nanoparticles diagnostics and big data, it has inevitably adopted the combination model of "health monitoring + wearable", which is a new new concept that entrepreneurs in Silicon Valley favor. But Google is a big name after all. Great attention as well as considerable controversies have been aroused as soon as the Google project was announced. Now let's explain Google X's nanoparticle watch, or "nanoparticle Fitbit".

201

Andy Conrad, the head of the Google X Life Science Lab (in 2015, Google's life science unit was established separately as Verily after Google's reorganization.), once used a vivid example in a news interview: the current health and disease monitoring methods we are using now are like a person who wants to understand the culture of Paris, but only flies to Paris once a year by helicopter. The current monitoring methods are not real-time or continuous, but intermittently time-effective and conceivable, so it is likely that the collection of important information for early disease diagnosis is omitted.

As early as the 1960s, the filmmakers have described the stunning cone shape of a 21st-centuary nanoparticle watch. Conrad likens the health watch to be developed by Google X to Spock's artifact portable handheld scientific analyzer Tricorder in the famous science fiction movie "Star Trek". Tricorder is convenient and easy to carry, so Spock often holds it in his hand. And it is so powerful because it has three functions including detecting and scanning signals, analyzing data as well as recording and storing signals. For Conrad, Tricorder uses functional nanoparticles as sensors to monitor the signals sent by cells in the body, then analyzes the signals by the watch, and finally displays and stores the results that are closely related to health.

Conrad spent his whole life doing research related to health monitoring. He worked at LabCorp, the largest laboratory testing company in the United States. His career path was changed by a conversation with Google X's founder Brin and the team of brilliant scientists he led. This genius adventurer was deeply attracted by the scientific research environment with solid financial strength where talented people gathered. So he decided to join this mysterious organization to change the world with his intelligent experience and superb technology, and then made the magic Tricoder a reality. Conrad further explained the technical principles and some details of the nanoparticle health monitoring watch. Surface-functionalized nanoparticles can identify molecular markers on the surface of disease cells, and distinguish diseased cells from healthy cells.

Circulating Tumor Cells are the main targets of Nanoparticles. They are a type of cancer cells that are isolated from the primary tumor and entered the blood circulation. They are the seeds of cancer transmission as they can reach other parts of the body and organs with the flow of blood, induce the growth of new cancer cells, and thus cause the metastasis and spread of cancer in the body. Circulating Tumor Cells have been observed and discovered as early as 1869 by Thomas Ashworth. A century later, in 1990, Liberti and Terstappen confirmed that Circulating Tumor Cells exist in the development of early and middle diseases, and then people started to realize the importance of such cells. But because the number of Circulating Tumor Cells is very small, and only 0.01% of them can cause cancer metastasis and spread, it is difficult to detect such cells. Until the last 5 years, with the improvement of detection technology, Circulating Tumor Cells have gradually become a research focus. The academic community has also created a new term "Liquid Biopsy" using Circulating Tumor Cells as a disease diagnosis to distinguish it from traditional tissue biopsy techniques.

The nanoparticles that Conrad planed to use are surface-functionalized magnetic iron oxide nanoparticles. Iron oxide nanoparticles were chosen because they are FDA-approved imaging contrast media in the human body. Oral nanoparticle tablets enter the blood circulation system and follow the blood flow to anywhere in the body to capture and identify diseased cells. The watchband worn on the wrist created a magnetic field that brings the nanoparicles together. After that, the watch will ask those collected "secret nanoparticle services": "What have you seen? Have you located the cancer?" Then the analysis equipment in the ring will analyze the information collected by the nanoparticles to get the current health report. Next, the magnetic field on the watch was removed, and those "secret nanoparticle services" will return to the blood to perform the next task.

The Google X Lab has always been mysterious. The nanoparticles "Health Monitoring Watch Program" is no expectation. Before Conrad publicizing this program which sounds like a plot in the fiction movie at the press conference, few people know Google X's program which is great and combines the most popular technologies in Silicon Valley. In order to perform this program which looks like what in a sci-fi movie, Google X has convened a group of brilliant scientists in the field of disease diagnosis. Like other surreal cutting-edge ideas, nanoparticles watches have also been questioned. Some experts in academia have reservations about whether Google X can implement this program. Experts in Case Western Reserve believe that Google X's hypothetical nanopill will be in trouble as soon as it enters the human body. Nanoparticles or functional groups on the surface of the particles are likely to be destroyed by the stomach and gastrointestinal tact before entering the blood circulation.

Scientists of Google X have obviously noticed this problem, and are actively seeking solutions. Although they do not announce any specific plan, Google X has secretly partnered with Entrega, a Massachusetts startup. Entrega has developed a small pitch containing nanoparticles, which will be encapsulated. After being decomposed by digestive enzymes in the body, the small patches will adhere to the intestinal wall, and then the nanoparticles will be gradually released from them and successfully enter the bloodstream.

Although Entrega does not disclose its formula, academia has published articles on degradable capsules made from a mixture of carbopol, pectin and Carboxymethylcelllulose sodium. There is a protective layer of ethyl cellulose on the capsules, which prevents the capsules and pitches containing nanoparticles from being prematurely resolved by the enzymes of the digestive system. It is likely that Entrega has used similar techniques and formulations.

The wearable watch that can convene "secret nanaparticle service" as Conrad described to the public is just an idea by now. However, Google X has already been looking for expert guidance. Professor Steven Conolly of Berkeley, California, was invited to have a report on imaging of magnetic nanoparticles in mice at Google X. At present, Conolly's laboratory is able to detect nanoparticles with a mouse-sized scanner, but the FDA has not yet approved the imaging experiments of magnetic nonoparticles in human body. News reports that Philips has started research on a scanner for magnetic nanoparticle imaging in human body at its research and development center in Hamburg, Germany. However, Professor Conolly rigorously pointed out that experiments in the early stage are necessary for the confirmation of the accuracy and reliability of magnetic nanoparticles, even with the FDA approval to use them in human body. But Bajaji, a Google X chemist specializing in nanoparticles, claims that his team has proven that magnetic nanoparticles can accurately target and identify the target in vitro experiments, and now they are preparing for the experiments in simulated artificial limbs.

Chad Mirkin, a powerful professor at the University of Northwestern University in the United States, also expressed reservations about Google X's ambitious project. Everyone in the field of nanobiomedicine knows Mirkin, and he is one of President Obama's scientific advisers. Mirkin has set up three nanotechnology companies, one of which named Nanosphere has successfully commercialized the technology of disease detection using gold nanoparticles as probes. Gold nanoparticles can quickly detect infections from blood, saliva and urine. Mirkin's laboratory has already published over 2,000 academic papers on this technology. And the gold nanoparticle probe has been industrialized, which can supply thousands of hospitals in the United States at any time. However, very few hospitals are currently willing to buy this probe. Mirkin also said that it is great and courageous for Google to invest vigorously in the field of life science which is not Google's expertise. Google X's nanoparticle Fitbit project currently seems to require investments of many years, and no one can be sure that this project will eventually succeed. The challenges for development cannot be overlooked. It is nervous as well as exciting because of the uncertain future of nanoparticle watch. And it is possible that one day in the future this Google watch is on our wrist.

The Defense Advanced Research Project Agency (DARPA) also has a project similar to Google X's nanoparticle project. In addition to strongly supporting the development of unmanned vehicles, DARPA has also invested a lot of funds in recent years to support scientific research projects of nanoparticle probes for real-time health monitoring instruments. DARPA's idea is to let hydrosol particles enter the soldiers' bodies by intravenous injection or tablets. The hydrodol contains 5 to 6 different fluorescent nanoparticles. The surfaces of these nanoparticles have been functionalized, and different ligands can recognize different biomolecular markers in the body. The nanoparticles are also bound to a fluorescent quencher. The quencher can absorb The energy released after the fluorescent molecular is excited, so that the fluorescence cannot be released normally. That is, before the nanoparticle is combined with a disease-related marker, the fluorescence of nanoparticle is annihilated by the quencher, thus it does not emit light. However, when the nanoparticles and the related markers are combined, the fluorescent quencher is disconnected from the surface of the nanoparticle, so that the energy after the electron excitation can be released normally, and fluorescence that can be detected are also produced. DARPA wants to target the products of human metabolism, many of which are early sign of disease. Each fluorescent signal corresponds to a specific metabolite. A comprehensive analysis report of soldiers' daily health can be made through collecting and analyzing the fluorescent signals in their bodies with in vitro fluorescence detectors. If the content of a certain metabolite exceeds certain limits, it is very likely to be an early signal of the disease, and corresponding preventive or therapeutic measures should be taken. DARPA's confidentiality work could be the best in the United States, therefore, the actual progress of this project is almost unknown to the public.

Other Interesting Inventions: Nanoparticle Hydrosol Smoking Cessation Patches and Nanoneedle Insulin Patches

Another interesting technology about nanoparticle medical devices is the smoking cessation patch. This technology was developed by Matt Phar of the University of Illinois at Urbana-Champaign, and its prototype has now been developed. Quitting smoking is a painful process for smokers and has a high recurrence rate. Many ex-smokers made a promise to quit at the beginning, but later they either gave up lack of persistence or failed to strictly follow the course of treatment.

Nanoparticle hydrosol patches are mainly used by doctors to control those who want to quit smoking by taking medicine. The patch contains a large number of hydrosol nanoparticles with smoking cessation drugs, circuit equipment and small mechanical devices. Circuit equipment is mainly used for remote control and data upload, while the small mechanical equipment is mainly used to promote the release of drugs from hydrosol nanoparticles. The patient puts a smoking cessation patch on his arm, and the circuit inside the patch allows the doctor to operate and control the patch through a computer. At the appointed time, the doctor can remotely operate the patch to make the mechanical device inside the patch act, squeeze or quantitatively release the enzyme, rupture or decompose the hydrodol, and release the internal drug from the nanoparticles to penetrate the patch layer to the skin.

By using a smoking cessation patch, even if a patient forgets to take medicine, his or her doctor can still effectively control and monitor the patient, so that the treatment can normally continue. Each dose and time will be uploaded to the control device and recorded through the circuit equipment. At the same time, developers are also considering connecting data to smartphones, so that patients can know and record their daily medication usage through their mobile phones. Professor Pharr who developed this technology gave a related report at an academic conference in San Francisco in April 2015 and showed pictures of its prototype. Professor Pharr revealed that at present his team is actively negotiating with manufacturers to achieve mass production of the prototypes. Meanwhile, they are also working with investors to achieve the commercialization of smoking cessation patches in the next one or two years.

Zhu Gu is a Chinese professor at the University of North Carolina at Chapel Hill, and the nanoneedle insulin patch was developed by him and his laboratory. Gu's grandmother died of diabetes, so he has been working to develop more effective insulin injection technology against diabetes. Diabetic patients need to monitor blood glucose levels everyday and inject insulin multiple times a day, which are very cumbersome processes, and even if some patients have an insulin syringe pump in their body, it is likely to be overdosed or underdosed.

The nanoneedle patch developed by Professor Gu is only the size of a fingernail, and there are more than 100 nanoneedles in the patch. When the patient puts the patch on the skin, the nanoneedle will penetrate into the blood vessel and he or she will feel a temporary tingling. The nanoneddle is filled with "small pockets" containing insulin and enzymes. When the blood glucose concentration is too high, glucose in the blood will penetrate through the pores of the "small pockets". Enzymes in the "small pockets" convert glucose to acidic substances, break down the "small pockets", and inject insulin into the blood vessels through nanoneddles. The design of "small pockets" allows the whole insulin to be released slowly within an hour.

Professor Gu has tested nanoneedle insulin patches on five mice. The data show that the patch can successfully control blood glucose levels in mice for up to nine hours. Professor Gu has already begun doing experiments on pigs because the skin of pigs is more closer to human skin. His ultimate goal is to develop a medical patch that allows diabetic patients to change the patch every two or three days, and to regulate their blood sugar levels simply and accurately without any pain. Because of this invention, Professor Gu won the prestigious award of MIT 35 innovators under 35 in 2015.

From watches that can depict the real-time physical conditions of human body, to patches that intelligently control accurate drug release, nanoparticles have made outstanding achievements in the development and application of medical equipment. Nanoparticles are just like the omnipotent "secret services" with power to do anything. With the continuous innovation and development of nanotechnology, medical equipment are developing to be miniature, intelligent, and controllable. The goal of the future development of nanoparticle medical equipment is to be smaller, faster, as well as more accurate and intelligent. This century may be the era of the nanomedical revolution. Let us embrace the great technological innovation brought by those small nanoparticles with a throbbing heart and innovative mind.

单词释义：

1. 纳米颗粒	nanoparticle
2. 纳米磁性氧化铁纳米颗粒	magnetic iron oxide nanoparticle
3. 金纳米颗粒	gold nanoparticle
4. 荧光纳米颗粒	fluorescent nanoparticle
5. 纳米水溶胶颗粒戒烟贴片	nanoparticle hydrosol smoking cessation patches
6. 纳米针胰岛素贴片	nanoneedle insulin patches
7. 分子标记物	biomarker
8. 抗体	antibody
9. 契合寡聚物	aptamer
10. 配合基	ligand
11. 循环肿瘤细胞	circulating tumor cells

12. 原发肿瘤	primary tumor
13. 荧光湮灭物	fluorescent quencher
14. 卡波普	carbopol
15. 果胶	pectin
16. 羧甲基纤维素钠	carboxymethylcellulose
17. 乙基纤维素	ethyl cellulose
18. 液体切片	liquid biopsy
19. 显像试验	imaging experiment
20. 大内密探	secret service

译文解析：

1. 被动语态

汉语多用主动句，且多无主语句，尤其常见于科技文本中。而英语恰恰相反，一般英文科技文本中作者为了使文风更加严谨，往往大量使用被动语态以及 it 作形式主语的句型。因此在我们翻译中文句段里的句子，尤其是无主语句子时，可以根据实际情况以及中英语言习惯的差异进行适当的转变。

原文：将收集到的数据显示在腕表的电子显示屏上。

译文：The collected data will be presented on the electronic display of the watch.

分析：这句话的主语未知，我们可以将中文中的无主句译为英文的被动句，更符合原文意思。

原文：纳米颗粒的表面可以通过化学修饰加入各种化学基因。这些化学基因可以和抗体、契合寡聚物、配合基的化学基因反应，形成牢固的化学键。

译文：The surface of nanoparticles can be chemically modified to add various chemical groups, which can react with the chemical groups of antibodies, aptamers and ligands to form strong chemical bonds. In this way, these ligands that recognize molecular markers on the surface of cancer cells are distributed on the surface of the nanoparticles.

分析：在这个例子中，将"通过化学修饰"这一方式状语翻译成被动可以使句子更简洁，增强连贯性。除此之外，在这个例子中，我也将两个句子用定语从句的形式合为一句，这也也使得译文更通顺、简洁。

2. 逻辑关系

汉语中，句子各成分之间不通过语言形式手段连接，而是通过语言内在的逻辑关系和叙事的时间顺序来表达语法意义和逻辑意义。汉语是意合语言，运用一种隐性连接，注重语义连贯，句子各成分之间不用或少用关联词，句子结构松散。而英文句子是显性连接的,即用我们看得见的语言形式手段来连接词语或句子。这种构句特点往往注重句子形式,注重以形显意。因此,英语会大量使用关系词和连接词,来确保句子逻辑关系正确。这也就要求我们在进行汉英翻译时注意发现句子内暗含的逻辑关系,并运用适当的关系词来连接前后文。

原文: Tricorder 方便易携,(因此)时常被斯波克先生拿在手中,(2)功能强大,可以同时具有信号探测扫描数据分析和记录存储信号的三重功能。

译文: Tricorder is convention and easy to carry, so Spock often hold it in his hand. And it is so powerful because it has three functions including detecting and scanning signals, analyzing data as well as recording and storing signals.

分析: 这个例子的句意中有一个隐含的因果关系:因其方便易携,所以时常被拿在手中,在进行汉英翻译时我们应当将这一隐含的因果关系翻译出来,这样的译文更严谨。除此之外,原文中的这一个长句其实由两个意群组成,中文习惯使用逗号,而英文的句子更注重完整性,因此我将这个长句分成了两个短句,并加上连接词"and",使得前后联系更紧密。

原文: 纳米颗粒表面的高聚物可以防止纳米颗粒在血液循环过程中黏附在一起,(从而)实现纳米颗粒在人体内长时间的循环,完成收集情报信息的使命。

译文: The high polymer on the surface of the nanoparticles can prevent the nanoparticles from sticking together during the blood circulation, so that the long-term circulation of nanoparticles in our bodies can be realized and information can be collected.

分析: 这个例子中隐含了一个递进的关系,因此翻译时我在前后两句之间添加了一个"so that"短语,这样可以使前后逻辑更清晰,句子更完整。

3. 词性转换。

英汉造句方式上的差异必然导致了词类使用频率的不同。汉语多使用动词等来表达,英语则多使用名词和介词来表达。因为英语句子中的谓语动词及其重要且只有一个,因此我们要善于利用词性转换来使句子结构完整,表达意思清晰。切忌完全对照中文的词性来翻译。

原文: 它们能帮助纳米颗粒从茫茫细胞海洋中一眼识别出病变细胞。

译文: Nanoparticles can easily identify diseased cells from the vast ocean of cells with their help.

分析: 这个例子将动词"帮助"转换成了名词短语"with their help",因为科技文本多使用名词,更能体现其客观性、准确性。

4. 长难句

当原文句子较长时,可以根据各分句传达的信息群进行适当断句或合并。一般情况下,若各分句的主语各不相同,且各个主语之间没有紧密的内在联系时,最好划分开来,而不是生搬硬套地译为一个冗长的整句,这样会反而使句子主语指代不明。适当断句不仅使文章逻辑更清晰,还能提高其可读性。

原文: Entrega 开发了一种微小的含有纳米颗粒的贴片,这些小贴片被封装在胶囊内,(2)在被体内的消化酶分解后,小贴片就会附着在肠壁上,在几个小时内逐渐释放出纳米颗粒,使它们能够顺利进入血液循环。

译文: Entrega has developed a small pitch containing nanoparticles, which will be encapsulated. (2) After being decomposed by digestive enzymes in the body, the small patches will adhere to the intestinal wall, then the nanoparticles will be gradually released from them and successfully enter the bloodstream.

分析: 这个例子中的汉语长句其实由两个意群组成,因此将其翻译成英文时,我们可以将其分成两个句子。同时,第 2 句中也有一个暗含的先后顺序,在翻译时可以添加一个"then",使句子逻辑更清晰。

第十二章　心脏修复

用脂肪细胞修补一颗受伤的心：从脂肪组织分离并体外培养能够获得脂肪基质/干细胞，利用适当的生长因子和蛋白质调制成的"鸡尾酒"，可以诱导脂肪基质/干细胞分化成为心肌细胞和血管内皮细胞。

原　因

癌症已成为当今人类健康的最大威胁，人体的很多器官都有可以罹患癌症，我们听说过脑癌、喉癌、食道癌、胃癌、肝癌、胰腺癌，但是好像从来没有听说过心癌，难道心脏就不得癌症？ 要讲清楚这个问题，首先我们来看看什么是癌症。癌症是恶性肿瘤的一类统称，是一类生长异常、能够浸润（invade）或扩散（spread）到身体其他组织的细胞所引起的疾病。人们常说的肿瘤其实并不等于癌症。

良性肿瘤细胞与身体其他正常组织有明确的界限，不会扩散到身体其他部位，生长繁殖速度相对于恶性肿瘤更慢，细胞的分化程度也更高，比较接近正常的细胞。但这并不代表良性肿瘤是无害的，这类细胞仍然会释放出对身体其他组织（比如内分泌组织肿瘤可能会分泌过量激素）或者对神经系统有损害的物质。同时，肿瘤组织本身对身体其他组织也有压迫作用，可能引起组织的缺血坏死和器质性损伤。很多种良性肿瘤非常有可能发展成为恶性肿瘤。所以医生通常建议手术切除良性肿瘤。由于恶性肿瘤细胞自带生长因子，功能性受体和增殖基因表达增高，因此恶性肿瘤细胞具备干细胞（stem cell）特性的潜能，由于干细胞可以"制造"细胞，因此，具有这一特性使得恶性肿瘤细胞也可以源源不断地制造出更多的肿瘤细胞。另一方面，恶性肿瘤细胞与周围的组织没有明显的界线，很容易入侵或扩散到其他组织。因此不论是手术、化疗或者放疗都很难去除这些癌细胞，即使切除也很难彻底清除，极易复发。

为什么很少听说心脏会发癌症呢？ 一是因为心肌细胞是一种"终末分化细胞（terminal differentiated）"，人出生后就不再分裂增殖。二是因为心脏中的血流速度非常快，身体其他部位的入侵性或扩散性的癌细胞很少能转移到心脏中。但这并不代表心脏不会罹患癌症，虽然心脏的主要组成细胞是心肌细胞，但是心脏中有很多血管，血管容易受癌细胞入侵,血管肉瘤是心脏恶性肿瘤(癌症)中常见的一种,还有一种横纹肌肉瘤多发在婴幼儿身上。总体来说,心脏的恶性肿瘤(癌症)发病率比例还是非常低的,但是心肌梗塞的发病率在全球持续增长,是危及人类健康的一大疾病。

2011年世界卫生组织统计,显示每年因心肌梗塞死亡的人数超过900万人,位居全球十大死因榜单第二。目前,美国每年心肌梗塞的发病人数在150万人左右,我国的心肌梗塞发病率也呈上升的趋势,每年新增50万病人。比如著名相声演员侯耀文,还有朝鲜前主席金正日,灵魂音乐教父詹姆斯布朗(James Brown)。心肌梗塞或急性心肌梗塞,俗称心脏病,是由于部分心脏组织无法得到足够的血液供给,导致不可逆的心肌损伤而引起的。人体的很多其他细胞,比如皮肤、肝脏等,都可以持续分裂,但是心肌细胞在人出生后就停止分裂增殖,也就是说,人一生的心肌细胞数目是一定的,因此无法再分裂增殖的心肌细胞不会受到癌细胞的影响,这使得心脏发生癌症的几率非常小;另一方面,由于心肌细胞不再分裂增殖,一旦心肌受损,可真是件要命的事。那么受损的心肌真的无法修复吗?

创　　造

随着医学的进步,心脏病患者可以通过急救度过危险期,但度过危险期不等于万事大吉,至少有三分之一度过危险期的患者的心脏会越来越虚弱,这在医学上称为"心力衰竭"。心力衰竭患者,可以通过药物、心率调节器或植入型去颤器等治疗;但严重心力衰竭者,在这些治疗方法效果不佳的情况下,最后的方法只有心脏移植或者人工心脏移植。

2014年,全球心脏移植总共2174例,远远无法满足患者需求(想象一下漫长的等待队伍)。而人工心脏真的就是心力衰竭患者的救星了吗? 1982年,西雅图的一位心衰患者巴尼 克拉克(Barney Clark)第一次接受了人工心脏移植,存活了112天。1985年,美国印第安纳州的一位名叫Bill Schroeder的患者在接受人工心脏移植后存活了620天,于1986年死亡,创下了手术后存活时间最长的纪录。他们移植的名为"杰维克 −7(Javik-7)"的人工心脏,它极其复杂,需要在体内植入装置,用导线、管子与体外笨重的设备相连,患者只能躺在床上。而且,这类人工心脏只能暂时使用,病人仍需进行心脏移植手术。

2013年12月,世界首例由Carmat研发的永久性人工心脏移植手术,在法国乔治蓬皮杜医院进行。这是世界首例具备可以避免人体排异、以电池为动力,可人工智能调节供血节奏和远程监测的人工心脏,理论上使用可长达5年。然而2014年3月,乔治蓬皮杜医院通报了首例心脏移植手术患者去世的消息。患者依靠这颗人工心脏仅仅存活了75天。人工心脏最大隐患就是脑血管阻塞并由此引发中风瘫痪,"长期性的成功"还有待进一步验证。一些科学家们认为完全用仪器代替心脏没有必要,应该将重点放在研究如何帮助患病的心脏恢复功能。也许有一种方法,可以修补那颗受伤的心。

修复一颗能够跳动的心脏需要做哪些事情呢?首先要解决的是材料。用猪器官代替人器官进行移植的研究甚嚣尘上,且不说人们思想上能否接受自己的胸腔里跳动的是一颗猪心,单只看不同物种器官差异之大,还是同源性材料更合理,比如自体干细胞。科学飞速发展的今天,我们已经能够从人体获得一定数量的干细胞,自体干细胞的优点是来源广泛,且容易获得(抽血抽骨髓分离后都可以获得),并且不会产生自身排异反应。可以说,自体干细胞是修复心脏的"五彩石"!但是它依旧有缺点和局限性:数量有限、非常脆弱、增殖能力极强、易引发癌症。

目前体外培养心肌细胞来说也是完全可行的,可是并非培养出健康的心肌细胞就可以拥有健康的心脏,心脏是一个有着精密细胞结构和复杂血管分布的器官,心肌细胞除了需要有规则地排列生长成特定的精密结构,细胞之间还需要建立起物理性和神经性的联系,有了这种联系,细胞之间可以传导电信号,心肌才能够收缩。如果没有这种联系,那些细胞并不能被称为心脏,就像有些实验室声称可以在实验室制造大脑,其实不过是培养出了一堆神经细胞。

科学家们尝试了用各种方法来帮助细胞形成期望的器官。比如利用生物 3D 打印技术制造器官,这听起来很科幻,但其实从 20 世纪 90 年代起科学家们就开始实验了。最初是一些结构和功能都较为简单的器官,比如膀胱和气管,肾脏和心脏这样复杂的器官还有待研究。

2014 年,美国维克弗里斯特大学再生医学研究所的安东尼·阿塔拉(Anthony Atala),通过 3D 打印技术直接用活细胞打印出肾脏。但是这种方法打印的肾脏缺少血管和肾小管这样的内部通道,无法真正用于移植;接着,针对这问题,美国宾夕法尼亚大学的 Jordan Miller 和他的团队以及麻省理工学院的研究团队提出了一项解决方案,Miller 先用可溶性的糖打印出"糖血管和糖肾小管",然后把整个血管外包上细胞外间质和可以形成血管的内皮细胞,最后冲洗掉糖。之后细胞开始生长,形成强有力的血管,甚至在一些大血管周围自发的长出完美的微血管。这个发育接近成熟的器官能够被身体接受,自发进行细节的调整,具备完整的功能。但是 3D 打印心脏,目前来说还是相当有难度,现有的技术打印出的最小物体只能够达到毫米级别,而心脏中最细的血管宽度仅为微米级别(约 11000 毫米)。心脏中错综复杂的微小的血管网络正是确保器官健康的关键。

是金子就会发光的

2001 年,科学家们将两种干细胞应用到研究:胚胎干细胞和成体干细胞。胚胎干细胞的优点在于可以根据需要改造成任何一类细胞,用于器官或者是组织的再生,比如肝细胞、神经元或者是心肌细胞。过去使用的人类胚胎干细胞通常取自生殖中心的剩余受精卵,但是胚胎干细胞的使用在美国一直伴随伦理争议,甚至有宗教组织认为这是谋杀。作为回应,美国政府 2001 年 8 月,对人类胚胎干细胞的使用进行了限制,目前只有有限几种已经在实验室培养的胚胎干细胞仍可以在研究中使用。

胚胎干细胞可以无限扩增是因为含有一种抗衰老蛋白端粒酶,癌细胞也拥有这样的端粒酶,因而可以不断增殖,获得永生。虽然胚胎干细胞分化为成熟细胞后就不再含有这种端粒酶,也就失去了永生的能力。但是,难免有那么小部分仍具备永生能力的胚胎干细胞有形成肿瘤细胞的可能。而且,所有从胚胎干细胞获得的用于重塑的细胞都含有原来受精卵的遗传物质,因此接受由人类胚胎干细胞改造的组织或器官的病人,他们可能也需要接受免疫抑制治疗来防止排异,而感染的风险也随之增加。

既然胚胎干细胞的研究难以开展,科学家们转而把重点放到成体干细胞上。成体干细胞优点是,它们不具备永生的能力,因此不太可能形成肿瘤;它们不是从胚胎获得的,所以也不受伦理束缚;并且可以从病人本身获得,降低了免疫排异。但是成体干细胞也并非完美。与胚胎干细胞相比,它们更成熟,因此在可定向诱导分化的细胞种类上有一定限制。比如,取自骨髓的血液成体干细胞可以分化出红细胞和白细胞但是不能分化成心肌细胞。

举例来说,间充质干细胞是从骨髓中找到的另外一种成体干细胞,这类干细胞可以分化成骨、脂肪和软骨,但没有证据显示可以分化成为心肌细胞。另外,成体干细胞治疗还有一个重要的应用限制,就是我们的身体只有有限的成体干细胞,如果每次我们都取一些成体干细胞用于治疗,将面临影响它们本职工作的风险。譬如,即使来自骨髓的干细胞可以分化成心肌细胞,但如果从骨髓中分离出大量干细胞用于治疗心脏病,可能会影响到骨髓的首要工作—制造血细胞,这样做显然是得不偿失。

有科学家大胆地提出:为什么不用脂肪呢? 2001 年,加利福尼亚大学洛杉矶分校(UCLA)的细胞生物学家发表了一篇关于脂肪干细胞的文章。整形外科医生和科学家们分析了从吸脂手术中获得的脂肪,并从中获得了大量的成体干细胞! 这是一个非常重大的发现,毕竟成体干细胞的缺乏是阻碍器官再生影响力的重要原因。如果真的有那么一种简单的方法,从脂肪中就可以获得成体干细胞,那就不需要冒着损耗珍贵的骨髓,或者是损耗更稀少的其他成体干细胞资源的危险。因为获取脂肪干细胞听起来相对简单,脂肪就分布在我们的皮肤下面,然而真的如此简单吗? 脂肪干细胞从间充质干细胞分化而来,进一步可以分化成脂肪细胞。脂肪细胞主要分为给身体储存营养物质的白色脂肪细胞和给身体供暖的棕色脂肪细胞两类,另外还有一类叫作米色脂肪。

要如何从一团脂肪中获得干细胞呢? UCLA 的科研人员通过吸脂手术取得病人

的白色脂肪,除去成熟的脂肪细胞,这一步非常简单,因为脂肪的密度比水小,会悬浮在培养液上被滤去。初步处理后的脂肪细胞包含了未成熟脂肪细胞、内皮细胞,还有疤痕组织脂肪细胞。而脂肪干细胞很可能就隐藏在这些未成熟的脂肪细胞里! 为了验证这一猜想,研究人员先尝试用脂肪造骨和软骨。将初步处理后的脂肪置于模拟的体内环境中,也就是说用和体内同样的生长因子和蛋白来诱导骨髓中间充质干细胞分化为骨和软骨。这一策略成功了,初步处理过的脂肪细胞在与体内相同的生长因子刺激下,分化成为骨和软骨细胞。与一般成体细胞不同,这些初步处理后的脂肪细胞几周内数量就可以从一百万扩增为几千万,远远超过修补组织的需求。因此仅一次吸脂手术获得的脂肪细胞就足够制造出几层骨和软骨,这对广大骨折、骨质疏松和慢性关节炎的患者来说是一个绝好的福音。

　　将脂肪变成骨或软骨的"魔术"非常成功,但是,如何把脂肪变成修复心脏可用的细胞呢? 比如心肌细胞或者是形成血管内壁的内皮细胞。在正常成年哺乳动物的心脏中主要是心肌细胞、内皮细胞以及其他极少的一些辅助性功能细胞。内皮细胞是心血管系统的主要结构组成,是血管张力和附近细胞生长的动态调节器。内皮细胞可以为心肌细胞提供血氧和营养物质,还可以引导和促进心肌细胞发育、收缩、损伤后再生。各细胞间的相互协调作用对心脏的发育至关重要。

　　如何用脂肪修复一颗受损心脏呢? 首先,将吸脂手术分离得到的脂肪经过初步处理,去除成熟脂肪;接着调一杯"鸡尾酒",类似于脂肪干细胞分化成为脂肪细胞一样,心肌干细胞分化成心肌细胞也需要多种生长因子和蛋白质的参与,幸运的是心血管领域的研究人员已经揭示了这一"鸡尾酒"配方。但是这些因子用于脂肪变心肌还是第一次,因此需要科学家们不断调试配方的浓度、成分比例和处理时间的长短等。历经重重困难,最终,把脂肪细胞变成能够有节奏收缩的心肌细胞,就像真正的心肌细胞一样!

　　之前提到了,目前的 3D 打印还不能够用来制造心脏,因为心脏不是一个简单的中空结构的器官,心脏中有错综复杂的血管网络,而脂肪细胞除了可以被诱导生成心肌细胞,还可以生成血管的内皮细胞,用于修复心脏病造成的心脏损伤,它简直是为了修补心脏而生的。但是科学家们并不满足,不断探索更简单的方法。2003 年,印第安纳大学的研究人员设计了一个巧妙的实验,将吸脂手术获得的脂肪细胞经过初步处理后,加到在特殊介质上培养的内皮细胞中,进一步共同培养。结果血管样细管的生长增加了好几倍。而且,与内皮细胞单独形成的极细的细管不同,这些初步处理后的脂肪细胞生成了一种与血管非常相似的粗结构。

　　因为这类初步处理后的脂肪细胞中有多种与促进血管成活、再生、生长相关的因子。多种因子协同作用,不但激活了内皮细胞,也使得内皮细胞更具抗压能力。为突出这些脂肪细胞对于血管生成的"基质"或者是滋养功能的体现,研究人员正式命名这些细胞为"脂肪基质细胞"（adipose stromal/Stem cells ASCs）。

ASCs 有着不可思议的应用于心脏疾病治疗的天然优势，它并不盲目供给生长因子，而是通过细胞对于体内氧气量的探测来精确调控制造生长因子。心脏受血栓困扰的病人遭受着心脏和四肢肌肉组织缺氧，而在低氧环境中，ASCs 能够为血管生成提供双倍甚至是三倍的必需因子，这一功能简直是为了心脏病人量身定制的。研究人员在小白鼠上进行手术实验，首先阻断小白鼠腿部末端的血液供给，然后随机选择其中一半数量小白鼠接受人类 ASCs 腿部肌肉注射。与未接受 ASCs 注射的小白鼠相比，接受注射的小鼠更迅速地通过新生血管恢复了供血。

"逆生长"

科学家们除了摸索出用 ASCs 治疗心脏疾病，勤奋的他们通过不断探索，想要用成熟脂肪"制造"脂肪干细胞。还记得上一节中我们提到吸脂手术获得的脂肪需要去除掉大量的成熟脂肪细胞吗？科学家们觉得这些成熟的脂肪细胞就这么被废弃是件很可惜的事，于是期望通过"去分化"的手段变废为宝。"去分化"是一个相对于"分化"的过程，"分化"简单来说就是干细胞（以造血干细胞为例）变成特定细胞（比如血小板、T 细胞、巨噬细胞等）的过程；"去分化"指特定功能的细胞退回接近干细胞的状态。再打个不是特别恰当，但是很直观的比方，"分化"就好比是小蝌蚪变成青蛙，"去分化"就好比是青蛙变回小蝌蚪。在生物界存在的组织再生现象，很多都需要"去分化"参与。

科学家们发现脂肪细胞就具备"去分化"的能力。他们通过一种特殊的"天花板"体外培养方法，可以诱导成熟的脂肪细胞退回纤维细胞状的细胞状态，这类细胞表达干细胞特异性基因，弱表达或者不再表达成熟脂肪细胞的特性基因。

经分析，这些细胞开始表达脂肪干细胞特异性基因，包括 Yamanaka 转录因子。2012 年诺奖得主 Yamanaka 先生，凭借表皮细胞诱导出多能干细胞，轰动了整个科研界。他带领团队从胚胎干细胞特异性表达的基因中筛选出 24 个代表，对表皮细胞进行基因工程改造，使其可以表达这些基因，通过细胞学实验选择出改造后具有干细胞特性的表皮细胞，从而筛选得到 4 个具有诱导表皮细胞成为多能性干细胞特性的基因，称为 Yamanaka 因子。而与成熟脂肪细胞相关的基因开始渐渐下调，甚至不表达。这一结果预示着"变废为宝"的推测拥有极大可能性。最大程度利用分离的脂肪细胞，诱导其成为脂肪干细胞，进一步应用于心血管细胞的生成，进行血管再生和心肌修复。在谈"脂"色变的当下，脂肪总算被正了回名，关键时候可是救命的宝贝。

临床试验和治疗：理想和现实

继 2001 年 UCLA 的研究人员发布了他们的研究成果后,脂肪干细胞领域火热起来。科学家们最大的成就就是这一研究成果被转化成临床试验和治疗方法。现阶段,心血管疾病的病人可以接受自己的 ASCs 的移植,希望可以帮助血管再生或者是增强心脏功能。这些治疗也是有风险的,即便是用来自自身的 ASCs 注射到心脏中,也始终是"异物"。因此,虽然不会引起通常移植所引发的免疫排斥反应,但如果注射进心脏的细胞没有找到适合的支架,就可能会死亡,从而引起极具破坏力的炎症反应。另外, ASCs 在体内存活时间基本不超过一周,因此需要进行反复注射,这样重复的注射也会增加风险。

理想很丰满,现实很骨感,科研还是需要继续。同时科学家们也在思索,如果试验失败还能得到什么？可能关于 ASCs 再生机制的详细揭示将会是最大收获,包括:如何找出最有效的 ASCs？如何实现细胞和需要细胞的心脏区域之间的靶向输送？怎样才能确保细胞在体内维持和整合组织,从而保证正确蛋白的持续释放？如果能实现这样精准的治疗,应该能够挽救许多由于心脏缺血引起的组织损伤。

同时,这一研究给广大科研人员展示了更广阔的方向:成体干细胞可以被诱导分化为特定细胞,比如骨细胞和软骨细胞,或许可以用于慢性关节痛的治疗。而各种特定基质细胞可以用于血管重建,帮助器官缺血的病人减轻痛苦。脂肪细胞对于人造器官的研究也有重要帮助,因为血管生成是器官生成的首要保证,而脂肪细胞具有血管生成的能力,使其在人造器官研究中占据重要地位。发掘人类脂肪在再生医学领域的作用,这对人类有不可估算的价值和机会,有可能改变现代医学的面貌。

参考译文:

Repair of Heart

Repairing an injured heart with adipocyte or stem cells that can be isolated from adipose tissue and cultivated in vitro are able to be made into cocktail' using growth factor and protein. This kind of cocktail' can induce adipocyte or stem cells to differentiate into cardiac muscle cells and vascular endothelial cells.

Cause

Cancer has become the biggest threat to human health today. Many human organs can

grow cancer. We have heard brain cancer, throat cancer, esophageal cancer, stomach cancer, liver cancer and pancreatic cancer, but we have never heard about heart cancer. Does the heart have cancer? To get this straight, let us first look at what cancer is. Cancer is a general designation of malignant tumor. It is a disease that is caused by a kind of abnormally growing cells which can invade or spread to other organizations. It is often said that tumor is not equal to cancer.

Benign tumor cells differentiate from other normal tissues of body. They will not spread to other parts of body, and grow and multiply more slowly than malignant tumors. Their differentiation level is also high, which is closely to normal cells. But that does not mean benign tumors are harmless. These cells can still release substances that can cause damage to other parts of the body, or nervous system, for example tumors of endocrine tissue may secrete excessive hormones, or damage the nervous system. At the same time, tumor tissue itself also has a compressive effect on other tissues of the body, which may cause ischemic necrosis and organic damage. Many types of benign tumors are very likely to develop into malignancies. So doctors often recommend surgery to remove benign tumors.

Because malignant tumor cells have growth factors, functional receptor gene expression and proliferation become higher, so malignant tumor cells own potential characteristics of stem cells. Because stem cells can manufacture cells, this feature allows malignant cells to continuously create more tumor cells. Malignant cells, On the other hand, have no clear boundaries with surrounding tissues and are easy to invade or spread to other tissues. Therefore, no matter it is surgery, chemotherapy or radiotherapy, it is difficult to remove these cancer cells, even the resection is also very difficult to completely remove, prone to recurrence.

Why is it so rare to hear about heart cancer? One reason is that cardiac muscle cell is a kind of terminal differentiated cell that does not divide and multiply after birth. The second reason is that blood flow in the heart is so fast. Invasive or diffusible cancer cells from other parts of the body rarely reach the heart. However, this does not mean that the heart does not suffer from cancer. Although cardiac muscle cell is the main component of cardiac cells, the heart contains many blood vulnerable to invade by cancer cells. Hemangiosarcoma is a common type of cardiac malignant tumor, and there is a type of rhabdomyosarcoma that often occurs in infants and young children.

Overall, the incidence rate of heart malignant cancer is still very low, but the incidence of myocardial infarction continues to increase worldwide, which is a major threat to human health.

According to the world health organization in 2011, more than 9 million people die from heart attack each year, making it the second leading case of death globally. At present, the annual number of myocardial infarction in the United States is about 1. 5 million, and the incidence of myocardial infarction in China is also on the rise, with an annual increase of 500000 patients. Such as Hou Yaowen, a famous comedian, and the former north Korean President Kim jong ll and James Brown, godfather of soul music.

Myocardial infarction or acute myocardial infarction, commonly know as a heart disease, is because the part of the heart tissue can not have enough blood supply, leading to irreversible myocardial damage. A lot of other cells of body, such as skin, liver, etc. can continue to split, but myocardial cell stops division and proliferation after birth. That is to say, the number of myocardial cells in people life is certain, so cardiac muscle cell that can not split and proliferate will not be influenced by cancer cell, which makes a small percentage of cardiac cancer. On the other hand, since cardiac muscle cells no longer divide and proliferate, damage to the heart muscle can be deadly. So is the damaged heart muscle really beyond repair?

Creation

With the progress of medicine, patients with heart disease can pass the dangerous period through emergency treatment, but passing the dangerous period does not mean everything is fine, the heart of at least one third of the patients who have passed the dangerous period will be increasingly weaker, which is called "heart failure" in medicine. Patients with heart failure can be treated with drugs, heart rate regulators or implanted defibrillators. But for people with severe heart failure, when these treatments don't work as well, the only alternative is a heart transplant or an artificial heart transplant.

In 2014, there were 2,174 heart transplants worldwide, far less than patients' needs (imagine the long waiting lines). And is artificial heart really the savior for heart failure patients? In 1982, Barney Clark, a heart failure patient in Seattle, received his first artificial heart transplant and survived 112 days. In 1985, Bill Schroeder, a patient in Indiana, survived 620 days after receiving an artificial heart transplant. Their artificial heart, called the javik-7, is so complex that it requires implants, wires and tubes connected to a cumbersome device outside the body, and the patient has to lie in bed. Moreover, such artificial hearts can only be used temporarily and patients still need heart transplants.

In December 2013, the world's first permanent artificial heart transplant, developed by Carmat, was performed at Georges Pompidou hospital in France. This is the world's first

artificial heart that can avoid rejection, run on by batteries, adjust the rhythm of blood supply by artificial intelligence and can be monitored remotely. It could theoretically be used for up to five years. But in March 2014, the Georges Pompidou hospital reported the death of the first heart transplant patient. The patient survived only 75 days with the artificial heart. "Long-term success" of the artificial heart remains to be tested. Some scientists think it is unnecessary to replace the heart completely with instruments, but they should focus on how to help the diseased heart restore its function. Maybe there is a way to mend that broken heart.

What does it take to fix a beating heart? The first thing to deal with is the material. There is a growing research on the replacement of human organs with pig organs. Apart from whether people can accept that what beats in their chest is a pig heart. In the view of big differences among various species, it is more reasonable to use homologous materials, such as autologous stem cells. With the rapid development of science today, we have been able to obtain a certain number of stem cells from human body. The advantage of autologous stem cells is that they are widely sourced and easy to obtain (even after the separation of blood extraction and bone marrow extraction) without self-rejection. It can be said that autologous stem cells are "colorful stones" to repair the heart! But it still has drawbacks and limitations: limited in number, very fragile, highly proliferating and prone to cancer.

Now it was completely feasible to cultivate myocardial cells, but cultivating a healthy heart muscle cells does not mean that patients will have a healthy heart, the heart is a organ with precise cellular structure and complex blood vessels distribution. Myocardial cells not only have the need to be regularly arranged into specifically precise structure. But among cells, it is necessary to establish a link between physical and neurological. With this link, it is possible to transmit electrical signals among cells, thus heart muscle can contract. Without that connection, those cells can't be called hearts, just as some labs that claim to be able to make brains in the lab have simply grown a bunch of nerve cells.

Once you have the appropriate material, you need to think about how to transplant it to the site where it needs to be transplanted. Ideally, the patch connects to the surrounding healthy tissue, transmits electrical signals, and coordinates the contraction of cardiac muscle cells. But in reality the transplanted stem cells does not often have the good "relations" with the surrounding healthy tissue. Disappointingly, transplanted stem cells did not differentiate into myocardial cells to repair the increasingly thin ventricular, but the fiber cells without contraction ability try to repair infarction, which make up a sacar that does not have the ability of pulse (beating), leading to more severe myocardial infarction become more severe.

Scientists have tried various ways to help cells form the desired organ. Using bio-3d printing to create organs, for example, sounds like science fiction, but scientists have been experimenting since the 1990s. Initially, the organs were simple in structure and function, such as the bladder and trachea, while the organs as complex as the kidney and heart have yet to be studied.

In 2014, Anthony Atala of the Regenerative Iedicine Institute at Wake Forest university in the United States used 3D printing to print kidneys directly from living cells. But the printed kidneys lack internal channels such as blood vessels and tubules, they cannot be used for transplanting; Then, aiming at this problem, Jordan Miller and his team in the university of Pennsylvania and a team of researchers at the Massachusetts institute of technology have proposed a solution, Miller printed out the first "blood sugar and sugar renal tubular" using soluble sugar, then the whole vascular outsourcing could be formed on the outer cell mass and vascular endothelial cells, the sugar was finally rinsed. Then the cells began to grow, forming powerful blood vessels and even spontaneously growing perfect capillaries around some of the larger vessels. The organ that is nearly mature could be accepted by the body, making detailed adjustments spontaneously, and was fully functional.

But the skill of 3d-printed hearts is still difficult, with existing technology producing only the smallest objects on the millimeter scale, while the smallest blood vessels in the heart are only on the micron scale (about 11,000 millimeters). The intricate network of tiny blood vessels in the heart is the key to organ health.

Gold will Shine Everywhere

In 2001, scientists applied two types of stem cells to research: embryonic stem cells and adult stem cells. The advantage of embryonic stem cells is that they can be made into any type of cell needed to regenerate organs or tissues, such as liver cells, neurons or heart cells. Human embryonic stem cells used in the past were usually derived from the remaining fertilized eggs of reproductive centers, but the use of embryonic stem cells in the United States has been fraught with ethical controversy, even with religious groups calling it murder. In response, the U.S. government restricted the use of human embryonic stem cells in August 2001, and only a limited number of embryonic stem cells that have been grown in laboratories can still be used in research.

Embryonic stem cells can proliferate indefinitely because they contain telomerase, an anti-ageing protein that cancer cells also possess, allowing them to proliferate and live

forever. Although embryonic stem cells no longer contain the telomerase when they become mature, they lose the ability to live forever. But it is hard to avoid the small number of embryonic stem cells that remain immortal are likely to form tumors. Moreover, all cells derived from embryonic stem cells for remodeling contain genetic material from the original fertilized egg, so patients who receive tissue or organs modified from human embryonic stem cells may also need immunosuppressive therapy to prevent rejection, which increases the risk of infection.

Since embryonic stem cell research is difficult, scientists are focusing on adult stem cells. The advantage of adult stem cells is that they do not have the capacity to live forever and are therefore unlikely to form tumors: they are not derived from embryos and are therefore not ethically bound; And it can be obtained from the patient itself, reducing immune rejection. But adult stem cells are not perfect. Compared with embryonic stem cells, they are more mature, so there are limits on the types of cells that can be directed to induce differentiation. For example, blood adult stem cells derived from bone marrow can differentiate into red and white blood cells but not into cardiac muscle cells.

Mesenchymal stem cells, for example, are another type of adult stem cell found in bone marrow that can differentiate into bone, fat and cartilage, but there is no evidence showing that they can differentiate into cardiomyocytes. Another important limitation of adult stem cell therapy is that our bodies only have a limited number of adult stem cells, and if we take some of them every time for treatment, we will risk disrupting their job. For example, even if stem cells from the bone marrow can differentiate into cardiac muscle cells, isolating large numbers of stem cells from the bone marrow to treat heart disease could interfere with the bone marrow's primary job of making blood cells.

Some scientists boldly put forward the question: Why not use fat? In 2001, cell biologists at the university of California, Los Angeles (UCLA) published a paper about fat stem cells. Plastic surgeons and scientists analyzed the fat from liposuction and got a lot of adult stem cells from it! This is a very important discovery, after all, the lack of adult stem cells is an important reason to hinder organ regeneration. If there was a simple way to get adult stem cells from fat, there would be no risk of wasting precious bone marrow, or even rarer resources of other adult stem cells. Because getting fat stem cells sounds relatively simple. Fat is distributed under our skin, but is it really that simple?

Fat stem cells derived from mesenchymal stem cells can further differentiate into fat cells. Fat cells are divided into two types: white adipocyte, which stores nutrients in the body, and brown adipocyte, which heats the body. There is also a type called beige adipocyte.

How do you get stem cells from a lump of fat? The UCLA researchers used liposuction to take white fat from patients and remove mature fat cells, which is a simple step because the fat is less dense than water and will be suspended in a nuntrient solution and filtered out. The preliminarily treated fat cells included immature fat cells, endothelial cells, and scar tissue fat cells. And fat stem cells are probably hiding in these immature fat cells! To test this hypothesis, the researchers first tried to build bone and cartilage using fat. The preliminarily treated fat was placed in a simulated vivo environment, meaning that the same growth factors and proteins were used to induce the differentiation of bone marrow mesenchymal stem cells into bone and cartilage. The strategy worked, and the treated preliminarily fat cells differentiated into bone and cartilage cells stimulated by the same growth factors in vivo. Unlike normal adult cells, these treated fat cells can grow from a million to tens of millions in just a few weeks, far exceeding the need to repair tissue. So a single liposuction procedure can produce enough fat cells to produce several layers of bone and cartilage, which is a boon for extensive patients with fractures, osteoporosis and chronic arthritis.

The "magic" of turning fat into bone or cartilage has been very successful, but how to turn fat into cells that can be used to repair the heart? Such as heart muscle cells or endothelial cells that form the lining of blood vessels. The normal adult mammalian heart is dominated by cardiomyocytes, endothelial cells, and a few other helper cells. Endothelial cells are the major structural components of the cardiovascular system and are dynamic regulators of vascular tension and nearby cell growth. Endothelial cells provide blood oxygen and nutrients to cardiomyocytes, as well as guide and promote cardiomyocyte development, contraction, and regeneration after injury. The coordination of the cells is essential for the development of the heart.

How to repair a damaged heart with fat? First, the fat separated by liposuction operation was treated preliminarily to remove the mature fat. Just as fat stem cells turn into fat cells, heart muscle stem cells turn into heart muscle cells using a variety of growth factors and proteins. Fortunately, cardiovascular researchers have revealed the recipe for this "cocktail". But this is the first time that these factors have been used in adipogenic myocardium, so scientists have had to adjust the formula's concentration, composition and treatment time. Eventually, after much difficulty, the fat cells were transformed into cardiomyocytes that could contract rhythmically, just like real cardiomyocytes! Mentioned before, the current 3D printing also cannot be used to make the heart, because the heart is

not a simple hollow organs of the structure, there are intricate network of blood vessels in the heart, and fat cells besides can be generated induced myocardial cells, can also generate vascular endothelial cells which are used to repair heart damage caused by heart disease, it is produced to repair the heart. But scientists are not satisfied, and keep searching for simpler ways. In 2003, researchers at indiana university devised a clever experiment in which fat cells from liposuction, after preliminary treatment, were added to endothelial cells grown on a special medium for further co-culture. As a result, the growth of the blood-vesse-like tubes increased several times. And, unlike the tiny tubes that endothelial cells form alone, the treated fat cells produced a crude structure very similar to blood vessels.

Because there are many factors related to promoting vascular viability, regeneration and growth in these preliminary treated fat cells. The synergistic action of various factors not only activates the endothelial cells, but also makes the endothelial cells more resistant to stress. To highlight these fat cells for angiogenesis "matrix" or is a reflection of nourishing function, the researchers formally named these cells for "adipose stromal cells".

ASCs has the incredible natural advantage of being used in the treatment of heart disease. Instead of blindly supplying growth factors, it uses cells to detect the amount of oxygen in the body to precisely regulate the production of growth factors. Patients suffering from heart clots suffer from a lack of oxygen in their heart and limb muscles. In a hypoxic environment, ASCs can provide twice or even three times the necessary factors for angiogenesis, a function tailored for heart patients. The researchers performed the surgery on mice, first by blocking blood supply to the ends of their legs, and then randomly selecting half of them to receive an intramuscular injection of human ASCs into their legs. Compared with mice that did not receive ASCs, the mice that received the ASCs were able to restore blood supply more quickly through the new vessels.

"Reverse Growth"

In addition to figuring out how to treat heart disease with ASCs, scientists worked diligently to "make" fat stem cells from mature fat. Remember in the last video when we talked about liposuction fat you need to get rid of a lot of mature fat cells? The scientists thought it was a pity that these mature fat cells were discarded, so they hoped to make them valuable by dedifferentiation.

"Dedifferentiation" is a process different from "differentiation". "differentiation" simply means that stem cells (take hematopoietic stem cells as an example) become specific cells (such as platelets, T cells, macrophages, etc.). Dedifferentiation refers to the return of cells with a specific function to a state similar to stem cells. It's not quite appropriate, but it's pretty intuitive. Differentiation is like a tadpole turning into a frog, and dedifferentiation is like a frog turning back into a tadpole. Many of the tissue regeneration phenomena that exist in the biological world require the participation of "dedifferentiation".

For example, earthworms cut in two can grow into two individual earthworms, or salamanders can grow new tails and limbs when their original organs are damaged or lost. The mechanism of "dedifferentiation" allows a large number of differentiated cells to return to a state similar to that of stem cells when stimulated by the outside world, regain the ability to reproduce and differentiate, and according to the needs of the body, produce a large number of cells with specific functions, so as to grow new organs.

Scientists have discovered that fat cells have the ability to dedifferentiate. They used a special "ceiling" in vitro culture method to induce mature fat cells to fall back into a fibro-like state, expressing stem-cell-specific genes and either weakly or no longer expressing the characteristic genes of mature fat cells.

"Ceiling" culture method takes advantage of the mature fat cells suspended in the characteristics of the cultures of the upper, as described in the section, mature fat cells were isolated from adipose tissue, in cell culture bottles, filled with culture, fat cells are sent to near the culture bottle "ceiling" position, stick on wall of the culture bottle with growth period, mature fat cells began to stick wall to grow, gradually show the structure of the fiber cells.

After analysis, these cells began to express fat-stem-cell-specific genes, including the Yamanaka transcription factor. Mr Yamanaka, who won the Nobel Prize in 2012, stunned the scientific world when he induced pluripotent stem cells from skin cells. He led the team from embryonic stem cells into specific expression genes selected 24 representative, to genetically engineered epidermal cells, make its can express these genes, selected by cytological experiments modified epidermal cells with stem cell properties, thus screening four with induction of epidermal cells into pluripotent stem cell properties gene, called Dr Yamanaka factor. Genes associated with mature fat cells began to be down-regulated or even not expressed.

This result indicates that the speculation of "turning waste into treasure" has a great possibility. To make the best use of the isolated fat cells, induce them to become fat stem cells, and further apply them to the generation of cardiovascular cells for angiogenesis and myocardial repair. At present, people feel frightful when talking about "fat", fat was finally back to the name, and saving people at the key time .

Clinical Trials and Treatments: Ideal and Reality

After UCLA researchers published their findings in 2001, the field of fat stem cells took off. The biggest achievement for scientists is that this research has been translated into clinical trials and treatments. At present, patients with cardiovascular disease can receive their own ASCs transplants, hoping to help regenerate blood vessels or improve heart function.

These treatments are also risky. Even when injected into the heart with ASCs from the body, they are still "foreign". So, while it won't cause the immune rejection normally associated with transplants, if the cells which are injected into the heart can't find a suitable scaffold, they can die, leading to a destructive inflammatory response. In addition, ASCs rarely survive more than a week in the body, so repeated injections are needed, which increases the risk.

The ideal is plump and the reality is bony, scientific research need to continue. At the same time, scientists are wondering, what if the experiment fails? Perhaps the biggest gain would be a detailed explanation of the regenerative mechanism of ASCs, including: how to find the most effective ASCs? How do you achieve targeted delivery between the cell and the heart region that needs the cell? How do you ensure that cells maintain and integrate tissue in the body so that the right proteins are released continuously? If such a precise treatment can be achieved, it should be able to save a lot of tissue damage caused by heart ischemia.

At the same time, the study shows that adult stem cells can be induced to differentiate into specific cells, such as bone cells and chondrocytes, and may be used to treat chronic joint pain. A variety of specific stromal cells can be used to rebuild blood vessels and help alleviate pain in patients with organ ischemia. Fat cells are also important to the study of artificial organs, because angiogenesis is the primary guarantee of organ growth, and the ability of fat cells to generate blood vessels makes them occupy an important position in the study of artificial organs. Exploring the role of human fat in the field of regenerative medicine has an inestimable value and opportunity for human beings and may take on a new look of modern medicine.

单词释义：

1. 脂肪细胞   adipocyte
2. 脂肪组织   adipose tissue

3. 心肌细胞	cardiac muscle cells	
4. 血管内皮细胞	vascular endothelial cell	
5. 分化	differentiate	
6. 食道癌	esophageal cancer	
7. 胰腺癌	pancreatic cancer	
8. 恶性肿瘤	malignant cancer	
9. 良性肿瘤细胞	benign tumor cells	
10. 缺血坏死	ischemic necrosis	
11. 器质性损伤	organic damage	
12. 分泌组织	secretory tissue	
13. 化疗	chemotherapy	
14. 放疗	radiotherapy	
15. 血管肉瘤	hemangiosarcoma	
16. 横纹肌肉瘤	rhabdomyosarcoma	
17. 终末分化细胞	teminal differentiated cells	
18. 心肌梗塞	myocardial infarction	
19. 心肌损伤	myocardial damage	
20. 分裂增殖	split and proliferate	
21. 心力衰竭	heart failure	
22. 心率调节器	heart rate regulators	
23. 植入型去颤器	implanted defibrillators	
24. 自体干细胞	autologous stem cells	
25. 同源性材料	homologous materials	
26. 自体干细胞	autologous stem cells	
27. 心肌细胞	myocardial cells	
28. 自身排异反应	self-rejection	
29. 心肌梗塞	myocardial infarction	
30. 膀胱和气管	the bladder and trachea	
31. 肾小管	tubules	
32. 可溶性的糖	soluble sugar	
33. 血管的内皮细胞	vascular endothelial cells	
34. 微血管	capillaries	
35. 胚胎干细胞	embryonic stem cells	
36. 成体干细胞	adult stem cells	

37. 剩余受精卵　　　　　remaining fertilized eggs

38. 端粒酶　　　　　　　telomerase

39. 免疫抑制治疗　　　　immunosuppressive therapy

40. 骨髓　　　　　　　　bone marrow

41. 成体干细胞　　　　　adult stem cells

42. 间充质干细胞　　　　Mesenchymal stem cells

43. 软骨　　　　　　　　cartilage

44. 脂肪干细胞　　　　　fat stem cells

45. 吸脂手术　　　　　　liposuction

46. 米色脂肪　　　　　　beige adipocyte

47. 培养液　　　　　　　nutrient solution

48. 内皮细胞　　　　　　endothelial cells

49. 疤痕组织脂肪细胞　　scar tissue fat cells

50. 软骨　　　　　　　　cartilage

51. 骨折　　　　　　　　fractures

52. 骨质疏松　　　　　　osteoporosis

53. 慢性关节炎　　　　　chronic arthritis

54. 辅助性功能细胞　　　helper cells

55. 心血管系统　　　　　cardiovascular system

56. 脂肪变心肌　　　　　adipogenic myocardium

57. 血管生成　　　　　　angiogenesis

58. 脂肪基质细胞　　　　adipose stromal/Stem cells ASCs

59. 血栓　　　　　　　　heart clots

60. 去分化　　　　　　　dedifferentiation

61. 血小板　　　　　　　Platelets

62. 巨噬细胞　　　　　　macrophages

63. 干细胞特异性基因　　stem-cell-specific genes

64. 培养液　　　　　　　culture

65. 转录因子　　　　　　transcription factor

66. 多能干细胞　　　　　pluripotent stem cells

67. 细胞学　　　　　　　cytological

68. 血管再生　　　　　　angiogenesis

69. 炎症反应　　　　　　inflammatory response

70. 靶向输送　　　　　　targeted delivery

译文解析:

1. 语态的转换。

科技英语中的被动语态使用极为广泛,在汉译英时,多把主动变被动。运用客观事物的角度,事实方法或特征,避免用第一第二人称。

原文: 用脂肪细胞修补一颗受伤的心:从脂肪组织分离并体外培养能够获得脂肪基质/干细胞,利用适当的生长因子和蛋白质调制成的"鸡尾酒"……

译文: Repair an injured heart with adipocyte or stem cells that can be isolated from adipose tissue and cultivated in vitro are able to be made into 'cocktail' using growth factor and protein.

分析: 原文的"从脂肪组织分离"译为"can be isolated from adipose tissue";调制成的"鸡尾酒"译为"are able to be made into cocktail"。

2. 句子的逻辑结构。

在进行长句翻译时,要弄懂句子结构,按照逻辑进行翻译。

原文: 要癌症是恶性肿瘤的一类统称,是一类生长异常、能够浸润(invade)或扩散(spread)到身体其他组织的细胞所引起的疾病。

译文: It is a disease that is caused by a kind of abnormally growing cells which can invade or spread to other organizations.

分析: 原文中,是一类后面是对前文癌症的解释说明,为使结构清晰易懂,就可以采用定语从句,把原文中分开的句子合并到一起,更符合英语的文本结构。

3. 连接词的使用。

英语属于形合语言,句与句之间的关系要用连接词表达出来

原文: 良性肿瘤细胞与身体其他正常组织有明确的界限,不会扩散到身体其他部位,生长繁殖速度相对于恶性肿瘤更慢,细胞的分化程度也更高,比较接近正常的细胞。

译文: Benign tumor cells differentiate from other normal tissues of body. They will not spread to other parts of body, and grow and multiply more slowly than malignant tumors. Their differentiation level is also high, which is closely to normal cells.

分析: 原文"比较接近正常的细胞"的主语是前面的"良性肿瘤细胞",在中文中,没有体现,但读者都知道主语是什么,是因为中文是意合语言,是通过语境体现出来的,但是英语是形合语言,要多用连接词,所以在译时,用了一个非限制性定语从句来体现此处的主语以及逻辑关系。

4. It is 句型的使用。

用 It is 句型来平衡句子结构,避免主语冗长,句子结构偏颇

原文: 因此不论是手术、化疗或者放疗都很难去除这些癌细胞,即使切除也很难彻底清除,极易复发。

译文: Therefore, no matter it is surgery, chemotherapy or radiotherapy, it is difficult to remove these cancer cells, even the resection is also very difficult to completely remove, prone to recurrence.

分析: 原文中主语是"手术、化疗或者放疗",为了避免主语过长,翻译时把主语都放在谓语动词前,整个句子变得头重脚轻,所以此处在处理时运用了 it is 句型用以平衡句子结构。

5. 语序的转换。

在中文中,定语一般放在被修饰词的前面,在科技英语中,定语后置的情况居多,定语尽量放在所修饰词的后面,符合译入语的语言习惯。

原文: 二是因为心脏中的血流速度非常快,身体其他部位的入侵性或扩散性的癌细胞很少能转移到心脏中。

译文: The second reason is that blood flow in the heart is so fast. Invasive or diffusible cancer cells from other parts of the body rarely reach the heart.

分析: 原文中"身体其他部位的"是"癌细胞"的定语,放在了所修饰词的前面,译文中,"from other parts of the body"放在了"Invasive or diffusible cancer cells"的后面,符合译入语的语言习惯,实现了功能对等。

6. 句子的整合。

科技英语大量运用长句,当原文是含有逻辑关系的分句时,在翻译时用连接词将它们整合成长句。

原文: 总体来说,心脏的恶性肿瘤(癌症)发病率比例还是非常低的,但是心肌梗塞的发病率在全球持续增长,是危及人类健康的一大疾病。

译文: Overall, the incidence rate of heart malignant cancer is still very low, but the incidence of myocardial infarction continues to increase worldwide, which is a major threat to human health.

分析: 原文有转折的关系,也存在限定关系,所以译文用了"but"和"which"引导的非限制性定语从句,把其整合成一个长句,更符合科技英语的文本特点,增强其客观性。

7. 句子的意译。

由于英汉两种语言文字之间所存在的悬殊差异,在实际翻译过程中我们很难做到词字上的完全对应,因此,为了准确地传达出原文的信息,译者往往需要对译文做一些增添或删减。

原文：2011 年世界卫生组织统计,显示每年因心肌梗塞死亡的人数超过 900 万人,位居全球十大死因榜单第二。

译文：According to the world health organization in 2011, more than 9 million people die from heart attack each year, making it the second leading case of death globally.

分析：原文想表达的是心肌梗塞位居全球十大死因榜单第二,所以在翻译时,把省略掉的代指心肌梗塞的 it 进行增译,补充出来,让译入语读者对内涵更加明白了解,这符合译入语的表达习惯,实现了功能对等。

8. 非谓语动词的使用。

科技英语翻译中,广泛使用非谓语动词,动词不定式、分词、动名词。

原文：心肌梗塞或急性心肌梗塞,俗称心脏病,是由于部分心脏组织无法得到足够的血液供给,导致不可逆的心肌损伤而引起的。

译文：Myocardial infarction or acute myocardial infarction, commonly know as a heart disease, is due to the part of the heart tissue can not have enough blood supply, leading to irreversible myocardial damage.

分析：原文中的"导致"就运用了非谓语动词现在分词表示主动,在翻译时译为"leading",符合科技英语的翻译习惯。

9. 翻译中的名词化。

在科技英语中,广泛使用名词化结构,即用名词短语表示动作,用名词短语表示调价、原因、目的、时间等状语从句。

原文：体的很多其他细胞,比如皮肤、肝脏等,都可以持续分裂,但是心肌细胞在人出生后就停止分裂增殖。

译文：A lot of other cells of body, such as skin, liver, etc. can continue to split, but myocardial cell stops division and proliferation after birth.

分析：原文中的"停止分裂增殖",是动词,在翻译时,为符合科技英语客观的要求,译为了"stop division and proliferation"名词,符合科技英语的翻译原则。

10. 逻辑顺序的调整。

在科技英语长句翻译时,当原文的逻辑顺序与译入语的逻辑顺序相同时,采用顺译法就可以。

原文：随着医学的进步,心脏病患者可以通过急救度过危险期,但度过危险期不等于万事大吉,至少有三分之一度过危险期的患者的心脏会越来越虚弱,这在医学上

称为"心力衰竭"。心力衰竭患者，可以通过药物、心率调节器或植入型去颤器等治疗；但严重心力衰竭者，在这些治疗方法效果不佳的情况下，最后的方法只有心脏移植或者人工心脏移植。

译文： With the progress of medicine, patients with heart disease can pass the dangerous period through emergency treatment, but passing the dangerous period does not mean everything is fine, at least one third of the patients who have passed the dangerous period will be increasingly weak heart, which is called "heart failure" in medicine.

分析： 原文的逻辑是递进的，首先阐释可以通过急救度过危险期，但又说不等于万事大吉，然后摆出事实，最后对其下定义，原文的逻辑与译入语的逻辑顺序一致，所以采用顺译法就可以。

11. 合译法。

科技英语翻译时可采用合译法，把分句合译成一个长句。

原文： 另外，成体干细胞治疗还有一个重要的应用限制，就是我们的身体只有有限的成体干细胞，如果每次我们都取一些成体干细胞用于治疗，将面临影响它们本职工作的风险。

译文： Another important limitation of adult stem cell therapy is that our bodies only have a limited number of adult stem cells, and if we take some of them every time for treatment, we will risk disrupting their job.

分析： 原文的分句，运用拆分重组，重新整合翻译的顺序，整改后第一句的主语变为"Another important limitation of adult stem cell therapy"，这是科技英语中长句翻译时常用的方法，对原文整句话进行重组再造，尽量符合译入语的语言习惯。

12. 增译法。

使用增译法把原文内含的逻辑结构译出来，因为英语是形合语言，所以必要的连接词是不可以像中文一样省略的。

原文： 一些科学家们认为完全用仪器代替心脏没有必要，应该将重点放在研究如何帮助患病的心脏恢复功能。

译文： Some scientists think it is unnecessary to replace the heart completely with instruments, but should focus on how to help the diseased heart restore its function.

分析： 原文两个分句中有转折关系，在中文中，依靠语境，读者可以得出，在翻译中就要把"but"的转折翻译出来，这样才符合译入语的表达习惯。

13. 分译法。

当原文有多个分句，逻辑不够清晰时，翻译时就应采用分译法，将原文逻辑重新归纳处理。

原文：科学飞速发展的今天，我们已经能够从人体获得一定数量的干细胞，自体干细胞的优点是来源广泛，且容易获得（抽血抽骨髓分离后都可以获得），并且不会产生自身排异反应。

译文：With the rapid development of science today, we have been able to obtain a certain number of stem cells from human body. The advantage of autologous stem cells is that they are widely sourced and easy to obtain（even after the separation of blood extraction and bone marrow extraction）without self-rejection.

分析：原文全是用逗号隔开，逻辑不够清晰，如果翻译完，还是这么多分句，容易让以英文读者混淆，并且科技英语要保证译文的逻辑清楚易懂，所以在翻译时，分成了两句话，这样的逻辑更清楚。

14. 名词化结构。

科技英语中多实用名词化结构，增强科技英语的客观性。

原文：2003 年，印第安纳大学的研究人员设计了一个巧妙的实验，将吸脂手术获得的脂肪细胞经过初步处理后，加到在特殊介质上培养的内皮细胞中，进一步共同培养。

译文：In 2003, researchers at indiana university devised a clever experiment in which fat cells from liposuction, after preliminary treatment, were added to endothelial cells grown on a special medium for further co-culture.

分析：原文中的"将吸脂手术获得的脂肪细胞经过初步处理后"，翻译成"after preliminary treatment"，将动词结构译为名词化结构，符合科技英语中的客观性原则。

15. 顺译法。

长句的翻译，当原文的逻辑顺序与译入语的逻辑顺序一致，采用顺译法

原文：而且，所有从胚胎干细胞获得的用于重塑的细胞都含有原来受精卵的遗传物质，因此接受由人类胚胎干细胞改造的组织或器官的病人，他们可能也需要接受免疫抑制治疗来防止排异，而感染的风险也随之增加。

译文：Moreover, all cells derived from embryonic stem cells for remodeling contain genetic material from the original fertilized egg, so patients who receive tissue or organs modified from human embryonic stem cells may also need immunosuppressive therapy to prevent rejection, which increases the risk of infection.

分析：原文中前置定语翻译为后置定语，"derived from"做"all cells"的后置定语，同样原文中的"接受由人类胚胎干细胞改造的组织或器官"前置定语在翻译时后置，由于原文的逻辑顺序与译入语的逻辑顺序一致，采用了顺译法，不需要改变句子的顺序。

16. 定语的处理。

科技英语中定语从句的灵活应用。

原文：要如何从一团脂肪中获得干细胞呢？UCLA 的科研人员通过吸脂手术取得病人的白色脂肪,除去成熟的脂肪细胞,这一步非常简单,因为脂肪的密度比水小,会悬浮在培养液上被滤去。

译文：How do you get stem cells from a lump of fat? The UCLA researchers used liposuction to take white fat from patients and remove mature fat cells, which is a simple step because the fat is less dense than water and will be suspended in a nuntrient solution and filtered out.

分析：中文是意合语言,所以逻辑关系都与蕴含在句与句之间,但是英语是形合语言,句与句之间的逻辑关系要用连接词明确的表现出来,这句话中间存在限定关系,所以加一个非限定性定语从句就更能清晰地表达出它们之间的逻辑关系,符合科技英语逻辑清晰的文本特点。

17. 变序法。

在翻译科技英语长句时,当原文逻辑顺序与译入语逻辑顺序不一致时,采用变序法进行长句翻译。

原文：历经重重困难,最终,把脂肪细胞变成能够有节奏收缩的心肌细胞,就像真正的心肌细胞一样!

译文：Eventually, after much difficulty, the fat cells were transformed into cardiomyocytes that could contract rhythmically, just like real cardiomyocytes!

分析：前文中的"最终"是整个句子的大前提,所以在翻译时提前在整个句子之前,"Eventually"提前,并且原文中的"历经重重困难"翻译成"after much difficulty"名词化结构,符合科技英语中的客观性特点。

18. 固定搭配的使用。

科技英语翻译时常用许多固定搭配,某些固定搭配的灵活使用。

原文：因为这类初步处理后的脂肪细胞中有多种与促进血管成活、再生、生长相关的因子。

译文：Because there are many factors related to promoting vascular viability, regeneration and growth in these preliminary treated fat cells.

分析：原文中陈述客观事实,翻译时就可以用"there be"句型表存在,很多句型的灵活运用对科技英语的翻译起重要作用。

第十三章 人体增强

外骨骼、动力机甲，人与智能机器走向"合体"

什么是机器人外骨骼——从科幻到现实

人因为身体的先天限制因素，没法像猎豹一样快和敏捷，更没法像蚂蚁那样扛比自身重好几倍的东西。小时候观看的动画片中，"圣斗士"们穿上"圣衣"后，战斗力激增，能够挫败敌手，维护安宁。不妨设想，有没有类似的装置，人穿上后能变得更强更敏捷，甚至使普通士兵变成神勇无敌的"圣斗士"呢？许多电影里就有类似的情节，比如在影片《极乐空间》（Elysium）中，男主角 Max 本已因受强辐射而身体虚弱，装上外骨骼后却能够和反面劲敌 Kruger 肉搏激战；再如影片《明日边缘》（Edge of Tomorrow）中，汤姆·克鲁斯作为人类士兵身穿单兵机甲与外星生物大战；电影《阿凡达》（Avatar）更构想了体型巨大、可让人坐入其中操控的 AMP 战斗机甲。还有，差点忘了说钢铁侠套装，但是这款装备过于科幻，既能飞又能发射手炮，胸口还有一个一辈子不用交电费的小型核反应堆提供能源。相比之下，还是前面几个比较接地气。

够了够了，别提这么多虚幻的电影了，现实中类似的技术到底发展得怎样呢？别急，咱们先介绍下基本概念。其实影片中出现的，穿在人身上的装置叫做机器人外骨骼，它能通过机械系统为人助力，其结构酷似节肢动物（如螃蟹）的坚硬外壳（学名为外骨骼，即骨头长在肉外面），而且在技术上属于机器人的范畴，因此得名。其中偏军用的装置有时也叫动力装甲。而尺寸较大、功能更强的，尤其是人可以坐在里面操控的称为机甲。机器人外骨骼目前主要应用于医疗康复、救援、工程作业以及军事等方面。

机器人外骨骼系统一般包括机械结构、传感部分、动力与传动部分、能源部分和控制部分。机械结构为整个系统提供结实的支撑，并通过绑带或其他方式固连在人身上来分担承重以及提供发力的基础。下半身型外骨骼与人身体固连部位主要是腰部和腿部；全身型外骨骼的固连部位除了腿部和腰部，还包括上肢和躯干。传感器和信号处理电路构成了传感部分，以采集人体运动趋势、位姿与力量等信息，为控制部分提供判断依据。动力与传动部分一般由电机、液压元件或气动元件提供驱动力或力矩,再通过传动元件传至机械结构,从而使外骨骼做出动作。多数外骨骼系统会采用电池提供总能源,但现有的电池几乎都不足以维持系统长时间高负荷工作,又不可能过分增大电池容积(过重,且外骨骼有尺寸限制),因此有些外骨骼会采用燃油和小型内燃机提供能源和原始动力。

控制部分的核心是微型电脑与控制软件,它能综合传感部分传来的信息,按照人的意图指挥动力传动部分。下面我们来看看当前世界范围内几个极具代表性的产品。

日本 HAL 系列康复/作业用外骨骼

众所周知,日本的机器人行业非常发达,机器人外骨骼技术也不在话下。其中最有代表性的当属日本筑波大学和日本科技公司"赛百达因"(Cyberdyne)联合开发的HAL 系列外骨骼。它有两个主要的版本:下半身型 HAL-3 和全身型 HAL-5。其功能定位是辅助行动受障碍的人士,或者助力强体力作业(比如救援工作需要搬开重物)等。

最早的原型是由现为日本筑波大学教授的 Yoshiyuki Sankai 提出的。早在 1989年,他获得机器人学博士学位后就开始了设计工作。他先用了 3 年时间整理绘制了人体控制腿部动作的神经网络,之后又用了 4 年时间制作了一部硬件原型机。它由电机提供动力,并通过电池供电。早期版本的重量很大,光是电池就有 22 公斤,需要 2 名助手帮助才能穿上,而且要连接至外部电脑,因此很不实用。最新的型号在重量方面有了很大改善,整套 HAL-5 才重 10 公斤,而且电池和电脑被集成环绕在腰间。HAL系列外骨骼的控制方式最有意思,不过在深入展开之前,咱们还是先了解下人是如何控制身体运动的吧。

当人想让身体做出动作的时候,脑部会产生控制信号,并通过运动神经传送至相应肌群,从而控制肌肉和骨骼的运动。这些神经信号多少会扩散到皮肤表面,形成表面肌电信号。虽然很微弱,但仍能被电子电路检测到。HAL 系列外骨骼通过表贴在人皮肤的传感器采集这些信号,控制外骨骼做出和人相同的动作,从而为人的行动助力。对于身患残疾或肢体运动障碍的使用者,这是很巧妙的办法。HAL 系列外骨骼目前已经在医疗机构大量使用,取得了一系列巨大的成功,于 2012 年 12 月获得国际医疗器械设计制造标准认可(ISO 13485),又于 2013 年 2 月获得国际安全性证书(世界第一款获此认证的动力外骨骼),并于同年 8 月获得 EC 证书获准在欧洲进行医疗应用(同类医疗用机器人中获准的第一款)。

日本 T52 Enryu 工程/救援机甲

日本还有个身躯庞大的机甲——T52 Enryu。它高 3.5 米,宽 2.4 米,重达 5 吨。两个胳膊各 6 米长,总共能抬起 1 吨重的负荷。强大的力量来源于液压驱动,而能源是柴油。它可由人坐在里面直接操控,也可远程遥控(装有摄像头辅助)。它于 2004 年由日本机器人公司 TMSUK 主要开发,设计目的是用于灾难救援,如地震、海啸和车祸等,由于其远程可控性,尤其适合代替人进入危险的环境。它还能操作工具切割金属等材质,破开车门解救被困人员。2006 年,T52 Enryu 在长冈技术科学大学接受测试中成功从雪堆上举起一辆汽车。

美国硅谷 BLEEX 军事/安防用外骨骼

美国硅谷是高科技的聚集地,在机器人外骨骼方面也有相当杰出的成就。加州大学伯克利分校人体工程与机器人实验室开发的"伯克利下肢外骨骼"(BLEEX)可谓是目前已公开的、在军事应用方面技术最领先的外骨骼系统。2000 年,它被美国国防高级研究计划局(英文简称 DARPA)看中并资助。该项目主要用于士兵、森林消防与应急救援人员,帮助他们长时间背负沉重的武器、通信设备和物资。这些苛刻的应用场合,要求外骨骼系统能提供很强的力量和较长的工作时间、保证机械和控制可靠,重量要轻并且要符合人体工学才能保证动作敏捷和长时间穿戴舒适。

第一台实验样机由双腿动力外骨骼、动力/能源单元和可背负各种物品的与框架构成。为做到力量强劲,BLEEX 采用液压驱动,并由燃油作为主要能源;同时电控部分仍由电池供电(官方资料称其为混合动力)。为保证在野外使用可靠,当燃油耗尽时,腿部外骨骼可轻易拆下,余下部分可像普通背包一样继续使用。2014 年 11 月,第一台实验样机成功亮相,试穿者身背重物却只感觉像几磅重,并能较灵活地蹲、跨、走、跑,跨过或俯身钻过障碍,以及上下坡。

BLEEX 的控制方式是一大亮点。传统检测表面肌电信号的方式比较适合身患残疾或具有肢体运动障碍的使用者,但其最大的问题在于传感器需要和皮肤密切接触,而且信号采集并不总是可靠(比如流汗状态下,传感器就没法紧贴皮肤;而且会改变信号通路的阻抗,信号检测就会不准确),显然不适用于军用这类对可靠性要求较高的场合。因此 BLEEX 另辟蹊径,采用力反馈的方式:当人腿部开始产生动作的时候,这个力量会带动腿部外骨骼一起产生相同的运动趋势,装在外骨骼上的传感器会敏感地捕捉到这个趋势并驱动外骨骼做出顺应这个力的动作,从而增强力量。不过此方法也不完美,因为要求穿戴者先做出动作趋势,外骨骼才能跟着加强这个动作,当穿戴者

做出快速或者高难度的动作时就会有阻碍或滞后感,而且也不适用于截瘫截肢的患者使用。

Bra-Santos Dumont 外骨骼让巴西截瘫少年在世界杯上开球

2014 年 6 月 12 日,圣保罗举行的巴西世界杯开幕式上,一名瘫痪少年在名为"Bra-Santos Dumont"的脑控外骨骼的帮助下开出了激动人心的第一球。这款脑控外骨骼是国际"再次行走计划(Walk Again Project)"的一个研究成果,由杜克大学教授 Miguel Nicolelis 领导。其灵感来自于 Miguel Nicolelis 教授的团队在 2013 年进行的一项实用且有趣的实验,他们开发出一套算法,能帮助恒河猴控制两只虚拟手臂。这款脑控外骨骼系列通过穿戴者佩戴的特殊"帽子"接收脑电波信号,通过装有动力装置的机械结构支撑这名少年的双腿,并帮助他的腿部运动。研究小组为外骨骼安装了一系列传感器,负责将触感、温度和力量等信息反馈给穿戴者,穿戴者能够感知是在何种表面行走。Miguel Ni-colelis 教授在接受法国媒体采访时表示:"外骨骼由大脑活动控制,并将信息反馈给穿戴者,这还是第一次。"

值得一提是,相比于前面提到的检测表面肌电信号和基于力反馈的两种控制方式,该外骨骼的控制方式特别适用于佩戴者身体已经截瘫或失去下肢的情况。这是因为穿戴者已经无法产生动作以提供力反馈,(截瘫患者)也没法形成表面肌电信号。但它也有缺陷,目前能够识别的脑神经信号是很有限的,而且难以保证信号检测准确。此外,这种方式需要将电极植入头皮或脑内,具有一定的创伤性。

中国自主研发的认知外骨骼机器人 1 号

中国在机器人外骨骼领域也占有重要的一席。在中科院常州先进制造技术研究所,有一款外骨骼在研发调试阶段。它名为 EXOP-1(认知外骨骼机器人 1 号),目前只有下半身,主要结构由航空铝制成。双腿的髋关节、膝关节和踝关节各有 1 个电机驱动(共 6 个),共装有 22 个传感器以及一个控制器,并由电池提供能源,在人腰部和腿部共有 9 处固定带。该外骨骼自身总重约 20 公斤,计划承重 70 公斤。它的控制方式和 BLEEX 很接近,都是基于力反馈预判人的动作趋势。中国正在研发外骨骼项目的机构中,已公开报道过的,还包括中科院合肥智能机械研究所和解放军南京总医院。

俄罗斯和以色列的医疗康复用外骨骼产品

医疗用外骨骼能为行走不便的老年人、下肢残疾或瘫痪人士带来重新行走的希望。这方面,除了先前提到的日本 HAL 系列外骨骼,俄罗斯的 ExoAtlet 康复用外骨骼也已进入临床测试并准备商业化。以色列 ReWalk Robotics 公司同样看准了这块市场,更是已经成功在纳斯达克上市。相比于军事和救援应用,这类医疗康复用外骨骼的机械结构和控制原理更简单,易于批量生产,而且价格更易被消费者接受,能满足重要需求且需求群体基数大,既接地气又不失技术壁垒,很受创投资本青睐。

目前存在的问题与未来发展趋势

前面提到的这些机器人外骨骼和机甲,它们的动力装置无非是电机、液压或气动元件等。这些动力元件要想产生足够大的力量,尺寸也必须做得较大,自重也跟着变大了,而且工作时还常有难以忍受的噪音。现有可用于机器人的各种电池的续航力也还很低,设计时也需平衡容量、尺寸和重量等因素。这些核心元件的问题,是当前所有实体机器人系统都要面临的,而且属于产业级别的瓶颈,在瓶颈解决之前,目前各家外骨骼产品的主要看点集中在系统集成优化程度(外观、尺寸、重量、续航力、价格、可靠性、力量与敏捷度之间的平衡)以及控制方式上的创新。

这个瓶颈同样也制约了机器人外骨骼在军事方面的实际应用。几年前就听说战斗民族俄罗斯在开发炫酷的单兵外骨骼系统,但至今也没看到确切消息。不过可以肯定的是,由于巨大的军事应用潜力,大国们不会在这个领域甘于落后。当硬件的技术瓶颈和成本逐步降低后,军用外骨骼会逐渐成为现实。民用级机器人外骨骼也有望进一步大规模商业化,尤其是在医疗康复方面。老年人的行走助力是一个正在逐步扩大的市场,会有越来越多的人有能力购买并愿意使用机器人外骨骼。

外骨骼还可用于虚拟现实交互、动作类游戏。穿戴者能够通过外骨骼感受到触感、力量感,从而体验近乎现实世界的反馈。举例来说,在模拟射箭游戏中,当玩家模仿拉弓的动作时,手臂和背部的外骨骼会施加一定的阻力,让玩家体验弓的张力。类似的游戏场景还有打高尔夫、模拟拳击对战等。甚至可以通过外骨骼主动做出动作来帮助玩家或运动员学习正确的动作。若外骨骼系统的成本能够大幅降低,并且虚拟现实技术进入成熟阶段,这二者结合形成的娱乐体验场景将是一个极为广阔的市场。

参考译文:

Human Body Enhancement

Exoskeleton, powered mecha—the integration of human and intelligent machines

What is the robotic exoskeleton—from science fiction to the reality

Because of the innate limitations of human body, human cannot be as fast and agile as a cheetah, nor can they carry things several times heavier than themselves like ants. In the cartoons that we watched as children, when the "Saint fighters" put their "suits" on, their fighting capacity surged, and they can defeat the enemy and maintain peace. Now imagine if there are similar suits that can make people stronger and more agile after wearing them, or even make an ordinary soldier an invincible "Saint fighter"? There are similar plots in many films, for instance, in the film Elysium, the chief actor Max had been weakened by intense radiation, but after being equipped with an exoskeleton, he was able to fight with Kruger, his vigorous opponent; and in the film Edge of Tomorrow, Tom Cruise, as a human soldier, was able to fight with alien creatures after wearing an individual mecha; also in the film Avatar, an enormous AMP battle mecha was designed, and it even could be manipulated by people sitting inside. Moreover, the Iron Man suit should also be addressed, but it appeared much whimsically. Because the Iron Man can fly and launch the hand-cannon when he puts the suit on, and there is also a small nuclear reactor for supplying energy on the chest of his suit which he never need to pay electricity bills. Therefore, through comparison, the first few types are relatively realizable.

It is enough to talk about so many illusory films, then how about the development of similar technologies in reality? Don't worry, let me introduce the basic concepts firstly. Actually, the device worn on the human body that appeared in the film is called the robotic exoskeleton, which can empower people through mechanical systems. Its structure bears a strong resemblance to the hard shell of arthropods (such as crabs) (its scientific name is exoskeleton, which means the bone grows outside), and the device technically belongs to the category of robots, hence it gets its name. Among them, sometimes the military devices are called powered armor. While those devices with larger size and more powerful functions are called mecha, which can even be manipulated by people sitting inside. At present, the robotic exoskeletons are primarily used in such fields as medical rehabilitation, rescue, engineering operations, and military.

Generally, the robotic exoskeleton system include mechanical structure, sensing part, power and transmission part, energy section and control segment. The mechanical structure provides a solid support for the entire system, and is fixed at the human body through bandages or other means to share the weight and provide the basis for exertion of strength. The lower extremity exoskeleton is mainly fixed at the waist and legs of people; while the systematic exoskeleton is fixed at not only the waist and legs, but also the upper limbs and the trunk. The sensor part consists of sensor and signal-processing circuit, which is used to collect such information as human motion trend, posture and strength to provide judgement data for the control segment. Driving force or torque/moment of force are usually provided by the electric machine, hydraulic component or pneumatic component in the power and transmission part, and then the force is transmitted to the mechanical structure through the driving element, thereby making the exoskeleton to move. The total energy of most exoskeleton systems is provided with batteries, but the existing batteries are hardly enough to support the high-load operation of the systems for a long time. And it is impossible to increase the battery volumes excessively (for the weight and size limitations of the exoskeleton), therefore, fuel oil and mini-type internal combustion engines are used to provide energy and raw power for some exoskeletons.

The key part of the control segment remains microcomputer and controlling software, which can integrate the information from the sensing part and direct the power and transmission part in accordance with the human intention. Now let's take a look at some of the most representative products in the world.

Representative products of the mechanical exoskeleton Japanese HAL series exoskeleton for rehabilitation / operation

As we all know, the Japanese robot industry is extremely developed, not to mention its robotic exoskeleton technology. Among them the most representative one is the HAL series exoskeleton that is jointly developed by the University of Tsukuba and the Japanese scientific and technical corporation Cyberdyne. This kind of exoskeleton contains two major versions: the lower extremity-type HAL-3 and the systematic HAL-5. Its functional orientation is to help the handicapped people or to assist physical exertion (for example, heavy objects need to be removed during rescue operation).

The earliest prototype of the exoskeleton was devised by Yoshiyuki Sankai, who is now a professor of the University of Tsukuba in Japan. As early as 1989, he began his design work after receiving his Doctor's degree in robotics. He spent three years drawing the neural network diagram of human body controlling legs movements, and then spent four years making a hardware prototype, which was powered by the electric machine and batteries. The early version of the exoskeleton was very heavy and just its batteries had a weight of 22kg. It is unpractical because it could be put on only with the help of two assistants, and also need to be connected with an external computer. The latest model has

improved a lot in terms of weight, and the whole HAL-5 weighs only 10kg with batteries and computer around the waist part. Although the controlling method of HAL series exoskeleton is extremely interesting, firstly let's learn how human control body movements before further proceeding.

When people want to make actions, their brain will generate controlling signals, which are transmitted to the corresponding muscle groups through motor nerves, so as to control the movement of muscles and bones. These nerve signals may somewhat spread to the skin surface, forming surface electromyographic (EMG) signals. Although very weak, it can still be detected by electronic circuit. HAL series exoskeleton collect these signals through the sensors attached to human skin, and then control the exoskeleton to perform the same movements as human being, thereby assisting human to make actions. It is a brilliant idea for users with disabilities or limb movement disorders. At present, HAL series exoskeletons have been widely used in medical institutions and achieved great achievements. It was recognized by the international standards for design and manufacture of medical devices (ISO 13485) in December 2012, and it also received the international safety certificate in February 2013 (It is the first kind of powered exoskeleton that obtained this certification in the world). Also, in August of the same year, HAL series exoskeletons were approved for medical application in Europe after receiving EC certificate (it is the first approved therapeutic robot of their kind).

Japanese T52 Enryu Engineering / Rescue Mecha

In Japan, there is also a huge mecha called T52 Enryu. It is 3.5 meters high, 2.4 meters wide and weighs 5 tons. Each arm is 6 meters long, which can lift a ton of things in total. Its great power is hydraulically driven with diesel oil as the energy source. It can be directly manipulated by people sitting inside or can be remotely controlled by people (it is equipped with a camera as an auxiliary). It was mainly developed by the Japanese robot company TMSUK in 2004 and was designed for disaster relief, such as earthquakes, tsunamis and traffic accidents. Thanks to its long-range controllability, it is especially suitable for dangerous environment instead of people. It can also cut metal and other materials by using tools, and then break the vehicle door to rescue the trapped people. In 2006, T52 Enryu successfully lifted a car from a snowdrift in a test at Nagaoka University of science and technology.

BLEEX Military / Security Exoskeleton in Silicon Valley

The Silicon Valley of the United States is a cluster of high technology, where an outstanding achievement has been made in robotic exoskeleton. The "Berkeley Lower Extremity Exoskeleton" (BLEEX), developed by Human Engineering and Robotic Laboratory of the University of California, Berkeley, is the most advanced exoskeleton system that is already publicized for military applications. In 2000, it was favored and funded by the US Defense Advanced Research Projects Agency (DARPA). The project is mainly used for soldiers, forest firefighters and emergency rescue workers to help them carry heavy weapons, communication equipment and supplies for a long time. In these demanding applications, exoskeleton system is required to provide strong strength and long working hours, as well as to ensure reliable machinery and control, light weight and its ergonomic design, thereby making quickness in one's movement and long-time wearing comfort.

The first experimental prototype consists of a lower-extremity powered exoskeleton, the power/energy unit and the frame that can carry various items. In order to achieve strong power, BLEEX is hydraulically driven with fuel oil as the main energy source. Meanwhile, the electric control part is still powered by batteries (officially known as hybrid power). To ensure its reliable use in the field, the lower-extremity exoskeleton can be easily removed when the fuel oil is exhausted, and the remained part can continue to be used as an ordinary backpack. The first experimental prototype successfully appeared in November 2014. People who wore the device only felt a few pounds of weight when carrying a heavy load, and they could flexibly squat, stride, walk, run, step over or lean over the obstacles, as well as climb and descend the slopes.

The control method of BLEEX is one of its great highlights. The traditional method of detecting surface electromyographic (EMG) signals is more suitable for users with disabilities or limb movement disorders. However, the biggest problem is that the sensor needs to be in close contact with the skin so that the acquisition of signal is not always reliable (for example, the sensor cannot be in close contact with the skin when someone is sweating; and the impedance of the signal channel can also be changed, thereby causing an inaccurate signal detection). Therefore, it is obviously not suitable for military applications which require high reliability. Then BLEEX took a new approach and adopted the method of force feedback: When the human legs began to move, driven by the force it produced, the lower-extremity exoskeleton would also generate the same movement trend. Then the

sensor installed on it would sensitively catch the trend and drive it to make the same action, so as to enhance the strength. But this method is also not good enough for the reason that the exoskeleton could follow the action only when the wearer made movement first. Hence the wearer would feel to be impeded when making fast or difficult movements, and it is also not suitable for paraplegic patients or amputees.

Bra-Santos Dumont Exoskeleton Makes a Brazilian Paraplegic Teenager to Kick off at the World Cup

On June 12, 2014, at the opening ceremony of the World Cup Brazil held in Sao Paulo, a paralyzed teenager scored the first exciting goal with the help of mind-controlled exoskeleton called "Bra-Santos Dumont". This mind-controlled exoskeleton was developed by the international "Walk Again Project", which was led by Miguel Nicolelis, a professor of Duke University. Inspired by a practical and interesting experiment conducted by Professor Miguel Nicolelis and his team in 2013, they developed an algorithm to help the rhesus monkeys control two virtual arms. The mind-controlled exoskeleton series can receive brainwave signals through a special "hat" worn by the teenager, and support the teenager's legs through a mechanical structure equipped with a power device, and then helps him move his legs. The research group installed a series of sensors on the exoskeleton that can report back such information as touch, temperature and strength to the wearer, who is able to perceive on the kind of surface they are walking. "It is the first time that the exoskeleton is controlled by brain activity and feeds back information to the wearer." Professor Miguel Nicolelis said in an interview with the French media.

It is worth mentioning that the control method of this mind-controlled exoskeleton is particularly suitable for paraplegic patients or amputees compared with another two control methods mentioned above. The reason is that paraplegic patients or amputees are no longer able to provide force feedback by making movements, nor can they make the surface electromyographic (EMG) signals. But it also has some defects. The brain nerve signals that can be recognized currently are very limited, and it is also difficult to ensure the accuracy of signal detection. Moreover, electrodes need to be implanted into the scalp or brain, which is somewhat traumatic.

China's Self-developed Cognitive Exoskeleton Robot Ⅰ

China also plays an important role in the field of exoskeleton robot. At the Changzhou

Institute of Advanced Manufacturing Technology of the Chinese Academy of Sciences, a kind of exoskeleton is in the research and debugging phase. It is called EXOP-1 (which is the cognitive exoskeleton robot 1). There is only the lower part at present, which is primarily made of aviation aluminum. The hip joint, knee joint and ankle joint of both legs are respectively driven by a motor (6 in total). 22 sensors and one controller are installed in total, with batteries for energy supply. There are 9 fixing bands at the waist and legs of human. The exoskeleton itself weighs about 20 kg and is planned to bear 70 kg. Its control method is similar with that of BLEEX, which are both based on force feedback to predict the trend of human action. Many institutions that are researching and developing exoskeleton projects in China have already been publicized, including the Hefei Institute of Intelligent Machinery of the Chinese Academy of Sciences and Nanjing General Hospital of the People's Liberation Army.

Exoskeleton Products for Medical Rehabilitation in Russia and Israel

Exoskeleton products for medical treatment can bring the hope of walking again for the elderly who have trouble walking, the disabled or paralyzed people. In this regard, in addition to the previously mentioned Japanese HAL series exoskeletons, Russian ExoAtlet exoskeleton for rehabilitation has also entered clinical tests and is ready for commercialization. The Israeli company ReWalk robotics also seizes the opportunity and has successfully listed on NASDAQ. Compared with military and rescue applications, the mechanical structure and control principle of this exoskeleton for medical rehabilitation are simpler, and easier to be produced largely. Also, the price is more acceptable for consumers. It can meet vital needs and has a large base of demanding groups. It is realizable, and also has technical barriers, and thus is favored by the venture capital.

Current problems and future development trends

The power devices of robotic exoskeletons and mecha mentioned above are nothing more than electric machine, hydraulic pressure or pneumatic components. In order to generate enough power, these power components must be large in size, with their own weight also increasing. And the noise accompanied is often unbearable when they are working. At present, the endurance of various batteries available for robots also remained in a low level, and it is also necessary to balance such factors as capacity, size and weight in design. The bottlenecks of these core components, which belong to an industrial level, are currently faced by all current physical robotic systems. Before the bottlenecks are solved, various exoskeleton products mainly focused on the innovation of both system integration optimization (balance among appearance, size, weight, endurance, price, reliability, strength and agility) and control methods.

This bottleneck also limits the practical application of robotic exoskeleton in military. A few years years ago, Russia, a nation famous for fighting, was heard to develop a fascinating exoskeleton system for individual soldiers, but the exact information has not been known yet. However, it is certain that the great powers will not be willing to lag behind in this field due to its huge potential in military application. The military exoskeleton will gradually become a reality with the technical bottlenecks and costs of hardware gradually reduced. The civil-type robotic exoskeletons are also expected to be further commercialized largely, particularly in medical rehabilitation. Providing assistance for the elderly to walk is a growing market, and more and more people will be able to buy and have desires to use robotic exoskeleton.

Exoskeleton can also be used in virtual reality interaction and action games. The wearers can feel the touch and strength through the exoskeleton, so as to experience the feedback of the near-real world. For example, in the simulated archery game, when the player imitates the action of pulling a bow, the exoskeleton of the arms and back will exert certain resistance, allowing the player to experience the tension of the bow. Similar game scenes include playing golf, simulating boxing and so on. It is even possible to help players or athletes to learn the correct actions by actively making actions through exoskeletons. If the cost of exoskeleton system can be greatly reduced, and virtual reality technology also enters a mature stage, then the entertainment experience scene combined by the two will form an extremely broad market.

单词释义：

1. 外骨骼　　　　　　　　exoskeleton
2. 机器人外骨骼　　　　　robotic exoskeleton
3. 机甲　　　　　　　　　mecha
4. 下半身型外骨骼　　　　the lower extremity exoskeleton
5. 电机　　　　　　　　　the electric machine
6. 液压元件　　　　　　　hydraulic component
7. 气动元件　　　　　　　pneumatic component
8. 力矩　　　　　　　　　torque/moment of force

9. 信号处理电路　　　　　　　signal-processing circuit

10. 内燃机　　　　　　　　　internal combustion engines

11. 表面肌电信号　　　　　　　surface electromyographic（EMG）signals

12. 美国国防高级研究计划局　　the US Defense Advanced Research Projects

　　　　　　　　　　　　　　Agency（DARPA）

13. 脑控外骨骼　　　　　　　　mind-controlled exoskeleton

14. 动力装甲　　　　　　　　　powered armor

15. 微型电脑　　　　　　　　　microcomputer

16. 运动神经　　　　　　　　　motor nerves

17. 神经信号　　　　　　　　　nerve signals

18. 肌群　　　　　　　　　　　muscle groups

19. 航空铝　　　　　　　　　　aviation aluminum

20. 手炮　　　　　　　　　　　hand-cannon

译文解析：

1. 善用被动语态。

原文：为保证在野外使用可靠,当燃油耗尽时,腿部外骨骼可轻易拆下,余下部分可像普通背包一样继续使用。（无主语）

译文：To ensure its reliable use in the field, the lower-extremity exoskeleton can be easily removed when the fuel oil is exhausted, and the remained part can continue to be used as an ordinary backpack.

分析：英语中的谓语至少有三分之一的句子使用被动语态,科技英语的表达方式主要是叙事说理,追求准确严密,客观性比较强；而被动语态能更客观、准确地描述事物的发展和变化,强调的是受动者。因而,科技英语在突出强调所要论证和说明的问题时,需要广泛使用被动语态,这样也能够避免使用主动语态的句子主观性带来的弊端。在汉语中使用被动句的频率远低于英语使用被动语态的频率,表示被动的手段也十分有限,因而在汉译英的过程中,需要对其进行语态转换处理,译为英语的被动句,这样既能够在深层意义关系上再现原句信息,也遵循了英语常用的表达方式。

2. 词性要转换。

原文：现有可用于机器人的各种电池的续航力也还很低,设计时也需平衡容量、尺寸和重量等因素。

译文：At present, the endurance of various batteries available for robots also remained a low level, and it is also necessary to balance such factors as capacity, size and

weight in design.

分析：中文属于动态性语言,善用动词,而英文属于静态性语言,善用名词与介词,那么我们在翻译的过程中,有时需要根据译文语言的习惯进行词性转换,如把原文中的动词转换为名词,把原文中的副词转换为介词等,使其更符合目标语的行文习惯。科技英语要求行文简洁,语言准确客观,强调客观存在的事实,那么在汉译英的过程中,我们需要进行适当的词性转换,以使其符合科技英语简明客观的特点。

3. 增译法。

原文：而且会改变信号通路的阻抗,信号检测就会(增词"因此")不准确。

译文：The impedance of the signal channel can also be changed, thereby causing an inaccurate signal detection.

分析：汉语属于意合语言,句子各成分之间通过语言内在的逻辑关系来表达语法意义和逻辑意义,运用隐性连接,注重语义连贯,句子各成分之间不用或少用关联词,句子结构松散。

英语属于形合语言,多用关联词来连接词语或句子,注重句子形式,注重以形显意。因此,英语大量使用关系词和连接词。那么在汉译英的过程中,我们需要将汉语中隐藏的逻辑关系词明确表达出来,以使译文连贯流畅。

4. 科技英语长句译法 将汉语短句连成长句,保持句子的连贯性。

原文：其实影片中出现的,穿在人身上的装置叫做机器人外骨骼,它能通过机械系统为人助力,//其结构酷似节肢动物(如螃蟹)的坚硬外壳,而且在技术上属于机器人的范畴……

译文：Actually, the device worn on the human body that appeared in the film is called the robotic exoskeleton, which can empower people through mechanical systems. Its structure bears a strong resemblance to the hard shell of arthropods (such as crabs), and technically the device belongs to the category of robots…

分析：汉语注重意合,结构较松散,短句较多,且句与句之间多用逗号,句子各成分之间内部逻辑严密英语注重形合,结构较严谨,长句较多,注重句子形式,注重以形显意。汉语中,由许多并列分句构成的句型,特点是语义联系松散、分句之间缺少连接词语,逻辑关系隐藏在内部含义当中。因此在翻译这类汉语长句时,要注意两种语言的差异,理清汉语句子间的逻辑关系,注意断句,可用连词、定语从句、独立结构等把汉语短句连成英语长句,使译文更加连贯,更符合英语的表达习惯。

第十四章　脑计划

2013 年 4 月 2 日,时值美国首都华盛顿特区最热闹,最富生机的"樱花节"。在樱花如雪的白宫,获得连任的奥巴马政府,雄心勃勃地宣布启动了一个被誉为、可以媲美"阿波罗计划"(Project Apoll)的项目——人类脑计划。脑计划的全称是"通过推动前沿创新性神经技术进行脑研究的计划"(Brain Research through Advancing Innovative Neurotechnoligies,英文缩写 BRAIN INITIATIVE)。

"可上九天揽月,可下五洋捉鳖",互联网已经把这颗直径 1.2 万多公里的星球变成一个"地球村"。宅男们大可以和机器人女友 Siri"谈情说爱",可穿戴式智能设备改变了我们未来的生活方式。在这个黑科技席卷全球的年代,很难想象我们自己肩上"三磅以内,两耳之间"的大脑,还如同暗箱一般,对我们保留了如此多的神秘与未知。阿尔茨海默病、帕金森病依旧是威胁中老年人健康的两大顽症,而对于它们的发病机理和治疗方案,科学界莫衷一是。精神分裂症、自闭症、毒品成瘾的研究成果早已汗牛充栋,但完整的答案依然像散落在各处的拼图。仅是从零碎的图景中窥见这个巨大奥秘的一角,就足以令科学家兴奋不已。有读者会问,揭开大脑的奥秘与我有什么关系?会让我更聪明、更有钱,还是更健康长寿? 恭喜你,你问对问题了。脑科学(广义为神经科学)或许不能点石成金,也不能让你学会七十二变。但多了解些大脑的知识,不但可以让你学习更轻松,摆脱拖延症,还可以增加你在茶余饭后不少高大上的谈资,诸如:为什么有的人更爱"劈腿"? 为什么有的人到"双十一"会控制不住自己的购物欲望?

我们和爱因斯坦的大脑有几条街的距离？

20 世纪上半叶正是物理学发展如日中天之时。尼尔斯·波尔(Niels Henrik David Bohr)、维尔纳·海森堡(Werner Karl Heisenberg)、埃尔温·薛定锷(Erwin Schrodinger)等物理学大师辈出,相对论、量子力学等理论风起云涌。彼时,艾萨克·牛顿(Sir Issac Newton)早已远去,史蒂芬·霍金(Stephen William Hawking)尚未成名。阿尔伯特·爱因斯坦(Albert Einstein)是全世界人眼中"天才"的代名词。关于爱因斯

坦的 IQ,连带他童年的各类传说在坊间为人津津乐道。一时之间,早教方面也在"如何培养出下一个爱因斯坦"上大做文章。但毫无疑问,人们有一个共识:"爱因斯坦之所以比普通人聪明,是因为他的大脑和我们不一样。"

很显然,他的医生也是"爱因斯坦大脑"的狂热粉丝。1955 年,当爱因斯坦在美国普林斯顿大学医院去世后,病理医生托马斯·哈维(Thomas Stdtz Harvey)在 75 小时内"偷走"了爱因斯坦的大脑,在脑动脉中注入防腐剂,请一位朋友切成了 240 片并保存在福尔马林溶液中。当哈维医生带着这颗大脑做横穿美国大陆旅行时,可是被 FBI 特工跟踪保护 4000 公里。

尽管哈维医生为了科研的"偷天换日"后来取得了爱因斯坦儿子汉斯的谅解,但汉斯提出了严格的条件:对其父亲大脑的研究,必须发表于高水平的科学刊物上。几十年过去了,哈维也没什么像样的相关科研成果发表。令人尴尬的是,爱因斯坦的大脑重量只有 1230 克,比起普通人平均重量 1400 克的大脑尚且不如跟海豚、大象的巨脑比起来简直更是相距甚远。虽然从脑切片上观察到顶叶部位有许多山脊状和凹槽状结构,也就是传说中的"大脑沟回比较多",但是这和天才的智商之间依然缺乏直接联系。

时间推进到 1980 年,背负巨大压力的哈维开始把脑切片分发给世界各地的研究者,组织小伙伴们"一起来找茬"。众人拾柴火焰高。很快,加利福尼亚大学伯克利分校的玛丽安·戴蒙(1985 年)和哈维(1999 年)分别撰文说:爱因斯坦脑中的胶质细胞(尤其是星形胶质细胞),而不是负责计算和记忆的神经元细胞比常人多。根据当时学术界的共识,胶质细胞不过是对神经元细胞起辅助作用的小伙计,难当大任。于是,哈维他们的"胶质说"也被嗤之以鼻。后来,他和加拿大科学家又共同宣称:爱因斯坦的"脑洞"大(大脑的岛叶顶盖和外侧沟是空的),而且有图有真相。这与不明真相的群众所做的猜测在一定程度上吻合,可是由于"脑洞说"缺乏功能性研究支持,也遭到其他科学家的质疑。中国科学家在这场科研追星运动中也不甘人后。2013 年,华东师范大学相关研究成果称:连接爱因斯坦两个大脑半球的结构——胼胝体厚于常人,因此左右半脑的交流可能更高效。

此外,关于爱因斯坦大脑不同脑区,尤其是关于数学、语言和计算脑区的研究成果也层出不穷。然而,若要就"我们和爱因斯坦的大脑有多远的距离?"给一个标准答案,神经科学家们大概争论三天三夜也不会有结果。甚至有人认为这是个伪命题:"爱因斯坦的大脑和我们只是不同,未必更好。"毕竟,即使个人的成就可以量化,大脑的功能却极其复杂、缺乏单一量化标准。世上从不缺乏机智过人屠狗辈,也不鲜见生搬硬套状元郎。否则现代社会就不会在 IQ(智商)的基础上延伸出 EQ(情商)、SQ(灵感智商)等一系列衡量大脑功能的参数。那么正确的问题似乎应该是:完美的大脑长啥样?是如何工作的?

揭开大脑神秘面纱的先驱：从高尔基与卡哈尔说起

在"身体发肤受之父母"、身体完整性神圣不可侵犯的年代，人们认为心才是意识和感觉的器官。直到今天我们还是习惯说"心痛"和"心碎"而不是"脑子痛"。如16世纪的文艺复兴一般，现代神经科学也发源于地中海国家。解剖学把大脑和人的思想行为联系在一起后，1873年，意大利细胞学家卡米洛·高尔基（Camillo Golgi）首次通过脑切片加铬酸盐-硝酸银染色法描述了脑中两种形态截然不同的细胞：神经元（Neuron）和胶质（Glia）细胞。大家才注意到，原来煮在火锅里的猪脑花不是仅由一种细胞组成的。这种利用重金属渗透显示脑细胞的染色法几经改进，便形成了在学界经久不衰的"高尔基染色法"，并获得了1906年的诺贝尔生理学和医学奖。即使现在已经有了更快速观察各种神经元和胶质细胞形态，甚至直接观察动物脑中活动的神经元的影像学方法，但延续了一个多世纪的高尔基染色依然是显示神经元外形最经济便捷的方式。

在高尔基染色的启发下，西班牙解剖学教授拉蒙·伊·卡哈尔（Ramony Cajal）确立了更灵敏的还原硝酸银染色法，并发现神经纤维精细结构（轴突和树突）和神经末梢之间的物理接触（即神经元间的"突触"结构）。因此与高尔基共享了1906年的诺贝尔生理学和医学奖。卡哈尔在神经生物学界是一个高山仰止的存在，正如达尔文之于生物学，孟德尔之于遗传学。和接受过传统解剖训练和艺术修养的高尔基比起来，卡哈尔其实是个半路出家、自学成才的"叛逆"中年人。不过在这个孕育了毕加索、塞万提斯和高迪的国家，即使是实验台前的科学怪咖也从不缺乏艺术细胞。

卡哈尔不满足于把染色用于做报告时秀几张高逼格的图片。酷爱绘画却在父亲压力下进入医学院的他，在切脑片（猫脑和鸡脑）与染色中找回了青年时期的爱好。在实验室小小一隅，卡哈尔把染色的脑切片放在用自己私房钱买来的老式显微镜下，一边哼着歌一边画出了不同脑区的神经元和它们组成的网络。在照相技术离商业化还遥遥无期的19世纪末除了极少数拥有显微镜的"高富帅"之外，普罗大众是没有机会看到肉眼无法区分的细胞和亚细胞结构的庐山真面目的，因此他被完全排除在细胞生物学家队伍之外。因此卡哈尔的工作相当于给电视机出现之前的群众放映微观世界的电影，而他既是制片人摄影师，又是放映员。当笔者在美国国立健康研究院见到卡哈尔绘出的小脑和海马神经网络图真迹时，其兴奋程度不亚于纯颜控见到了完美版的帅哥和美女。用网友的话说就是"神经元的秘密花园，美爆了"！

诺奖肯定了卡哈尔工作的技术性成果，但卡哈尔对于神经科学的贡献却远不止于此。才华横溢的他是个玩跨界的高手，不仅是画家、运动员、科学家，还是个伟大的思想家。基于他在显微镜下观察到的神经元结构，他首次提出"神经元假说"，大胆挑战了当时的主流说法"网状神经假说"。网状神经假说认为，神经细胞之间相互贯通形成一张巨大的网络；细胞之间没有屏障，物质和信息可以自由传递。卡哈尔则认为：每个神经元都是独立的功能单位；细胞间接触却不联通；并且神经元是有极性的，即信息是单向传递的，由树突至胞体再到轴突。在未经功能性实验证实的当年，仅凭形态学证据（而且还不是特别充分的证据）就做出这样的假设，虽不是天马行空也颇需要想象力。在科学界，假说的提出是要讲证据的，卡哈尔的离经叛道惹恼了以高尔基为代表的圈内大佬们。两人在发表诺贝尔奖获奖演说时就针锋相对、各执一词，让首次由两人分享的生理学或医学奖颁发得异常尴尬。

时间（还有实践）是检验真理的标准。1894 年，英国生理学家查尔斯谢灵顿通过研究膝跳反射中神经肌接头的结构，支持了"接触但不联通"的观点，并将神经元之间这种连接结构称为"突触"。后来的研究都证明突触包含大量蛋白、脂质和"信号分子"（神经递质的极其精密复杂的结构）。谢灵顿膝跳反射实验还证实：神经元的信息传导有特定方向。接着科学家在神经纤维上记录到电活动，阐释了神经信息传递方式——电传导。网状神经说被潮水般的证据钉在耻辱柱上，很快被抛弃，卡哈尔深刻的洞察力再次为世人惊叹。但历史少以线性发展，科学之吊诡从来不会就此打住。就如物理学上光的"粒子性"和"波动性"争论一样，人们以为看到了大结局，不料片尾还有上帝准备的彩蛋。尽管，神经元间的信号传递由突触前细胞传向突触后细胞已经铁证如山，但越来越多的证据也指出：神经细胞间信息的传递并非完全单向，突触后细胞对突触前细胞释放的信号的强弱有调节作用，从而形成反馈环路。

经过一个世纪的发展，神经生物学虽然还只是生物学板块中的小鲜肉，然而在分子生物学、电生理学、药理学和迅速发展的影像学技术推动下已成为生命科学领域中的当红明星。美国《科学》杂志在庆祝创刊 125 周年之际公布了今后半个世纪最具挑战性的 125 个科学问题，其中狭义脑科学范畴内就有 15 个问题，占据八分之一席位。了解越多，问题也越多，探索和求知莫不如此。

连接组学：我思故我在

"我是谁?"是一个深奥的哲学命题，吕秀才用它兵不血刃杀死了姬无命。英国著名演化生物学家理查德·道金斯（Richard Dawkins）在《自私的基因》中给出这样的答案：生物体本质上不过是一组基因存于世上需要的暂时的盒子。这些基因为了长生不死，选择各种肉体（不同表型的组合），并淘汰劣等的队友（不利于生存的基因）、选择强大的伙伴（利于生存的基因），最终表现为物种的进化。这也就是说：我不是我，我是我的基因组！

克隆技术早已问世，如果撇开伦理问题，复制一个人的全部基因信息，就能得到完全相同的副本吗？逻辑强大的看官一定想到了：用纯生物学方法，从复制一个拥有相同基因的细胞开始，生长出的两个人应是高度类似却不一致的（参见同卵双胞胎）。一个人不能两次踏入同一条河流。随机发生在两个个体上任何微不足道的不同事件都足以塑造不同的记忆，乃至于改变他们的人生轨迹。

TED（Technology、Entertainment、Design，美国一家私有非营利机构）演讲中，麻省理工学院教授承现峻（Sebastian Seung）给出一个更精确的解读："我是我的连接组（connectome）。"自"基因组"一词问世以来，各种"组学"已经被好大喜功的科学家们"玩坏了"。单打独斗拼智商、能力在现代组学研究中已然捉襟见肘，整合人才和资源才是王道。"连接组"早在 2005 年就被提出，意指描绘脑内神经元间联系（全部突触的集合）的整体图谱。这不就是"脑计划"的另一个版本吗？若非奥巴马倾一国之力整合资源，单个实验室充其量只能小打小闹地在模式生物上做连接组学。要知道最低等的模式动物——1 毫米长的秀丽隐杆线虫，根本没有脑区，然而绘制它区区 302 个神经元的连接组直到 2012 年才宣告完成。

早在 2002 年，微软创始人之一保罗·艾伦斥巨资在他的私人研究所启动了"艾伦脑图谱工程"（Allen Brain Atlas），旨在用基因组学、神经解剖学来建立小鼠和人脑中基因表达的三维图谱。这一亿美元花得掷地有声。2006 年，首个小鼠大脑基因表达图谱公布。截至 2012 年，已有成年小鼠脑、发育中小鼠脑、成年人脑、发育中人脑、灵长类脑、小鼠脑连接图谱和小鼠脊椎图谱 7 种图谱公布。更重要的是，它的所有信息对公众免费开放。它就像一座公共阅览室，研究者和医生可以方便地上网查阅，这一工程为全世界数以万计的神经科学家们节约了至少十年的单独探索时间。

神经科学黑科技：你问我要去向何方，我指着大脑的方向

以分子水平为基础，自下而上的还原论方法在基础研究中高歌猛进，阐释了神经递质释放与神经信号传递的机制、神经可塑性（学习与记忆）的原理以及阿尔茨海默病、帕金森病为代表的脑疾病发病过程中的分子变化。前景看似一片光明。然而回到整体层面，分子水平的经验却常常不可重复。许多在培养皿中的细胞，甚至是模式生物上作用显著的神经类药物，却在临床实验中纷纷败下阵来。这不过是因为管中窥豹，仅见一斑。由于缺乏对神经系统的整体认识，很容易忽略其他重要的影响因素。大脑仍以一个千丝万缕打结的线团形式呈现在研究者面前。每个研究者手执一个线头各有独到见解，然而靠这样解开线团无异于缘木求鱼。"脑计划"正是要开启一种自上而下的整体方法论，让研究者抽丝剥茧地揭开大脑之谜。

面对这个重约三磅的物体，最大的挑战来自：1. 神经细胞种类多样化；2. 对特定功能的神经环路解剖结构和细胞组成不了解；3. 不能观察和控制活脑中神经细胞的活

动。我们需要什么样的法宝呢？脑计划的专家很快圈定了以下几个方向：标记——神经细胞种类（脑虹技术），示踪——神经环路（CLARITY），观察——高分辨率、高灵敏度和高通量（简称"三高"）成像技术（超分辨率显微技术冷冻电镜和激光片层扫描显微技术），控制——特定细胞类型神经环路调控（光遗传学）。

标记：脑虹深处

在"神经科学家宁可共用一把牙刷也不愿共享数据"的年代，两个哈佛大学教授杰夫·里奇曼（Jeff Lichtman）和约书亚·塞恩斯（Joshua Sanes）称得上是"情比金坚"了。两人从20世纪90年代在华盛顿大学圣路易斯校区开始的合作一直持续至今。塞恩斯是研究视网膜神经网络的翘楚，但视网膜研究在神经科学领域已颇为边缘化。许多人认为它不过是由视神经连接到大脑上的编外组织，根本算不上脑的一部分。然而视网膜上却集合了多种类型的神经元细胞（视杆细胞视锥细胞、双极细胞和神经节细胞），而且是最容易分离的完整局部神经网络。身为神经遗传学家，长期以来塞恩斯苦恼于不能同时标记多种类型的神经元。而里奇曼多年专注于神经发育中突触形成过程的研究，擅长影像学，被塞恩斯誉为"全世界做突触形成活体成像最牛专家"。两人一拍即合，开始了漫漫征程。

海洋生物在黑暗中发出荧光的秘密于20世纪60年代被日本人下村修破解，他从发光水母体内找到的绿色荧光蛋白，此项发现照亮了生物学家研究的征途。研究者很快通过遗传学手段将荧光蛋白的基因导入动物体内表达，就能轻松定位、观察目标细胞。里奇曼就是这方面的行家，但这次他挑战的是区分脑中密密麻麻的不同神经元的胞体，理清它们伸出的纠缠不清的线头（轴突和树突）。一种颜色明显不够用了。不过这个时候荧光蛋白家族中已新添好几位成员：红色、橙色、黄色以及青色荧光蛋白。

接下来进入塞恩斯最爱的遗传学游戏时间：他们把红、黄、青色的荧光蛋白基因，像珠子一样连成一串，每个基因之间加一个遗传重组位点和一个表达终止元件，这一串导入小鼠脑细胞中的荧光蛋白基因组合中只有排在第一个的基因能表达。体内一个叫Cre的重组蛋白酶随机选择两个重组原件，剪去中间片段，这样就得到不同的荧光蛋白组合：不发生重组时只有第一个红色荧光蛋白表达，细胞标记为红色；发生重组1时红色荧光蛋白基因丢失，暴露出的黄色荧光蛋白表达，细胞显示为黄色；发生重组2时，只剩青色荧光蛋白表达，细胞为青色。在基因"珠串"前加上细胞特异的基因表达开关（启动子），就可以标记特定的神经细胞。另外，还可以给"珠串"加上亚细胞定位的GPS系统，选择让荧光蛋白表达在胞体内（实心圆圈）或是只表达在细胞膜上（空心圆圈）或细胞核内（圈内有点）。紧接着里奇曼和塞恩斯给珠串打了个"补丁"加上橙色荧光蛋白，升级为1.1版本。这样就可以得到4种颜色的细胞了。

254

　　但"游戏上瘾"的人根本停不下升级的节奏。脑虹 2.0 版几乎同时推出：将四个不同颜色的荧光蛋白基因方向两两相对地串起来，基因对之间加入重组位点。要知道基因的表达和阅读文字一样是单方向的，方向相对的红、青基因对中，只有第一个正向红色蛋白的可以被读出。发生一次重组，基因对的方向就颠倒一下，原本反向的青色基因就会被读出。当绿黄、红青四色蛋白基因配对串起之后，同样有 4 种不同重组方式带来 4 种颜色的细胞。那么当把这些基因"珠串"扔到神经细胞中表达，可以得到多少种颜色的细胞呢？如果你说 4 种就"too young too simple"了。排列组合下？也不对！复杂的生物体怎么可能给出这么简单的答案！从一颗受精卵开始发育成的大脑经历了无数次有丝分裂，每一次分裂都可能发生一次重组，每一次重组都会表达一种不同颜色的荧光蛋白。所以转基因鼠成年脑中每个神经元显示的颜色都是四种荧光蛋白颜色的叠加，有的红色多些有些偏绿一点有的看上去是紫色（红青色叠加后效果）。理论上可产生的颜色有无数种！但是由于显微镜波长和肉眼分辨的极限，我们能观察并区分只有近100多种颜色。

　　电影《绿野仙踪》的主题曲《彩虹深处》许多人耳熟能详。2007 年，英国《自然》杂志重磅推出名为《脑虹深处》的报道。在里奇曼和塞恩斯合作的这项划时代技术中，科学家终于突破了"高尔基染色"颜色单一着色细胞少的瓶颈。在荧光显微镜下，小鼠脑中的神经元能像电视显像管一样，呈现出五彩缤纷的颜色。"脑虹"技术不仅在于炫丽，它的问世使科学家能够标记并长距离追踪动物的神经回路，而不再限于某个脑区，还能观察神经元是如何连接到神经网络中的。除了转基因小鼠，低等模式生物果蝇在这些实验上，遗传学操作使"脑虹"的应用更加便捷。后来里奇曼研究组改进了方法，用病毒感染模式研究生物大脑，并提高荧光蛋白表达效率，这样一来，我们在普通动物的脑中也可以看到美丽炫丽的"脑虹"了。虽然"脑虹"使得神经回路标记和区分不同细胞的局面大大改观，然而它的局限性也是显而易见的：仅限于研究表达了荧光蛋白的神经细胞，转基因动物或病毒感染率等客观条件至关重要；需要将许多脑切片叠加起来才能得到完整的神经网络图像，而这本身就是一个难题。

参考译文：

Brain Initiative

　　On April 2, 2013, the Washington D. C was celebrating the most bustling and vigorous Sakura Festival. In the White House, amid the cherry blossoms, the newly re-elected Obama Administration ambitiously announced a plan which rivals the Project Apoll-namely, the Brain Initiative, and the full name for it is Brain Research through Advancing

Innovative Neurotechnologies.

As a Chinese saying goes, "You can clasp the moon in the Ninth Heaven and seize turtles deep down in the Five Seas." Internet turns the Earth, a planet with a diameter over 12000 km, into a global village where Otakus could fall in love with their robot girlfriend and wearable smart devices change our lifestyles in the future. In the era that black technology was all the rage, it is hard to imagine that our brain, a 3-pound organ between our two ears, just like the dark box still keep so much mystery and unknowns to us. Alzheimer and Parkinson remains two intractable disease threatening the health of middle aged and elderly people. However, their pathogeny and treatment are still a matter of discussion. Abundant research results of schizophrenia, autism and drug addiction were made while the complete answers are still scattered around like pieces of puzzle. But a glimpse of the vast mystery in this fragment picture is enough to excite the scientists.

Some readers may wonder "What does unraveling the mystery of brain have to do with me? Will it make me smarter, richer or healthier?" Congratulations! You asked the right question. Brain Science (broadly defined as Neuroscience) may neither be the Midas Touch, nor will it teach you seventy-two kinds of changes. Instead, learning more about brain can help you study easily, get rid of procrastination, and increase a lot of interesting conversations going around the table, such as: Why do some people prefer "cheating"? Why do some people fail to control their shopping desire during "Double Eleven Festival"?

What's the Difference Between Our Brain and Einstein's?

The first half of the 20th century was the heyday of physics development. Masters of physics such as Niels Henrik David Bohr, Werner Karl Heisenberg, Erwin Schrodinger have come forth in succession. Theories like relativity and quantum mechanics raged as storm. At that time, Sir Isaac Newton was passed away and Stephen William Hawking was still a nobody. Albert Einstein is the synonym for "genius" in the eyes of people all over the world. His IQ, together with all kinds of legends about his childhood, is widely talked about. Therefore, early education also made a big fuss on "how to cultivate the next Einstein". But there is no doubt that "Einstein is smarter than ordinary people because his brain is different from ours."

Apparently, his doctor is also a big fan of "Einstein's Brain". In 1955, after Einstein died at Princeton University Hospital, Thomas Stdtz Harvey, a pathologist, "stolen" Einstein's brain and injected preservatives into the cerebral arteries within 75 hours. He asked a friend to cut it into240 pieces and kept them in formalin solution. When Dr.

Harvey took the brain to travel across the United States, he was tracked and protected by FBI agents for 4,000 kilometers.

Although Dr. Harvey later obtained the forgiveness of Hans-Einstein's son that he stole the brain for scientific research, Hans put forward a strict condition for him: The research on his father's brain must be published on high-level scientific journals. However, decades had passed and Harvey had not published any valuable research results about it. What's embarrassing was that Einstein's brain weighs only 1230 grams, far less than the average 1400 grams of human brain let alone the giant brains of dolphins and elephants. Although the parietal lobes of Einstein's brain, observed from those slices, have many ridge-like and groove-like structures, or in other words "has more brain sulci and gyri", it still had nothing to do with the intelligence quotient of genius.

As time went by, in 1980, Harvey, who was under great pressure, began to distribute brain slices to researchers all over the world, and organized his partners to "find fault together". The more the merrier. Soon, Marian Damon (1985) at the University of California, Berkeley, and Harvey (1999) respectively wrote that there were more glial cells (especially astrocytes) than neurons were responsible for computation and memory in Einstein's brain. According to the consensus of the academic circles at that time, glial cells were just a small part that played an auxiliary role in neuron cells. As a result, Harvey and their "colloid theory" were looked down by the others. Later, he and Canadian scientists jointly declared that Einstein's "brain hole" was larger than the normal (the insular tectum and lateral sulcus of the brain were empty), and pictures would tell the truth. To some extent, this is consistent with the conjecture made by the onlookers. However, due to the lack of functional research support, the "brain hole theory" remained as a question to other scientists. Chinese scientists are not willing to be left behind in this scientific research campaign. In 2013, research results from East China Normal University indicated that the corpus callosum, a structure connecting Einstein's two cerebral hemispheres, was thicker than that of the ordinary, thus the communication between the left and right hemispheres might be more efficient.

In addition, research results on different regions of Einstein's brain, especially on mathematics, language, and computational regions, were endless. However, even neuroscientists argue for three days and nights, it remains equivocal to answer the question of "how far are we from Einstein's brain?" Some even think this is a pseudo-proposition: "Einstein's brain is just different from us, not better than us." After all, even individual achievements can be quantified, the brain functions are extremely complex and lack a single

quantification standard. The fool can be resourceful while the wise can copy mechanically. Otherwise, modern society will not extend a series of parameters that measure brain function such as EQ (Emotional Quotient), SQ (Spiritual Intelligence Quotient) and so on.

Therefore, the right question seems to be asked is "What does a perfect brain look like and how does it work?"

Pioneers Who Unveiled the Mystery of Brain: Let's Start with Golgi and Cajal

In an age when body is sacred and inviolable, we believe that the heart is the organ to generate consciousness and feeling. We are still used to say "heart attack" or "heart break" instead of "brain attack". Like the Renaissance in the 16th century was originated in Mediterranean countries, so did the modern neuroscience. After linked the brain to human's thought and behavior by using anatomy ways, in 1873, the Italian cytologist Camillo Golgi used chromate-silver nitrate to stain brain sections and first described two different types of cells in our brain: neurons and glial cells. It was only then we noticed that the pig brain cooked in hot pot was composed of various kinds of cells. After several improvements, this method of staining brain cells by heavy metal penetrations became the well-known "Golgi staining", which won the Nobel Prize in Physiology and Medicine in 1906. Even though there was a more rapid way to observe the morphology of various neurons and glial cells, or directly observe the activity of neurons in the animal brain, Golgi staining, which had lasted for more than a century, is still the most economical and convenient way to show the shape of neurons.

Inspired by the Golgi staining, Ramon y Cajal, a Spanish professor of anatomy, created a more sensitive method-reducing silver nitrate staining and then discovered the physical contact between the fine structures of nerve fibers (neurites and dendrites) and nerve endings (the "synaptic" structures between neurons). As a result, he shared the Nobel Prize in Physiology and Medicine with Golgi. Cajal is to neurobiology what Darwin is to biology and Mendel to genetics. But different from Golgi who had learnt traditional anatomy and art, Cajal switched to this new profession in the midcourse of his life and was in fact a self-taught middle-aged "rebel". But in Spain, motherland for Picasso, Cervantes and Gowdy, even the geeks in laboratory knows something about art.

Cajal was unsatisfied with using staining to show a few high-end pictures when making reports. As a paintholic, he was admitted to medical school under the pressure of his father, but he finally found his hobbies of his youth by making brain sections (cat brain and

258

chickens brain) and staining. In a small corner of the lab, Cajal put the stained brain sections under an old-fashioned microscope bought with his own pocket money, humming and drawing the neurons and their networks in different brain regions. At the end of the 19th century, when photography was still a long way from commercialization, the general public, with the exception of "Gao Fu Shuai" (person who is tall, rich and handsome) who had microscopes, had no chance to see the real cellular and subcellular structures that the naked eye could not distinguish, and thus he was excluded from the team of cell biologists. Therefore, Cajal's job was equivalent to showing micro-world movies to the masses before the advent of television, and he was both a producer, a photographer and a projector. When I (the author) saw Cajal's drawings of neural networks in the cerebellum and hippocampus at the National Institutes of Health, I was as excited as if the face judger had seen the perfectly handsome man and beautiful woman. In the words of netizens, "the secret garden of neurons was brilliant!"

The Nobel Prize affirmed the technical achievements of Cajal's research, but his contribution to neuroscience went far beyond that. As a talented cross-border professional, Cajal is not only a painter, an athlete and a scientist, but also a great thinker. Based on the observed structure of neurons, he firstly proposed the "neuron hypothesis", boldly challenging the mainstream- "the reticular nerve hypothesis"- at that time. According to the reticular nerve hypothesis, nerve cells connect with each other to form a huge net; There is no barrier between cells thus materials and information can be transmitted within freely. But Cajal thought that each neuron was an isolated functional unit; cells contact but not connect with each other; and neurons have polarity which means that information flows along the neuron in one direction, namely, from dendrite to cell body and then to neurite. In the year without functional experiment, it was not easy to make such a hypothesis on the basis of morphological evidence (even the evidence was particularly sufficient) and this needed imagination. In the scientific circle, hypotheses are based on evidence, and Cajal's rebel against the mainstream angered some of the most prominent figures in the world, especially Golgi. They traded barbs in their Nobel speeches, making the Physiology or Medicine Awards shared by two scientists for the first time extremely embarrassing.

Time (and practice) is the test of truth. In 1894, Charles Sherrington, a British physiologist, supported the idea of "contact but not connect" by studying the structure of neuromuscular junction in knee jerk reflex, and named this connection structure between neurons "synapse". Subsequent studies have shown that synapses contain large amounts of proteins, lipids, and "signaling molecules" (extremely sophisticated structures of

neurotransmitters). The knee jerk reflex also approved the specific direction of information transmission between neurons. Scientists then recorded electrical activity on the nerve fibers, which explained how nerve information was transmitted, namely, electrical conduction. The reticular nerve theory was nailed on the stigma column by the flood of evidence, and was soon abandoned. The world was once again amazed by Cajal's profound insight. But history seldom develops linearly, and the paradox of science never dies. Like the "particle" and "wave" debates in physics, people believed they had seen the finale in neuroscience, but there was a prepared Easter egg at the end. Although it had been proved that the signal transmitted from presynaptic cells to postsynaptic cells, increasing evidence also showed that the information between nerve cells is not completely transmitted in one direction and that postsynaptic cells can regulate the strength of signals released by presynaptic cells to form a feedback loop.

After a century of development, neurobiology, although still an emerging subject in the biology, has become the hottest star in life science under the impetus of molecular biology, electrophysiology, pharmacology and rapidly-developed imageology. On the occasion of celebrating the 125th anniversary of its founding, Science magazine published the most challenging 125 scientific issues for the next half century which included 15 questions of narrow sense of brain science, occupying one eighth of the seats. The more you know, the more questions you have. That's how you explore and learn.

Connectome: I Think, Therefore I Am.

"Who am I?" It is a profound philosophical proposition. In his book The Selfish Gene, Richard Dawkins, a famous British evolutionary biologist, gave the answer: organisms are essentially nothing more than a temporary box for a set of genes to exist in the world. In order to be immortal, these genes choose various kinds of bodies (combinations of different phenotypes), eliminate inferior teammates (genes that are not conducive to survive), select powerful partners (genes that are conducive to survive), and finally lead to the evolution of species. That is to say: I am not who I really am. I am the genome of myself!

Cloning has been invented for a long time. But if you put aside the ethical issues and copy one's entire genetic information, can you get the exact same copy? A logical viewer must seek out the way: apply pure biological methods to copy a cell with the same gene and the two people grown from these cells should be highly similar but inconsistent (see identical twins). One cannot step into the same river twice. Any trivial but different events that occur randomly on two individuals is enough to shape different memories for them and then change their life.

Sebastian Seung, a professor at the Massachusetts Institute of Technology, gave a more precise answer in his TED speech (Technology, Entertainment, Design, a private non-profit organization in the United States): "I'm the connectome of myself." Since the advent of the term "genome", all kinds of "Omics" have been "played" by scientists who crave for great achievements and success. Intelligence and ability alone are not enough for the modern omics research. Integrating talents and resources is the way to go. The term "connectome" was first put forward in 2005, referring to an overall picture of the connections (the collection of all synapses) between neurons in the brain. Isn't it another version of the "Brain Initiative"? Without Obama's efforts to consolidate resources in America, a single laboratory can only do connected omics research on model organisms. You have to know that a 1-millimeter-long caenorhabditis elegans has no brain regions at all, but the mapping of its 302 neurons connectome was not completed until 2012.

Early in 2002, Paul Allen, one of the founders of Microsoft, launched the Allen Brain Atlas Project in his private research institute, aiming to build a three-dimensional map of gene expression in mice and human brain with the guidance of genomics and neuroanatomy. That 100 million dollars were well spent. In the year 2006, the first map of gene expression in the brain of mice was released and by 2012, seven maps had been published, including adult mice brain, developing mice brain, adult human brain, developing human brain, primate brain, mice brain connectome and mice spine. What's more, all of its information is free to the public. The Allen Brain Atlas Project is like a public library where researchers and doctors can easily access to the information, helping thousands of neuroscientists around the world save at least ten years to explore by themselves.

Black Neuroscience Technology: Ask Me Where to Go, and I Will Point to the Brain

Based on molecule, the bottom-up reductionist approach had made great progress in basic research, explaining the mechanism of how the neurotransmitter is released and neural signals are transmitted, the principle of neuroplasticity (learning and memorizing), and the molecular changes in the pathogenesis of brain diseases like Alzheimer's disease and Parkinson's disease. Future seems so bright. But the experience on molecular level is unrepeatable if we turn back to the overall level. Cells in petri dishes or even neurodrugs that work well on model organism, had failed in clinical trials because they only saw

segment of a whole. Without a holistic understanding of the nervous system, it's easy to ignore other important factors. The brain is presented to researchers as a tangled ball of threads. Each researcher holds a thread in his hand with his own idea, but untangling the thread in this way is like milking the bull. Therefore, the Brain Initiative created a top-down methodology that allows researchers to unravel the mysteries of our brain.

The biggest challenges in research this three-pound object come from: 1. variety of nerve cells type; 2. lack of understanding of the anatomy and cell composition in the specific functional neural loop; 3. inability to observe and control the activity of nerve cells in the living brain. What kind of method should we adopt? Experts of the Brain Initiative soon gave their answers: mark-types of nerve cells (brain rainbow technology); trace-neural circuits (CLARITY); observe-high resolution, high sensitivity and high flux ("three highs" for short) imaging technology (super-resolution microscopy frozen electron microscope and laser layer scanning microscopy), control-a specific cell type neural circuits control (optogenetics).

Mark: Deep Down in the Brain Rainbow

At a time when neuroscientists would rather share a toothbrush than data, the friendship two Harvard professors, Jeff Lichtman and Joshua Sanes, were "solid than gold.". They have been working together since the 1990s at St. Louis campus of Washington University. Sanes was an outstanding researcher in the study of retinal neural networks, but this research had been marginalized in neuroscience. Many people believed that it is just an extraneous tissue connected to the brain by the optic nerve, not a part of the brain at all. The retina, on the other hand, is home to various types of neurons (rod cells, cone cells, bipolar cells, and ganglion cells), and is the most complete regional neural network that can be easily separated. As a neurogeneticist, Sanes had long been frustrated by the inability to mark multiple types of neurons at the same time. But Lichtman, who had been focusing on the research of synaptic formation process in neurodevelopment for many years, was good at imaging and was praised by Sanes as "the world's best expert in imaging synaptic formation". These two experts hit it off with each other and started the long journey.

The secret that marine organisms can glow in the dark was solved by the Japanese Shimamura in the 1960s. He found out the green fluorescent protein from the glowing jellyfish, which illuminated the way for biologists. Researchers then genetically inserted the gene of the fluorescent protein into animals and waited for the expression, which allowed them to easily locate and observe target cells. Lichtman was an expert on this. But this time, his challenge was to distinguish the bodies of different neurons in the brain's dense network and sort out the tangled threads they connect with each other (neurite and dendrites). One color was absolutely not enough. By this time, however, several new members of the fluorescent protein family had been observed: red, orange, yellow, and cyan fluorescent proteins.

Next came Sanes' favorite genetic game: they connected red, yellow and blue fluorescent protein genes like beads on a string, and added a genetic recombination site and an expression termination element between each gene, thus only the first gene in this fluorescent protein gene combinations in mouse brain cells can be expressed. A recombinant protease named CRE randomly selects two recombination components and cuts out the middle fragment to obtain different combinations of fluorescent proteins: when there is no recombination, only the first red fluorescent protein is expressed, and the cell is marked with red; when recombination 1 occurs, the red fluorescent protein gene is lost, and the exposed yellow fluorescent protein is expressed, then the cell is marked with yellow; when recombination 2 occurs, only cyan fluorescent protein is expressed and the cell is cyan. By adding a specific gene expression switch (promoter) before the gene "bead string", specific nerve cells can be marked. In addition, the "bead string" can be added with a sub cell positioning GPS system, and the fluorescent protein can therefore be expressed in the cell body (solid circle) or only on the cell membrane (hollow circle) or in the nucleus (dot in the circle). Lichtman and Sanes then "patched" the gene beads with an orange fluorescent protein to upgraded them to version 1.1. In this way, we can obtain cells in four colors.

But game addicts never stop upgrading. Brain Rainbow Version 2.0 was introduced almost simultaneously: four fluorescent protein genes with different colors were linked in opposite directions with recombination sites added between the gene pairs. It is important to know that gene expression is a unidirectional action like reading. Only the first forward direction red protein being read out of the opposite pairs of red and blue genes. In the process of a recombination, the direction of gene pair will be reversed, and the original reverse cyan gene will be read out. When the green, yellow, red and cyan protein genes are paired together, the four different ways of recombinations can led to cells of four colors. how many colors of cells can you get when you introduce these gene beads into nerve cells? If your anwser is four, you are "too young too simple." Permutate and combinate the colors? Wrong! How can a complex biosome give us such a simple answer! The brain developed from an oosperm has undergone numerous mitosis and in each process a recombination may occur which will express a different color of fluorescent protein. As a result, the color displayed by each neuron in the adult brain of transgenic mice is the superposition of four kinds of fluorescent protein colors, some of which are more red, some of which are slightly green, some of which look purple (the effect of superposition of red and cyan). Theoretically, there are countless colors that can be produced! However, due to the limitation of the wavelength of the microscope and that of the naked eye to distinguish, we can only observe and distinguish about 100 kinds of colors.

Many people are familiar with the theme song "Over the Rainbow" of the movie "Wizard of Oz". In 2007, the British magazine Nature launched a report called "Deep in the Brain Rainbow". In the groundbreaking technique that Lichtman and Sanes collaborated to create, scientists finally broke through the bottleneck of "Golgi Staining" that can only stain few cells with single color. Under the fluorescence microscope, neurons in mice brains are as colorful as the television picture tubes. The "Brain Rainbow" technology is not only dazzling, but also enables scientists to mark and trace the neural circuits of animals in a long distance instead of being limited to a certain brain region, and to observe how neurons are connected to the neural network. In addition to the research on transgenic mice and lower-level model organism drosophila, genetic manipulation has made the application of "Brain Rainbow" more convenient. Later, Lichtman and his team improved their method by studying the biological brain with viral infection and increasing the efficiency of fluorescent protein expression. In this way, we could see the beautiful rainbow in the brains of ordinary animals. Although "Brain Rainbow" technology makes it easier to mark and distinguish different cells in neural circuits, its limitations are also obvious: it can only study neural cells which express fluorescent protein. Objective conditions like transgenic animals, virus infection rate and so on are of vital importance; It is necessary to stack many brain sections to get a complete neural network image, and this method in itself is hard to achieve.

单词释义：

1. 阿尔茨海默病   alzheimer disease(AD)
2. 帕金森病   parkinson's disease(PD)
3. 精神分裂症   schizophrenia
4. 自闭症   autism
5. 防腐剂   preservatives

6. 顶叶　　　　　　　parietal lobe

7. 胼胝体　　　　　　corpus Callosum

8. 轴突　　　　　　　neurite

9. 树突　　　　　　　dendrite

10. 突触　　　　　　　synapse

11. 铬酸盐-硝酸银　　chromate-silver nitrate

12. 高尔基染色法　　　golgi staining

13. 亚细胞结构　　　　subcellular

14. 网状神经假说　　　reticular nerve hypothesis

15. 神经递质　　　　　neurotransmitter

16. 基因组　　　　　　genome

17. 秀丽隐杆线虫　　　caenorhabditis elegans

18. 脑虹　　　　　　　brain Rainbow

19. 视网膜神经网络　　retinal neural network

20. 启动子　　　　　　promoter

译文解析：

1. 直译法

原文：可上九天揽月，可下五洋捉鳖。

译文：You can clasp the moon in the Ninth Heaven and seize turtles deep down in the Five Seas.

解析：这句话的翻译出自毛泽东的词《水调歌头·重上井冈山》，译者直接将原文的内容不加修饰的翻译出来，一个是能够原封不动的表达出全文的气势，还有一个就是能够感染读者。

2. 意译法

原文：世上从不缺乏机智过人屠狗辈，也不鲜见生搬硬套状元郎。

译文：The fool can be resourceful while the wise can copy mechanically.

解析：这句话是全文中最难翻译的，首先是因为他的中文原文就已经让人难以理解，甚至读不通顺。所以在翻译时用直译一个是不能完整地找到一一对应，还有一个就是显得译文十分僵硬。因此，在对这句话的译文进行处理时，我直接采用了意译的方法，现将原文用通俗的话翻译一遍，在进行翻译。

3. 减译法

原文：在"身体发肤受之父母"、身体完整性神圣不可侵犯的年代，人们认为心才

是意识和感觉的器官。

译文: In an age when body is sacred and inviolable, we believe that the heart is the organ to generate consciousness and feeling.

解析:这里运用了减译,原因是原文中身体发肤受之父母已经包含了身体神圣不可侵犯的意思了,所以删去哪一部分其实都可以,在这里我选择删去身体发肤受之父母那一句是因为在英语读者的思想中并没有中国古老的孝道文化,相反,他们更重视自己的权利,讲究个人主义,所以留下身体神圣不可侵犯反而更能引起读者共鸣。

4.增译与衔接连贯

原文: 汉斯提出了严格的条件:对其父亲大脑的研究,必须发表于高水平的科学刊物上。几十年过去了,哈维也没什么像样的相关科研成果发表。

译文: Hans put forward a strict condition for him: The research on his father's brain must be published on high-level scientific journals. However, decades had passed and Harvey had not published any valuable research results about it.

解析:原文中并没有出现任何转折连词,但是我在翻译的时候却将其包含的转折关系直接翻译了出来。上文说的是汉斯提出要求,但是下文却说得是哈维的失败,仔细体会会有一种期望落空,无奈的感情色彩。如果不进行增译,译文会显得较为突兀,这是因为英文比较注重逻辑,因此我在此进行了增译。

第十五章　深度学习

　　人工智能一直以来就是高科技行业中经久不衰的话题。"深蓝"电脑、百度大脑、微软的实时翻译系统、谷歌的无人驾驶汽车、亚马逊正在试验的能送快递的无人机,在这创新浪潮中,人工智能开始悄无声息地渗透进我们的生活。

　　在人工智能领域,深度学习(deep learning)是这一波高科技革命潮流中最璀璨的明珠。深度学习大大提高了计算机识别图像和语音的精度,在有些测试上还超过了人类的识别精度,比如 ImageNet 的图像识别测试、LFW 的人脸识别测试等。由于深度学习的卓越性能,各大高科技公司都在打造自己的研发团队,比如百度的深度学习研究院、谷歌的谷歌大脑团队、Facebook 的人工智能研究院等,还有无数的创业公司投入到深度学习的研发当中,创业时间不长就能取得很高的估值。那么为什么深度学习有如此强大的能力,甚至在某些领域超过人脑? 为什么高科技行业要投入如此多的资源在这个研究方向上? 想要把这个问题研究清楚,我们就要回到三十多年前,从深度学习的前身"人工神经网络"的诞生说起。

感知器:神经网络的第一次兴起和衰落

　　1958 年盛夏,美国华盛顿特区,生活像往常一样平静,政府办公区内西装革履的公务员来回穿梭着,博物馆里满是放假前来学习的孩子们(还有辛苦的家长)。可能很少有人知道,就在这个城市,一件人工智能领域里程碑式的事件就要发生。Frank Rosenblatt,康奈尔航空实验室(Cornell Aeronautical Laboratory)的科学家和资助他的美国海军一起在华盛顿举行了记者招待会,宣布拥有学习和认知功能的计算机——马克一号(Mark-I)的诞生。马克一号的理论基础是感知器算法(Perceptron),该算法现在已经成为人工智能领域的经典算法。通过对输入的数据进行分析,感知器算法可以根据所犯的错误调整自身的参数,从而达到学习的目的。

　　根据神经元算法,马克一号可以识别简单的字母图片,这在当时引起了科技界一片惊叹。当年 7 月 1 号的《纽约时报》报道:"海军军方向公众展示了一台可以讲话行走、观看、写作和自我认知的机器原型。"然而这仅仅是美好的理想,1969 年,麻省理工学院的两名科学家在他们合著的书中证明,单层神经元算法有很强的局限性,甚至无法学习到"异或"(exclusive or)表达式的规律。人工神经网络相关的研究因此停滞了10年之久,这被称为人工神经网络的第一个冬天。然而黑暗中却也蕴藏着希望,虽然单层的神经元算法能力有限,但如果将多层神经元连接起来,就可以创造出功能强大的多层神经网络,这为20世纪80年代神经网络研究的复兴打下了基础。实际上,现在火爆的深度学习中的"深度"一词,就是指神经网络层数的加深。

神经网络的结构

最简单的神经网络有三层组成:输入层、隐含层和输出层。

· 输入层:将输入的数据转化成一串数字;

· 隐含层:根据输入层的数字,计算出一组中间结果;

· 输出层:根据隐含层得到的中间结果,做最终的决策。

举个判断颜值的例子来说明这几层的作用:

· 输入层将他(她)的照片转化成一组数字,比如每个像素的 R/G/B 值(如果照片有 100 万像素,那么输入层总共有 300 万个数字);

· 隐含层计算出该长相的几个特点,比如身高、腿长、胖瘦、眼睛大小、鼻梁高低等;

· 输出层根据这些特点输出颜值的评分。

在上述过程中每一层神经元的输出取决于三个要素:

1)上一层神经元的输出;

2)层与层之间的连接:也就是神经网络结构图中的实线。如果不同层中的神经元之间有连接,那么这条连接左边(前一层)的神经元对连接右边(后一层)的神经元就有影响。影响的大小取决于这条连接的权重。权重越大,影响就越大。还是以判断颜值的应用为例:这个例子中,隐含层有一个神经元专门负责计算鼻梁的高低,那么这个神经元应该与输入层中描述鼻子附近像素的神经元建立高权重的连接,而与输入层其余的神经元只有微弱的连接关系;

3)激励函数:在神经网络工作时,每层神经元根据前一层的输出和对应的连接权重做加权求和后产生一个数值。而这个数值需要经过一个非线性变换后再传递到下一层去。这个非线性变换就是通过激励函数来实现的。实际应用中常见的函数包括 sigmoid 函数和 tanh 函数等。以 sigmoid 函数为例,激励函数可以理解为将任意一个输入数值转化成 0 ~ 100 的分数的过程。还以神经网络判断颜值为例,当输入图像非常极端时,通过激励函数产生的颜值分数也很极端:或是 0 分(惨不忍睹)或是 100 分(惊为天人)。当输入的图像是一张正常的人脸时,通过激励函数生成的分数就具有不确定性。这时,如果输入的脸过于平庸,神经网络对自己的决策就最不确定,给出的分数也最模糊(50 分,也就是颜值一般 ,不高不低)。

激励函数的作用类似于神经细胞的信息传导。信息传导的一个重要理论是"全有全无律"(All-or-none-law),就是说一个初始刺激,只要达到了阈电位,就能产生离子的流动,改变跨膜电位。而这个跨膜电位的改变能引起临近位置上细胞膜电位的改变,这就使得神经细胞的兴奋能沿着一定的路径传导下去。而跨膜电位改变的幅度只与初始刺激是否达到阈值电位有关,与具体的初始刺激的强度无关。这个过程与神经网络中激励函数的性质非常相似。

复杂的神经网络的隐含层一般有多个,隐含层数量越多,表达能力越强,可以解决的问题就更多。深度学习之所以优于早期的神经网络,主要是它可以从大规模数据中有效地训练出更多隐含层。那么深度学习模型中不同隐含层的关系是什么呢? 以图像为例,近期的研究显示,最初的隐含层,一般表示简单的图像特征,比如边缘、直角、曲线等。后面的隐含层可以表达更高级的语义特征,比如汽车轮胎、建筑物的窗户、动物的头部轮廓等。

现有公开的深度学习模型最多可以有 150 层网络,但无论结构多么复杂,只要知道了输入和网络的结构和参数,很容易根据之前的公式计算出最后的输出。具有挑战性的是如何根据已有的数据,优化计算出网络的参数(最重要的参数是神经元连接的权重),这就引出了一位人工智能领域的传奇人物 Geoff Hinton 和他发明的人工智能领域最伟大的算法(之一):后向传播(back propagation)。

神经网络的优化

20 世纪 80 年代的一天早上,加州大学圣迭戈分校心理学系的 David E. Rumelhart 教授来到办公室,看到神经网络算法在计算机上通宵运算输出的结果,露出了欣慰的笑容。因为结果表明,他和同事 Geoff Hinton 等人发明的后向传播算法,可以模拟布尔代数中的"异或"运算符以及更复杂的函数,这比 20 年前 Frank Rosenblatt 发明的感知器算法进了一大步。

后向传播算法解决的是多层神经网络的优化问题。神经网络的优化是基于一组训练数据的参数调整过程。这个过程类似老师教学生做题。不同的是在学校里老师教学的顺序一般是给出题目,再给解题思路,最后给出答案。而在科研人员优化神经网络时,会给它大量的题目和答案,让神经网络自己去寻找解题的方法。所谓"训练数据",也就是这大量的题目和答案的组合。比如上一节最后,识别汽车图片的例子中:汽车的图片就是题目,汽车的类型(跑车、卡车、越野车等类别)就是答案。大量的图片和对应的汽车类型就构成了训练数据。优化过程完成后,解题的方法就蕴含在神

经网络的各个隐含层之中。

在这个过程中,最初几层参数的优化最困难,因为它们与神经网络输出的函数关系最复杂。Rumelhart 和 Hinton 教授发明的后向传播算法,巧妙地利用了神经网络层与层之间的递推关系,从最后一层的参数开始,逐层向前优化。每一层参数的优化只和它后一层的参数有关,这就大大简化了需要表达的函数关系,有效地解决了多层神经网络的优化问题。

后向传播算法被发明后,神经网络在科技界又流行了一段时间。1987 年 9 月 15 日的《纽约时报》以"计算机学会了学习"为题报道当时神经网络的研究进展,列举了从大学到公司利用神经网络的例子,其中之一是一家金融公司利用神经网络对贷款申请的历史进行评估测试,测试的结果是可以使公司在该历史区间的利润增长近 30%,这很像是今天基于大数据的互联网金融的雏形。著名的科幻电影《终结者》系列中阿诺德施瓦辛格主演的机器人的一句台词就是"我的 CPU 是神经网络处理器,一个可以学习的电脑"。然而好景不长,神经网络虽然建模功能强大,但由于当时的数据量有限,神经网络模型难以解释等问题导致人们对神经网络的热情减弱,开始偏向更简单、易于解读的模型,比如支持向量机(Support Vector Machine)。支持向量机对不同种类样本"间隔最大化"(maximize margin)的思路在样本数量有限时非常有效,所以在缺乏大规模训练数据的时代受到了科研人员的广泛欢迎。在整个 20 世纪 90 年代和 21 世纪初,神经网络的研究陷入了低谷,在各个人工智能领域的前沿学术会议上很少能看到神经网络的论文。如果在开会时一个科学家跟其他参会人员说他(她)是做神经网络研究的,听到的人便会露出不解的表情,好像在说为什么要去研究这种过时的东西?可是,只要合适的时机到来,是金子总会发光。

21 世纪神经网络的复兴:深度学习横空出世

2013 年春天的多伦多,谷歌办公室坐落在繁华的金融区内,毗邻美丽的多伦多港。Hinton 教授经常到这里办公,将人工智能的新进展融于谷歌的产品中。虽然在这里办公的主要是负责市场和销售的同事,几乎没有做科研的员工,但在一些社交活动中,Hinton 还是会和大家自我介绍一下,比如这样:

同事甲:经常看您若有所思,或低头思考,或写程序,您是从事什么工作的呢?

Hinton:你是用安卓手机吧?

同事甲:啊,是的。

Hinton:里面的语音识别好用吗?

同事甲:还不错,语音搜索比较准确。

Hinton:嗯,这就是我做的。

在让多伦多同事崇拜（或不解）之外，Hinton 教授没有提到的是，他的公司（DNNResearch Inc.）之前刚刚卖给了谷歌，据传收购金额为上亿美元，而公司核心员工只有 Hinton 和他的两个学生，公司的核心技术就是深度学习的高效优化算法及其在计算机视觉、语音识别、自然语言理解中的应用。同一时期，除了 Hinton 教授，当年同时发明后向传播算法的 Lecun 教授被 Facebook 重金聘请为人工智能实验室的研究总监。各个大学的研究团队也纷纷转向深度学习相关的科研领域，仿佛没有在学术会议发过深度学习的文章就落后于时代了。那么深度学习相对于传统的神经网络做了哪些改进，足以受到学术界和业界的科研人员的集体追捧呢？

当神经网络遇上大数据

2011 年谷歌和斯坦福大学的科学家联合发表了一篇文章，利用谷歌的分布式系统来训练超大规模神经网络，把 Youtube 的视频输入到该神经网络中，在没有任何人为干预只有视频中的图片的情况下，该神经网络可以自行识别视频中的人脸、人体，甚至还有猫脸（Youtube 上有很多萌猫的视频）。这个神经网络有 10 亿个神经元连接，动用了 1000 台机器、16000 个 CPU，在 1000 万个图像上进行优化。这篇文章在当时引起了巨大的反响，一个原因是它精准的结果，神经网络可以自发地找到在视频中反复出现的物体。另一个原因是这项科研项目在神经网络规模上的突破。20 世纪 60 年代到 80 年代神经网络衰落的原因就是数据的缺乏和高性能计算资源的不足，导致无法在很大规模数据集上优化神经网络，随着 20 世纪末、21 世纪初互联网和分布式系统的兴起，大数据的相关技术日趋成熟，这对深度学习的发展起到了巨大的推广作用。2012 年，谷歌又在神经处理年会上发表了另外一篇文章，详细阐述了如何利用分布式系统来训练大规模神经网络。

除了分布式系统，专用硬件的出现也促进了深度神经网络的发展。比如显卡（GPU）、为了某种算法定制的现场可编程逻辑门阵列（FPGA）和专用集成电路（ASIC）。专用硬件提供了强大的并行计算性能，神经网络的训练速度达到了质的飞越。除了系统层面的性能改进，算法的改进也使得深度学习的实用性有所提高。下面我们就分几个方面来总结近期使深度学习成为机器学习主流方向的几个技术进展。

分布式模型

对于超大规模神经网络，一台机器很难处理一个模型中所有的参数。所以需要将一个模型分解成不同的部分，分布到多个机器中，这样的一组机器就是一个模型，称作"模型副本"（Model Replica）。

利用每一台机器只处理一小部分模型参数的思路,我们只需增加机器就可以产生很大规模的神经网络。唯一的困难是有的神经网络连接需要横跨不同的机器。这需要不同机器进行通信,而通信会带来网络带宽开销和计算上的延时。好在很多问题,比如图像处理,神经网络只需要"局部连接",也就是下一级的神经元只处理上一层中有限的几个神经元传递过来的信息。这就好像我们将一个图像分成几个子区域(比如蓝天的部分、地面的部分、建筑物的部分等),然后分别用神经网络模型中不同的神经元进行处理,得到不同子区域的处理结果后再汇总。这种机制大大减少了不同机器之间的神经元连接,使得不同机器间通信的成本降低,提高了分布式模型方法的计算效率。

并行数据处理

分布式模型算法解决了对超大模型参数的存储问题。而并行数据处理解决的是如何利用大数据对神经网络模型进行优化的问题。这里我们借用日本经典动漫《火影忍者》中的一个桥段来进行类比,以便大家理解。《火影忍者》的主角漩涡鸣人有一个重要的技能叫作"影分身",就是瞬间创造出多个自己的克隆体(类似于模型副本)。这个技能除了用于打怪兽之外,一个重要的用途是利用多个影分身的头脑进行快速学习。基本思路就是让每个影分身同时学习一项新的高级忍术,学习一段时间让影分身回归本体,将学习的经验整合,这样就可以大大加快学习的进度。并行数据处理就是利用类似的思路来加快神经网络的优化。在具体的优化算法方面,随机梯度下降(SGD)由于其简单易用的特性而广受欢迎。然而传统 SGD 方法很难并行化,因为该算法每次对训练数据集中的一条数据做处理,更新模型,然后处理下一条数据时以前一次处理的结果作为基础。这样的算法顺序性太强,虽然简单有效却很难利用多台机器加速计算。为了将不同的训练数据分布到不同的机器上同时计算,需要利用异步 SGD 的方法。

该算法的特点是每个"模型副本"只对输入数据的一部分进行优化,不同模型副本的优化结果通过一个"参数服务器"集群进行通信。每一个模型副本定期向参数服务器上传自己的优化结果,并下载最新的模型参数继续优化。如果说这里每个模型副本是一个影分身,那么参数服务器就是本体的大脑,在不断接受和归纳所有影分身学到的知识。这种异步的方式利用分布式系统来处理大规模的训练数据,使得我们可以用几万台机器来训练同一个模型,几个小时就可以完成以前需要计算几周的结果,因此利用大数据产生大规模神经网络的技术成为可能。

硬件加速

在计算机内部,除了通用计算单元 CPU 之外,还有用于图形显示的专用芯片,叫作 GPU。GPU 最早由英伟达在 1999 年发明,用于处理图像的变形、光照变化、渲染引擎等,主要用于游戏和电影特效。基本上在每一个 3D 游戏发烧友的电脑里都会有一块性能强劲的 GPU。GPU 的很多应用都涉及矩阵和向量运算,它的体系结构是为了快速并行的矩阵运算设计的。在单个机器上的并行计算一般有处理器上的多个核(core)来协作完成。比如现在一般的家用电脑有四核(core)就算是标准配置了,而 GPU 为了提高并行计算性能可以有成百上千个更小型且计算效率更高的核,比如英伟达在 2006 年推出的 GeForce8800 显卡有 128 个核;而 2013 年推出的 Tesla K40 有 2880 个核。另外,为了支持快速并行计算,GPU 的缓存系统也进行了专门的设计,减小了缓存的容量,提高了访问内存的带宽(每秒内存读写的数据量)。因此在矩阵计算这个特殊的领域,GPU 的优势不言而喻。因为神经网络在每一层之间的信息传递在数学上也可以表示成为矩阵运算,所以随着 GPU 性能的加强,科研人员就开始尝试利用 GPU 对神经网络的计算进行加速。比如 2012 年取得 ImageNet 图像识别测评第一名的 Hinton 教授的团队,在一台机器上利用了两个 GPU 训练同一个神经网络,取得当年最好的评测成绩。

深度学习的发展使英伟达公司看到了 GPU 在嵌入式自动化系统方面(譬如未来可以应用在可穿戴设备中、无人机、无人车)的潜力,于是将之作为公司重要的战略方向。这方面英伟达的最新产品就是 Jetson 系列的嵌入式芯片,该芯片只有信用卡大小,却有 CPU、GPU、内存等计算机的重要组件,可以在该芯片上实时运行深度学习算法。该芯片可以用于无人车、无人机等智能机器平台。可以预想英伟达在这方面的持续努力肯定可以带动深度学习的实际应用。除了英伟达之外,高通(Qualcomm)、IBM 等也在研究能够部署深度学习的芯片。将复杂的深度学习模型部署到低功耗低成本的芯片上将是芯片工业诞生下一个十亿美元公司的风口。

算法改进

除了硬件方面的提升,在软件方面一个重要的改进方向就是对优化出的模型进行压缩。比如将神经网络参数从一个浮点数压缩成一个数值范围有限的整数。或者设法减少每一层神经元的数量和不同层之间的连接数。这有两个好处。首先模型压缩可以提高模型的运算速度,这对很多需要处理实时数据的行业至关重要。比如在安防行业中,如果发现有可疑人员出现深度学习模型应该在第一时间做出响应,向监控部门发出警告。如果有一定的延时,可能会造成无法挽回的损失。其次,这种优化可以使得神经网络占用计算资源更小,更适合应用在计算资源有限的环境中,比如手机、机器人、智能手表等。除了速度上的改进,能够让超大规模的神经网络从大数据中学到有用的知识,而不是噪音,科研人员还提出了优化算法上的改进。比如有一种简单实用的方法叫作 Dropout 就是在训练每一层神经网络时,随机将一些神经元的输出强制为 0,也就是弃用(Drop)这个神经元。这种做法可以认为是将一个大型神经网络拆解成多个小规模神经网络,对它们的输出求平均。在求平均时,数据中的噪音便可以被有效降低。另外,更简单且非对称的激励函数(Rectified Linear Units)也被证明可以更快速有效地完成神经网络的学习。

开源战略

现在深度学习有很多开源软件可供选择,最受欢迎的是 Theano、Torch 和 Caffe。而 2015 年 11 月,谷歌也发布了自己深度学习系统 TensorFlow 的开源版本。TensorFlow 的特点是编程界面简单易用,模型开发出来后可以方便地部署到云端、手机、GPU 等多种计算平台上。由于 TensorFlow 采用了数据流图的结构,使得模型训练时不需要依赖单独的服务器(Parameter Server)来维护模型参数的状态,这使得 TensorFlow 的结构被大大简化了。作为谷歌内部最新一代的系统,TensorFlow 已经应用到多个产品部门中,所以开源 TensorFlow 的决定震惊了工业界和学术界。其实谷歌的这一步,也有制定行业标准的意味,为自己的云计算战略服务。最近几年很多重要的开源软件的设计思想都源自谷歌发表的论文,比如 Hadoop 和 HBase 分别来自于谷歌的 MapReduce 和 Big-Table。这本身倒是没什么问题,谷歌与同行分享心得本是好事。可是在云计算时代技术标准非常重要,假如所有研究人员都用 TensorFlow 做模型,那以后如果他们需要利用云计算扩展自己模型的规模,考虑到已有的系统和云平台的兼容性,自然首选谷歌的云服务。所以开源 TensorFlow 除了推动整个深度学习领域的进步,也是谷歌在商业拓展下的一步好棋。

关于未来的大胆设想

20XX 年的某一天,小明下班后坐上自家的无人驾驶汽车去赴一个饭局。虽然路上很堵,但小明仍然可以在座位上听音乐上网。网站上有一页写到历史上的汽车,小明看得饶有趣味,不知不觉就开到了饭馆。下车的时候小明在想,以前的人们真是不容易,要考驾照,要在这么拥堵的路上浪费珍贵的生命来驾驶,还要找停车位,买行车保险,这么麻烦还真不如不买车。下车之后,小明让汽车去附近做专车拉活挣钱,饭馆的机器人服务员将小明带到朋友所在包间吃饭。

没有人能准确地知道让以上场景变为现实还需要多长的时间,但无可否认的是深度学习已经在图像和语音识别等基础领域取得了重要的突破,而这些正是上述场景实现的基础。未来的深度学习一方面有可能在目前尚不成熟的自然语言理解上取得更大的突破,另一方面有可能会被部署到更多的平台中,使得用户可以方便地体验深度学习算法带来的便利。相信随着相关软件和硬件的不断完善,深度学习的模型会变得更快更有效,在某些任务上比人类做得更好,成为下一波信息产业革命的重要推动力量,带动交通、安防、工业制造、电子商务等行业的跨越式发展。

虽然目前国际上最领先的人工智能技术掌握在谷歌、微软、IBM 等公司中,但我国的人工智能技术与国际领先水平差距不大。一个重要的原因是在 21 世纪初的互联网革命中,我国涌现了大量优秀的互联网公司,它们的技术和数据的积累为深度学习技术研发提供了有利的基础。在国家层面,国务院在 2015 年发布了《关于积极推进"互联网+"行动的指导意见》,其中专门有一节论述"互联网+"大战略下人工智能的发展规划。相信在政府部门和企业的一起努力下,我国深度学习技术的研究可以后来居上,并应用到不同的行业中,推动国家整体科技实力的提升。

参考译文:

Deep Learning

AI has always been an enduring topic in the high-tech industry. AI has begun to penetrate our lives silently in the wave of innovation, for instance "Deep Blue" computer, Baidu Brain, Microsoft real-time translation system, Google unmanned vehicles and Amazon UAVs that can deliver express delivery during the course of testing.

DL (deep learning) catches people's eyes most in this wave of high-tech revolution in the field of artificial intelligence. DL has greatly improved the accuracy of computer recognition of images and speech, and in some tests it has surpassed the recognition accuracy of humans, such as ImageNet image recognition test, LFW face recognition test, etc. For the outstanding performance of DL, all major high-tech companies are building their own research and development teams, such as Baidu Deep Learning Research Institute, Google Brain Team, and Facebook institute of AI. At the same time, there are countless start-up companies engaged in the research and development of DL, and they can achieve high valuations within a short period of time. So what are the reasons for powerful DL, even exceeding human brains in some fields? Why are so many resources in high-tech

industry invested to this research direction? To deal with this problem, we have to talk about the birth of the predecessor of DL—— Artificial Neural Network back to more than thirty years ago.

Perceptron: the First Rise and Decline of Neural Network

In the midsummer of 1958, Washington D. C. , life was as quiet as usual—— in the government office area, civil servants in suits shuttled back and forth and museums was crowded with children (and parents working hard) who come to study on vacation. Probably few people knew that there would be a landmark event in AI field right here. Frank Rosenblatt, a scientist in Cornell Aeronautical Laboratory and US Navy who funded him announced that Mark-I, a computer with learning and cognitive functions was born. The theoretical foundation of Mark-I is perceptron algorithm, which has been classic in AI field. Perceptron algorithm can learn by analyzing data input and adjusting its own parameters according to the mistakes.

According to neuron algorithm, Mark-I could recognize simple letters and pictures, which caused a shock in the scientific and technological circles at that time. The New York Times July 1st of that year reported: "The Navy showed the public a machine prototype that could speak, walk, write and cognize itself. " However, this was just a sweet ideal. In 1969, two scientists in MTI demonstrated in their co-authored book that single-layer neuron algorithms have strong limitations, and they even can't learn the rules of exclusive or expression. Therefore, researches on Artificial Neural Networks (ANN) had stalled for 10 years, which was called the first winter of it. However, there were also hopes in the dilemma. Although single-layer neuron algorithms have limited capabilities, if multi-layer neurons are connected, a powerful multi-layer neural network can be created, which laid a foundation for the revival of neural network researches in the 1980s. In fact, the popular word "deep" in "deep learning" now refers to the deepening of the number of neural network layers.

Structure of the Neural Network

The simplest neural network consists of three layers: the input layer, the hidden layer, and the output layer.

· Input layer: convert the input data into a series of numbers;

· Hidden layer: According to numbers from the input layer, calculate a set of intermediate results;

　　· Output layer: Make final decisions based on the intermediate results obtained from the hidden layer.

　　Take an example of judging the face score to illustrate the role of these layers:

　　· The input layer converts the his photo of him or her into a set of numbers, such as the R/ G/ B value of each pixel (If the photo has 1 million pixels, then there would be 3 million numbers totally in the input layer.) ;

　　· The hidden layer calculates several characteristics of the appearance, such as height, leg length, fat level, eye size, nose height, etc. ;

　　· The output layer outputs the degree of the face score based on these characteristics.

　　The output of each layer of neurons in the above process depends on three factors:

　　1) The output of neurons from the previous layer;

　　2)The connection between layers: that is the solid line in the neural network structure diagram. If there is a connection between neurons in different layers, the neuron connected to the left (the previous layer) will have effects on the neuron connected to the right (the next layer). The magnitude of impact depends on the weight of this connection. The greater the weight, the greater the impact. Taking applications to judging face scores as examples, we can find: In this example, there is a neuron in the hidden layer specially responsible for calculating the height of the bridge of the nose; this neuron should establish a high-weight connection with neurons describing pixels near the nose in the input layer, while only having a weak connection with other neurons in the input layer.

　　3) Excitation function: When the neural network works, each layer of neurons performs a weighted sum based on the output of the previous layer and the corresponding connection weights to generate a value. This value needs to undergo a non-linear transformation before being transferred to the next layer. This non-linear transformation is achieved by the excitation function. Common functions in practical applications include the sigmoid function and the tanh function, etc. Taking the sigmoid function as an example, the excitation function can be understood as the process where any input value is converted into the score from 0 to 100. Take the neural network to judge the face score as an example; when the input image is very extreme, the face score generated by the excitation function is also very extreme: either 0 (horrible) or 100 (fair). When the input image is a common face, the score generated by the excitation function is uncertain. At the time, if the input face image is too plain, the neural network is not sure about the decision made by itself, so the score given is also the most vague (50, that means the face score is mediocre,

not high or low).

The function of the excitation function is similar to the information transmission of nerve cells. An important theory of information transmission is "All-or-none-law", which means, an initial stimulus, as long as it reaches the threshold potential, can generate the flow of ions to change the transmembrane potential. This can cause changes in the cell membrane potential at nearby locations, which allows the excitability of nerve cells to be transmitted along a certain path. The extent of the change in transmembrane potential is only related to whether the initial stimulus reaches the threshold potential and has nothing to do with the intensity of the specific initial stimulus. This process is very similar to the nature of the excitation function in neural networks.

Complex neural networks usually have multiple hidden layers, and if there are more hidden layers and stronger expressiveness, the more problems can be solved. The main reason why DL is better than earlier neural networks is that it can train more hidden layers from large-scale data effectively. So what is the relationship among different hidden layers in a DL model? Taking images as examples, a recent research shows that initial hidden layers are generally represent simple image features, such as edges, right angles, curves, etc. The hidden layers behind can express more advanced semantic features, such as car tires, windows of buildings, contours of animal heads, etc. .

The public existing DL models can have at most 150 layers of networks, but no matter how complex the structure is, as long as you know the structures and parameters of the networks and the input, it is easy to calculate the final output according to the previous formula. The challenge is how to optimize the network parameters based on the existing data (the most important parameter is the weight of neural connections), which led to a legend in AI field——Geoff Hinton and (one of) the greatest algorithms he invented in the field: back propagation.

Optimization of Neural Networks

One morning in the 1980s, Pro. David E. Rumelhart from the Department of Psychology at the University of California, San Diego came to the office and saw the output of the neural network algorithm's overnight calculation on a computer with a gratified smile, because the results showed that the back propagation algorithm invented by him and his colleague Geoff Hinton and others can simulate the "exclusive or" operator and more complex functions in Boolean algebra, which is a big step forward from the perceptron algorithm invented by Frank Rosenblatt 20 years ago.

Back propagation algorithm solved the problem of the optimization of multi-layer neural networks. The optimization of neural networks is based on the parameter adjustment process of a set of training data, which is similar to the teacher teaching students to solve problems. The difference is that the order of teacher teaching at school is generally to give the problem, then give the idea of solving the problem, and finally give the answer. When the researcher optimizes the neural network, a large number of questions and answers given, and him or her will let the neural network find the method of solving the problem by itself. The so-called "training data" is the combination of the massive questions and answers. For example, in the example of identifying a car picture at the end of the previous chapter: the picture of the car is the question, and the type of car (sports car, truck, off-road vehicle, etc.) is the answer. A lot of pictures and their corresponding car types constitute the training data. After the optimization process, the method of solving the problem is contained in each hidden layer of the neural network.

In this process, the optimization of the first few layers of parameters is the most difficult because they have the most complex functional relationship with the output of the neural network. The back-propagation algorithm invented by Pro. Rumelhart and Hinton skillfully takes advantage of the recurrence relations between the layers of the neural network to optimize it layer by layer, starting from the parameters of the last layer. The optimization of the parameters of each layer is only related to the parameters of the layer behind it, which greatly simplifies the functional relationships that need to be expressed and effectively solves the problem of the multi-layer optimization of neural networks.

After the invention of back propagation algorithm, neural networks became popular again in the scientific and technological circles for a while. The New York Times on September 15, 1987 reported on the research progress of neural networks at the time with the title "Computers Learned to Learn", citing examples of using neural networks from universities to companies, and one of which is a financial company using neural networks to evaluate and test the history of loan applications. The results of the test can increase the company's profit in this historical interval by nearly 30%, which is very similar to the prototype of Internet Finance based on big data today. The line of the robot acted by Arnold Schwarzenegger in the famous science fiction movie "Terminator" series is "My CPU is a neural network processor, a computer that can learn". However, every day is not Sunday. Although the modeling of neural networks is powerful, due to the limited data size then, problems such as the difficulty of interpreting the neural network model caused people's enthusiasm for neural networks to weaken, and they began to prefer simpler models which

are easier to be read, such as support vector machines. The thought of support vector machines maximizing margins of various samples gained wide popularity among researchers during the era when large-scale training data is lacked, because it was very effective when the size of samples was limited. Throughout the 1990s and early 2000s, the research on neural networks fell into its rock bottom, and papers on neural networks rarely appeared at cutting-edge academic conferences in various fields of AI. If a scientist tells other participants at the meeting that he or she was doing neural network researches, people who heard it would have a puzzled look, as if saying why study this outdated thing? However, his (her) time would come with an appropriate opportunity.

The Revival of Neural Networks in the 21st Century: the Unique Viable Birth of Deep Learning

In the spring of 2013 in Toronto, Google's office was located in a bustling financial district next to the beautiful Port of Toronto. Pro. Hinton often came to work here, incorporating new advances in AI into Google's products. Although colleagues responsible for marketing and sales worked here mainly, almost no staff responsible for scientific researches, Hinton will still introduce himself to everyone in some social activities, such as:

Colleague A: I often see you seem to be absorbed in thought, or lower the head to think, or writing programs. What kind of work do you do?

Hinton: Are you using an Android phone?

Colleague A: Ah, yes.

Hinton: Is the speech recognition inside easy to use?

Colleague A: Not bad, the voice search is accurate.

Hinton: Well, I made it.

In addition to being admired (or confused) by his colleagues in Toronto, what Pro. Hinton did not mention is that his company (DNNResearch Inc.) had just been sold to Google, and it was told to be worth hundreds of millions of dollars. Hinton and his two students are the company's core staff; the core technology is the efficient optimization algorithm for DL and its applications in computer vision, speech recognition, and natural language understanding. In the same period, in addition to Pro. Hinton, Pro. Lecun who invented backward propagation algorithm contemporarily was employed to direct the research of AI Lab at a high cost at the same time over the same period. The research teams of various universities had also turned to research fields related to DL, as if they would fall behind the times without publishing articles concerning to DL at academic conferences. And, what improvements does DL have over traditional neural networks that are enough to be sought after by academics and industry researchers?

Upon Neural Networks Meeting Big Data

According to a paper published jointly by Google and scientists in Stanford University in 2011, a neural network with super large scale which can be trained by the distributed system from Google can recognize human faces and human bodies, or even cat faces (there are a lot of videos about cute cats on Youtube) itself, only with pictures in videos and without any human intervention after videos on Youtube being input into it. This neural network had 1 billion neurons connected and used 1,000 computers and 16,000 CPUs to optimize on 10 million images. This article caused a strong reaction at the time. For on side, it owed to its accurate results. The neural network could spontaneously find objects repeatedly appearing in the video. For another side, it owed to the breakthrough of the research program in the scale of neural networks. The deficiency of data and high performance compute resources made neural networks can not be optimized on the basic of data sets with very large scale, resulting in the decline of neural networks from 1960s to 1980s. However, technology related to big data was becoming mature with the rise of Internet and distributed system at the end of 20th century and beginning of 21st century, which played an important role in the promotion of the development of DL. In 2012, Google also published another article in its annual meeting of neural processing, which elaborated on how to use distributed systems to train large-scale neural networks.

In addition to distributed systems, the emergence of specialized hardware had also promoted the development of deep neural networks, such as graphics cards (GPUs), field programmable gate arrays (FGPA) customized for certain algorithms and application-specific integrated circuits (ASIC). With the powerful parallel computing performance provided by the specialized hardwares, the training speed of the neural network had a qualitative leap. In addition to the performance improvements at the system level, the improvement of the algorithm had also improved the practicality of DL. Next, we summarize the technical progresses that has made DL the mainstream of machine learning in several aspects.

Distributed Model

For ultra-large-scale neural networks, it is difficult for a machine to process all the

parameters in a model. Therefore, a model needs to be decomposed into different parts and distributed to multiple machines. Such a group of machines is a model, called "Model Replica".

By the method that each machine processes only a fraction of model parameters, we can generate a large-scale neural network by simply adding machines. The only difficulty is that some neural network connections need to stretch across different machines. This requires different machines to communicate, and communication will result to network bandwidth overheads and computational delays. Fortunately, as for many problems, such as image processing, neural networks only require "local connections"; that means, the neurons in the next level only process the information transmitted by a limited number of neurons in the previous layer. It is as if we divide one image into several sub-regions (such as the blue sky part, the ground part, and the building part etc.), and then use different neurons in the neural network model to process them, and obtain the results of processing in different sub-regions, and then summarize them. This mechanism greatly reduces the neuron connections between different machines, the cost of communication between different machines and improved the computational efficiency of distributed model methods.

Parallel Data Processing

The distributed model algorithm solves the problem of storing large model parameters. However, what parallel data processing solves is the problem of how to use big data to optimize the neural network model. Here we spin a Japanese classic anime "Naruto" reference to make an analogy for everyone to understand. Naruto's main character, Uzumaki Naruto, has an important skill called "shadow avatars", which is to create multiple clones (similar to model copies) instantly. In addition to KOing enemies, another important use of this skill is to use the minds of multiple shadow avatars for fast learning. The basic idea is to let each shadow avatar learn a new advanced ninjutsu at the same time; after a period of time of learning, Naruto will let the shadow avatars return to the initial body and integrated experience learnt. The method can greatly speed up the progress of learning. Parallel data processing is to use similar ideas to accelerate the optimization of neural networks. In terms of specific optimization algorithms, stochastic gradient descent (SGD) is widely used because it is simple and easy to use. However, the traditional SGD method is difficult to parallelize because the algorithm processes one piece of data in the training data set at a time; refresh the models; then process the next data on the basis of the previous processing results. Such an algorithm is too sequential. Although it is simple and effective, it is hard to use multiple machines to accelerate the calculation. In order to distribute different training data to different machines at the same time, the asynchronous SGD method is needed.

The characteristic of this algorithm is that each "model copy" only optimizes a part of the input data. The optimization results of different model copies communicated through a "parameter server" cluster. Each model copy periodically uploads its optimization results to the parameter server and downloads the latest model parameters to continue its optimization. If each model copy here is a shadow avatar, the parameter server will be the brain of the initial body, and it will continuously accept and summarize the knowledge learned by all shadow avatars. This asynchronous way uses distribution system to process large-scale training data, so that we can use tens of thousands of machines to train the same model, and it can get the results that previously required several weeks of calculation in a few hours. Therefore, it makes the technology of generating large-scale neural networks by using big data possible.

Hardware Acceleration

Inside the computer, in addition to the general-purpose computation unit CPU, there are special-purpose chips for graphic display, called GPUs. GPUs were first invented by NVIDIA in 1999 and were used to process image deformation, lighting changes, rendering engines, etc., mainly used for special effects of games and movies. Basically, a powerful GPU is built in every 3D game enthusiast's computer. Many applications of GPUs involve matrix and vector operations, and its architecture is designed for fast parallel matrix operations. Parallel computing on a single machine is usually completed by multiple cores on the processor. For example, a home computer has four cores as standard configuration today, but there are hundreds of smaller and more computationally efficient cores in GPUs to improve parallel computing performance, such as GeForce 8800 graphics card in 2006 with 128 cores, and Tesla K40 launched in 2013 with 2880 cores, and they were both launched by NVIDIA. In addition, for fast parallel computing, the cache system of GPU has also been specifically designed to reduce the cache capacity and increase the bandwidth to access internal storage (the amount of data read and written in internal storage per second). Therefore, advantages of GPUs are self-evident in the special field——matrix calculations. Because the massage passing of neural networks in each layer can also be expressed mathematically as matrix operations, and as the performance of GPUs is enhanced, researchers begin to use GPUs to accelerate the calculation of neural networks.

For example, Pro. Hinton's team, who got the first place in ImageNet image recognition evaluation in 2012, used two GPUs to train the same neural network on one machine, and achieved the best evaluation results of the year.

The development of DL made NVIDIA see the potential of GPUs in embedded automation systems (such as wearable devices, unmanned aerial vehicles (UAV), and unmanned vehicles in the future), so it became an important strategic direction for the company. NVIDIA's latest product about this is the embedded chip of the Jetson series, which is only the size of a credit card, but has important components of the computer such as CPU, GPU, internal storage, etc., and DL algorithms can be run on the chip in real time. The chip can be used in intelligent machine platforms such as unmanned vehicles, UAVs, etc. It is envisioned that NVIDIA's continued efforts in this field will definitely drive the practical application of DL. In addition to NVIDIA, companies like Qualcomm and IBM are also developing chips capable of deploying deep learning. The deployment of complex deep learning models on low-power and low-cost chips will be the turning point in which the next billion-dollar company in the chip industry will be born.

Algorithm Improvement

In addition to the improvements in hardware, an important improvement direction in software is to compress the optimized model. For example, to compress neural network parameters from a floating-point number to an integer with a limited numerical range, or to try to reduce the number of neurons in each layer and the number of connections between different layers. There are two benefits. On one side, the compression of model can speed up the model's computing, and it is critical for many industries needing to process real-time data. For example, in the security industry, if anyone suspicious is found, DL models should make a response at the first time and issue a warning to the monitoring department. If there is a delay, it may cause irreparable losses. On the other side, this optimization can make the neural network occupy less computing resources and more suitable for applications with limited computing resources, such as mobile phones, robots, smart watches, etc. In addition to improvements on speed, useful knowledge can be learned by ultra-large-scale neural networks from big data rather than noise, and researchers have also proposed improvements on optimizating algorithms. For example, there is a simple and practical method called Dropout, which means when training each layer of neural networks, the output of some neurons is forced to be 0 randomly; that means the neuron is dropped. This approach can be considered as disassembling a large neural network into multiple small-scale neural networks and computing the average of their output. In the process of computing the average, the noise in the data can be effectively reduced. In addition, simpler and asymmetric rectified linear units have also been proved to complete learning neural networks more quickly and efficiently.

Open Source Strategy

There are many open source software to choose for DL; the most popular are Theano, Torch, and Caffe. In November 2015, Google also released an open source version of its own DL system——TensorFlow. TensorFlow was outstanding for its simple and easy-to-use programming interface; after the development, the model could be easily deployed to a variety of computing platforms such as the cloud, mobile phones, and GPUs. Using the structure of the data flow diagram, when training the model, TensorFlow did not need to rely on a separate server (Parameter Server) to maintain the status of model parameters, which made the structure of TensorFlow greatly simplified. As the latest generation of Google's internal system, TensorFlow had been applied to multiple product departments, so the decision to release an open source version of TensorFlow had shocked industrial and academic circles. In fact, the move by Google also gave off the hint to develop industry standards to serve its own cloud computing strategy. In recent years, many important open source software design ideas have been derived from papers published by Google, such as Hadoop from Google's MapReduce and HBase from Google's Big-Table. There is nothing wrong with it and it is good for Google to share the experience with its peers. But in the era of cloud computing, technical standards are very important. If all researchers use TensorFlow as a model, when they need to use cloud computing to expand the scale of their models, they would take the compatibility of existing systems and cloud platforms into consideration and prefer Google's cloud service naturally. Therefore, in addition to promoting the progress of the entire DL field, the open source version TensorFlow is also a good move for Google to develop its business.

Bold Visions for the Future

One day in 20XX, Xiao Ming will take his own autonomous vehicle to a dinner party after work. Despite the heavy traffic, Xiao Ming will still be able to listen to music in his seat and surf the Internet. There will be a page on the website about the cars previous, and it will be so interesting that he will drive to the restaurant insensibly. When getting out of

the car, Xiao Ming will be thinking that it was really not easy for people in the past, because they had to take a driving license exam, waste precious lifetime to drive on such a congested road, find parking spots and buy driving insurances; it was so troublesome that people would better not buy cars. After getting out of the car, Xiao Ming will ask the car to offer chauffeur-driven car-on-demand service nearby to earn money, and the robot waiter in the restaurant will take Xiao Ming to his friend's private dining room.

No one knows exactly how long it will take to make the above scenario a reality, but it is undeniable that DL has made major breakthroughs in basic areas such as image and speech recognition, and these are basics for the realization of the above scenario. On the one hand, DL in the future may achieve greater breakthroughs in the understanding of natural languages which are not yet mature; on the other hand, it may be deployed on more platforms, so that users can easily experience conveniences brought by DL algorithms. I believe that DL models will become faster and more efficient with the continuous improvements of related software and hardware, and do better than humans in certain tasks, becoming an important impetus for the next wave of information industry revolution to drive the leap-forward development of transportation, security, industrial manufacturing, e-commerce and other industries.

Although the most advanced artificial intelligence technologies in the world is currently mastered by companies such as Google, Microsoft, and IBM, the gap between China's AI technology and international advanced technology is not large. For it, there is an important reason that in the Internet revolution in the early 21st century, a large number of outstanding Internet companies had sprung up in China, and their accumulation of technology and data provided a stable foundation for the research and development of DL technology. At the country level, in 2015 the State Council issued the Guidelines on Actively Promoting the "Internet +" Action, in which there is a section to specially discuss the development plan of AI under the grand "Internet +" strategy. Therefore, it is believed that with the joint efforts of government departments and enterprises, the research on DL technology in China can surpass the formers as a latecomer and be applied to different industries to enhance overall national science and technology strength.

单词释义:

1. 激励函数　　　　driving function
2. 跨膜电位　　　　transmembrane potential
3. 隐含层　　　　　hidden layer

4. 递推关系　　　　recurrence relation

5. 专用硬件　　　　dedicated hardware

6. 并行数据处理　　parallelization of data processing

7. 矩阵运算　　　　matrix operation

8. 浮点数　　　　　floating number

9. 开源软件　　　　open source software

10. 兼容性　　　　 compatibility

译文解析：

1. 对新词、强文化性词汇及俗语的翻译。

原文：横空出世

译文：the unique viable birth

分析：横空出世是汉语中的成语，意为突然出现或（不被注意的人）突然做出引人注目的事情。查阅词义后，译者将其译为"the unique viable birth"，且变化了词性。

原文：在人工智能领域，深度学习（deep learning）是这一波高科技革命潮流中最璀璨的明珠。

译文：DL（deep learning）catches people's eyes most in this wave of high-tech revolution in the field of artificial intelligence.

分析："最璀璨的明珠"本就是汉语中的习惯用法，译者在这里并未进行直译，而是根据其本意"最瞩目的"将其进行意译，处理为"catches people's eyes most"，即英语中瞩目的、吸引目光的惯用表达。译者尽量使两种语言在翻译是时对等，使译文及最大可能符合译入语的特征及表达习惯。

2. 长句的翻译。

原文：2011 年谷歌和斯坦福大学的科学家联合发表了一篇文章，利用谷歌的分布式系统来训练超大规模神经网络，把 Youtube 的视频输入到该神经网络中，在没有任何人为干预只有视频中的图片的情况下，该神经网络可以自行识别视频中的人脸、人体，甚至还有猫脸（Youtube 上有很多萌猫的视频）。

译文：According to a paper published jointly by Google and scientists in Stanford University in 2011, a neural network with super large scale which can be trained by the distributed system from Google can recognize human faces and human bodies, or even cat faces (there are a lot of videos about cute cats on Youtube) itself, only with pictures in videos and without any human intervention after videos on Youtube being input into it.

分析：汉语原文为超长句,而在翻译的时候,译者仔细分析各个元素之间的关系,根据英语的特点——树状结构,重新进行了整合,进而得出译文。

3. 词性转换。

原文：具有挑战性的是如何根据已有的数据,优化计算出网络的参数(最重要的参数是神经元连接的权重),这就引出了一位人工智能领域的传奇人物 Geoff Hinton 和他发明的人工智能领域最伟大的算法(之一):后向传播(back propagation)。

译文：The challenge is how to optimize the network parameters based on the existing data (the most important parameter is the weight of neural connections), which led to a legend in AI field—— Geoff Hinton and (one of) the greatest algorithms he invented in the field: back propagation.

分析：这里将"具有挑战性的"由形容词处理为一个名词"challenge",转换词性,符合科技英语简洁的特征。

4. 增译法。

原文："深蓝"电脑、百度大脑、微软的实时翻译系统、谷歌的无人驾驶汽车、亚马逊正在试验的能送快递的无人机,在这创新浪潮中,人工智能开始悄无声息地渗透进我们的生活。

译文：AI has begun to penetrate our lives silently in the wave of innovation, for instance "Deep Blue" computer, Baidu Brain, Microsoft real-time translation system, Google unmanned vehicles and Amazon UAVs that can deliver express delivery during the course of testing.

分析：译文未按照原文本的顺序进行翻译,而是将重点提前,进行了重新整合,将举的例子放到后面,并增译了"for instance",使句段更加连贯。这样一来更符合英语重形合和前重点的特点,是译入语读者能更好地理解文意。

5. 科技英语中多用长句。

原文：20 世纪 60 年代到 80 年代神经网络衰落的原因就是数据的缺乏和高性能计算资源的不足,导致无法在很大规模数据集上优化神经网络,随着 20 世纪末、21 世纪初互联网和分布式系统的兴起,大数据的相关技术日趋成熟,这对深度学习的发展起到了巨大的推广作用。

译文：The deficiency of data and high performance compute resource made neural networks can not be optimized on the basic of data sets with very large scale, resulting in the decline of neural networks from 1960s to 1980s, however, technology related to big data was becoming mature with the rise of Internet and distributed system at the end of 20th century and beginning of 21st century, which played an important role in the promotion of the development of DL.

分析：原文虽然看似时汉语长句，然而其实是三个句子组合而成，所以译者对其因果、转折逻辑关系进行了整合，重新调整了语序，并增译了关联词"however"，使其逻辑关系更加明显，结构更加具象。同时译者也运用了定语从句，是各个因素能融合到一起，形成长句，这样更符合科技英语多长句的特征。

6. 被动语态的翻译。

善用被动语态，科技英语中常用被动语态，可以使译文更加客观

原文：复杂的神经网络的隐含层一般有多个，隐含层数量越多，表达能力越强，可以解决的问题就更多。

译文：Complex neural networks usually have multiple hidden layers, and if there are more hidden layers and stronger expressiveness, the more problems can be solved.

分析：原文中的"可以解决的问题就更多"本是主动语态，这符合汉语的表达习惯，可是译入语英语，尤其是科技英语常用物作主语的被动语态，这样能使文本更加客观，具有说服力。故译者将其处理为"the more problems can be solved"。

7. 定语从句的应用。

在翻译中可以灵活运用定语从句，以达到使译文能忠实地传递原文的内容。

原文：马克一号的理论基础是感知器算法（Perceptron），该算法现在已经成为人工智能领域的经典算法。

译文：The theoretical foundation of Mark-I is perceptron algorithm, which has been classic in AI field.

分析：汉语原文为两小句，译文根据这两小句中的相同的因素"算法"，通过定语从句将其整合为一句，使句子不过于冗长和啰嗦，符合科技英语简洁的特征。

8. 句子的整合。

在翻译时，为了使意思更加连贯，句式更加紧凑，可以将零散的汉语短句整合成长句。

原文：信息传导的一个重要理论是"全有全无律"（All-or-none-law），就是说一个初始刺激，只要达到了阈电位，就能产生离子的流动，改变跨膜电位。

译文：An important theory of information transmission is "All-or-none-law", which means, an initial stimulus, as long as it reaches the threshold potential, can generate the flow of ions to change the transmembrane potential.

分析：汉语原文本来虽然只有一个句号，而实际上却是多个小短句融合而成。

如果按其本来句型的话过于啰唆,于是译者将对各小句按逻辑关系及相同要素进行整合,增加"which means"和连词"as long as",将"就能"译为目的状语"to……"。

9. 注意科技英语的简洁性。

原文： 这样的算法顺序性太强,虽然简单有效却很难利用多台机器加速计算。

译文： Such an algorithm is too sequential. Although it is simple and effective, it is hard to use multiple machines to accelerate the calculation.

分析： 原文中的"顺序性太强"为名词加形容词的组合,本质上是主系表的结构。因此译者在翻译时沿用主系表结构,不过变换了主语,从"算法的顺序性"变为"算法"。"顺序性太强"则用一个副词加形容词的结构"too sequential"表达。这样做的目的是使译文更加简洁,不说废话,能符合科技英语的简洁特征。